VOLUME ONE HUNDRED AND FORTY FIVE

Current Topics in DEVELOPMENTAL BIOLOGY

Amphibian Models of Development and Disease

CURRENT TOPICS IN DEVELOPMENTAL BIOLOGY

"A meeting-ground for critical review and discussion of developmental processes"
A.A. Moscona and Alberto Monroy (Volume 1, 1966)

VOLUME ONE HUNDRED AND FORTY FIVE

CURRENT TOPICS IN DEVELOPMENTAL BIOLOGY

Amphibian Models of Development and Disease

Edited by

SERGEI Y. SOKOL

Department of Cell, Developmental and Regenerative Biology,
Icahn School of Medicine at Mount Sinai,
New York, NY, United States

Academic Press is an imprint of Elsevier
50 Hampshire Street, 5th Floor, Cambridge, MA 02139, United States
525 B Street, Suite 1650, San Diego, CA 92101, United States
The Boulevard, Langford Lane, Kidlington, Oxford OX5 1GB, United Kingdom
125 London Wall, London, EC2Y 5AS, United Kingdom

First edition 2021

Notices
Knowledge and best practice in this field are constantly changing. As new research and experience broaden our understanding, changes in research methods, professional practices, or medical treatment may become necessary.

Practitioners and researchers must always rely on their own experience and knowledge in evaluating and using any information, methods, compounds, or experiments described herein. In using such information or methods they should be mindful of their own safety and the safety of others, including parties for whom they have a professional responsibility.

To the fullest extent of the law, neither the Publisher nor the authors, contributors, or editors, assume any liability for any injury and/or damage to persons or property as a matter of products liability, negligence or otherwise, or from any use or operation of any methods, products, instructions, or ideas contained in the material herein.

ISBN: 978-0-12-816833-2
ISSN: 0070-2153

For information on all Academic Press publications
visit our website at https://www.elsevier.com/books-and-journals

Publisher: Zoe Kruze
Developmental Editor: Naiza Ermin Mendoza
Production Project Manager: Denny Mansingh
Cover Designer: Greg Harris

Typeset by SPi Global, India

Contents

Section III
Amphibian models for regeneration and disease

Contributors

Aparna B. Baxi
Department of Chemistry & Biochemistry, University of Maryland, College Park, College Park, MD; Department of Anatomy and Cell Biology, The George Washington University, Washington, DC, United States

Ira L. Blitz
Department of Developmental and Cell Biology, University of California, Irvine, CA, United States

Ken W.Y. Cho
Department of Developmental and Cell Biology, University of California, Irvine, CA, United States

Caitlin Collins
Department of Cell and Developmental Biology, Lurie Comprehensive Cancer Center, Northwestern University, Feinberg School of Medicine, Chicago, IL, United States

Eugenia M. del Pino
Escuela de Ciencias Biológicas, Pontificia Universidad Católica del Ecuador, Quito, Ecuador

Karen Echeverri
Marine Biological Laboratory, Eugene Bell Center for Regenerative Biology and Tissue Engineering, University of Chicago, Woods Hole, MA, United States

Nicole A. Edwards
Division of Developmental Biology, Center for Stem Cell and Organoid Medicine, Perinatal Institute, Cincinnati Children's Hospital Medical Center, Cincinnati, OH, United States

Claudia Marcela Arenas Gómez
Marine Biological Laboratory, Eugene Bell Center for Regenerative Biology and Tissue Engineering, University of Chicago, Woods Hole, MA, United States

Mustafa K. Khokha
Pediatric Genomics Discovery Program, Department of Pediatrics and Genetics, Yale University School of Medicine, New Haven, CT, United States

Valentyna Kostiuk
Pediatric Genomics Discovery Program, Department of Pediatrics and Genetics, Yale University School of Medicine, New Haven, CT, United States

Miho Matsuda
Department of Cell, Developmental and Regenerative Biology, Icahn School of Medicine at Mount Sinai, New York, NY, United States

Laura Medina-Cuadra
Université Paris-Saclay, Faculté des Sciences d'Orsay, CNRS UMR3347; Institut Curie, Université PSL, CNRS UMR3347, Inserm U1021, Signalisation radiobiologie et cancer, Orsay, France

Brian J. Mitchell
Department of Cell and Developmental Biology, Lurie Comprehensive Cancer Center, Northwestern University, Feinberg School of Medicine, Chicago, IL, United States

Anne H. Monsoro-Burq
Université Paris-Saclay, Faculté des Sciences d'Orsay, CNRS UMR3347; Institut Curie, Université PSL, CNRS UMR3347, Inserm U1021, Signalisation radiobiologie et cancer, Orsay; Institut Universitaire de France, Paris, France

Peter Nemes
Department of Chemistry & Biochemistry, University of Maryland, College Park, College Park, MD; Department of Anatomy and Cell Biology, The George Washington University, Washington, DC, United States

Leena R. Pade
Department of Chemistry & Biochemistry, University of Maryland, College Park, College Park, MD, United States

Sergei Y. Sokol
Department of Cell, Developmental and Regenerative Biology, Icahn School of Medicine at Mount Sinai, New York, NY, United States

Masanori Taira
Department of Biological Sciences, Faculty of Science and Engineering, Chuo University, Bunkyo-ku, Tokyo, Japan

Rosa Ventrella
Department of Cell and Developmental Biology, Lurie Comprehensive Cancer Center, Northwestern University, Feinberg School of Medicine, Chicago, IL, United States

Yuuri Yasuoka
Laboratory for Comprehensive Genomic Analysis, RIKEN Center for Integrative Medical Sciences, Yokohama, Japan

Aaron M. Zorn
Division of Developmental Biology, Center for Stem Cell and Organoid Medicine, Perinatal Institute, Cincinnati Children's Hospital Medical Center; Department of Pediatrics, University of Cincinnati College of Medicine, Cincinnati, OH, United States

Preface

Among various animal models, amphibians occupy a special place in driving basic biology research and paving way toward better understanding of human diseases and developing new therapies. The main impact of amphibian species as research organisms has been arguably on the fields of embryology and evolutionary biology, although their contributions to basic cell and molecular biology, regeneration, genetics, toxicology, and immunology are also evident. Key biological discoveries have been made using amphibian models. For example, Hilde Mangold and Hans Spemann have demonstrated the primary embryonic induction in salamander embryos. The regulation of the cell cycle by maturation-promoting factor, containing the cyclin-Cdk complex, has been first discovered in frog oocyte cytoplasm. This volume, although not comprehensive, testifies to the power of amphibians as one of the top experimental models for both basic biology and disease.

Amphibian models: Pros and cons

What makes the amphibians, especially *Xenopus*, so special? Amphibians usually lay large and abundant eggs that provide plenty of material for biochemical and embryological experiments. The embryos develop externally, allowing easy monitoring and even high-resolution live imaging. Frog embryos are inexpensive to culture, being raised in simple buffered solutions, yet they maintain many conserved features of human and mouse embryos. Amphibians also possess significant regenerative ability, as has been first described in the 18th century for a salamander limb. *Xenopus* has become a popular laboratory model after the discovery that it can be used for a robust human pregnancy test. John Gurdon has carried out nuclear transplantations to demonstrate that *Xenopus* embryonic cells can be reprogrammed. This discovery attracted further attention to *Xenopus* as a premier embryological model. On one hand, the above strengths made *Xenopus* embryos very suitable for studying early inductive interactions, gene regulatory networks, collective cell movements, organogenesis, genetics, and evolution. On the other hand, long generation time and the allotetraploid genome of *Xenopus laevis* reduce the attractiveness of this organism for classical genetic approaches. To overcome this disadvantage, the researchers started to use the related species, *Xenopus tropicalis*, which is diploid and acquires sexual maturity much faster than *laevis*. Current research using *Xenopus* has gained significant support from the National Institutes of Health through the Xenbase (www.xenbase.org) and *Xenopus* resource centers in the USA and in Europe.

Section I of the current volume of *CTDB* illustrates some basic biological questions that can be addressed using amphibian models. Chapter 1 written by Brian Mitchell and colleagues focuses on the simultaneous regulation of cell fate and behavior during the formation of multiciliated cells of the skin using advanced imaging techniques. How neuroepithelial cells polarize and activate contractile actomyosin complexes during neural plate closure is discussed by Matsuda and Sokol in Chapter 2. While the main knowledge of planar cell polarity (PCP) has come from *Drosophila* genetics, studies of the neural plate in *Xenopus* and other vertebrate models suggest new links between PCP proteins and actomyosin contractility. In Chapter 3, Edwards and Zorn review fundamental knowledge of endoderm development obtained using *Xenopus* embryos and provide examples of modeling relevant human disease. In Chapter 4, Eugenia del Pino details the common and divergent features of embryonic development of marsupial frogs as compared to *Xenopus* to give insights into the evolution of these species. Among notable adaptations of these animals to terrestrial life are embryo development inside the body of the mother ensuring high humidity and modified patterns of gastrulation.

Section II of the volume is devoted to systems biology approaches in amphibians. New systems biology approaches such as genome sequencing and analysis, quantitative transcriptomics, and proteomics revive the attractiveness of *Xenopus* for basic biology. In Chapter 5, Yasuoka and Taira describe insights obtained from genome sequencing on the evolution and function of LIM homeodomain proteins. The combination of LIM domains with a homeodomain in these proteins produced remarkable diversity of their functions in the assembly of transcriptional protein complexes and in modulating gene regulatory networks. In Chapter 6, Blitz and Cho review the existing knowledge of zygotic genome activation (ZGA) obtained with the transcriptomics approach. They present evidence to suggest that ZGA is not a one-step process, but several distinct events regulating transcriptional activation. Baxi et al. (Chapter 7) discuss the applicability of high-resolution mass spectrometry-based proteomics to *Xenopus* developmental neurobiology. Large protein amount in *Xenopus* early blastomeres allows to detect many constituents, including those that affect lineage decisions, such as geminin for neural cell fate. These results reveal high sensitivity of the approach and hold great promise for future studies using proteomics.

Finally, Section III focuses on the models that are relevant to regeneration and human disease. In Chapter 8, Karen Echeverri describes the continuing advantages of salamanders for studies of regeneration. Being the first model of regeneration reported by Spallanzani in 1768, newts remain very appealing due to efficient CRISPR-mediated gene editing and successful single-cell RNA sequencing. Two last chapters review the literature on *Xenopus* as a model for the discovery of new genes that are relevant to human disease. In Chapter 9, Kostiuk and Khokha describe how *Xenopus* helps study the contribution of patient mutations to complex phenotypes such as congenital heart disease and heterotaxy. Finally, in Chapter 10, Cuadra and Monsoro-Burq review emerging frog models for studying neural crest pathologies, including many congenital defects as well as melanoma and neuroblastoma cancers that are prone to metastases.

Looking ahead

The past several years witnessed the development of many new cutting-edge experimental approaches, such as single cell omics, powerful methods of big data analysis, or genome editing allowing to knock-in and knock-out DNA sequences of interest. In addition, light-sheet and super-resolution microscopy are now available to analyze dynamic developmental processes at high resolution. Due to these new approaches, it is anticipated that the many advantages of amphibians discussed in this volume would be enhanced further and allow these models to maintain their standing at the forefront of modern biology and biomedicine for years to come.

Acknowledgments

I am grateful to all the authors who recognize the power of amphibian models and contributed to this volume. I express my apologies to many of my colleagues who are using amphibian models and whose work has not been included in this book. I also acknowledge the funding from the National Institutes of Health.

Sergei Y. Sokol
Icahn School of Medicine at Mount Sinai

From early development to morphogenesis and tissue patterning

CHAPTER ONE

Building a ciliated epithelium: Transcriptional regulation and radial intercalation of multiciliated cells

Caitlin Collins†, Rosa Ventrella†, and Brian J. Mitchell*

Department of Cell and Developmental Biology, Lurie Comprehensive Cancer Center, Northwestern University, Feinberg School of Medicine, Chicago, IL, United States

*Corresponding author: e-mail address: brian-mitchell@northwestern.edu

Contents

Abstract

The epidermis of the *Xenopus* embryo has emerged as a powerful tool for studying the development of a ciliated epithelium. Interspersed throughout the epithelium are multiciliated cells (MCCs) with 100+ motile cilia that beat in a coordinated manner to generate fluid flow over the surface of the cell. MCCs are essential for various developmental processes and, furthermore, ciliary dysfunction is associated with numerous pathologies. Therefore, understanding the cellular mechanisms involved in establishing a ciliated epithelium are of particular interest. MCCs originate in the inner epithelial layer of *Xenopus* skin, where Notch signaling plays a critical role in determining which progenitors will adopt a ciliated cell fate. Then, activation of various transcriptional regulators, such as GemC1 and MCIDAS, initiate the MCC transcriptional program, resulting in

† These authors contributed equally to this work.

Current Topics in Developmental Biology, Volume 145
ISSN 0070-2153
https://doi.org/10.1016/bs.ctdb.2020.08.001

centriole amplification and the formation of motile cilia. Following specification and differentiation, MCCs undergo the process of radial intercalation, where cells apically migrate from the inner layer to the outer epithelial layer. This process involves the cooperation of various cytoskeletal networks, activation of various signaling molecules, and changes in cell-ECM and cell-cell adhesion. Coordination of these cellular processes is required for complete incorporation into the outer epithelial layer and generation of a functional ciliated epithelium. Here, we highlight recent advances made in understanding the transcriptional cascades required for MCC specification and differentiation and the coordination of cellular processes that facilitate radial intercalation. Proper regulation of these signaling pathways and processes are the foundation for developing a ciliated epithelium.

1. Introduction

Ciliated epithelia are important in a wide variety of physiological contexts, where they function to generate fluid flow over the surface of various tissues. In mammals, these epithelia are found in the airway to help mucus flow over the surface, the ependyma that line the brain ventricles to move cerebral spinal fluid, and the oviduct where they facilitate ovum transport. Fluid flow is generated by highly specialized multiciliated cells (MCCs) that contain 100+ cilia that beat in a coordinated and polarized manner to generate the directed fluid flow over the surface (Brooks & Wallingford, 2014; Choksi, Lauter, Swoboda, & Roy, 2014). Importantly, defects in the generation or maintenance of a ciliated epithelium are found in many pathologies, as ciliary dysfunction and ciliopathies are central to many diseases (Hildebrandt, Benzing, & Katsanis, 2011; Reiter & Leroux, 2017). Therefore, there is significant interest in understanding the molecular events and signaling pathways that regulate MCC specification, differentiation, and function.

The epidermis of the *Xenopus* embryo has emerged as a powerful model system for studying the development and maintenance of a ciliated epithelium (Walentek & Quigley, 2017; Werner & Mitchell, 2013). Although this epithelium is derived from the ectoderm in *Xenopus* rather than the endoderm, the transcriptional cascades that govern the formation of this tissue, as well as its morphogenesis, are highly conserved with mammalian lung tissue (Brooks & Wallingford, 2014; Walentek & Quigley, 2017). Recent single cell RNA sequencing in *Xenopus* has provided us with a powerful tool to trace MCC differentiation in the developing vertebrate embryo. This targeted tracing revealed that differentiation of MCCs begins as early as stage 11 of development when MCC genes including forkhead box protein

J1 (Foxj1), myb, forkhead box protein n4 (Foxn4), and regulatory factor X2 (Rfx2) become enriched in cells that will develop into ciliated epidermal progenitors (Angerilli, Smialowski, & Rupp, 2018; Briggs et al., 2018). Activation of various transcriptional pathways and signaling networks within ciliated cell precursors commits these cells to an MCC lineage and establishes a precise transcriptional cascade that drives MCC differentiation.

A significant body of work has contributed to our understanding of the process of MCC specification and differentiation, but it was not initially known where these processes occurred. The *Xenopus* skin is multilayered, leading to the possibility that MCCs are either (1) derived from cells in the most superficial layer of the ectoderm where they are found in their mature state, or (2) derived from a deeper layer, requiring migration to the outer epithelium. In order to test these possibilities, transplantation studies were performed in which labeled outer epithelial cells or labeled deep ectodermal cells (below the outer layer) were transplanted onto unlabeled embryos during mid-gastrulation. These classic transplantation and tracing studies revealed that MCC specification and differentiation occurs within a deeper layer of the *Xenopus* skin and is followed by a short, but directed, apical migration into the outer epidermis, a process termed radial intercalation (Drysdale & Elinson, 1992). Of note, the early transplantation studies revealed that another cell type, later identified as ionocytes (ICs), intercalates alongside of MCCs (Drysdale & Elinson, 1992; Quigley, Stubbs, & Kintner, 2011), and later studies revealed that a third cell type, small secretory cells (SSCs), also undergoes radial intercalation (Dubaissi et al., 2014).

Following the intercalation of multiple cell populations, the mature ciliated epithelium is comprised of the following cell types: secretory cells that release mucus (e.g., Goblet cells in mammals and mucus-secreting cells in *Xenopus*) and antimicrobial substances (e.g., Club cells in mammals and SSCs in *Xenopus*), MCCs that generate fluid flow and distribute the released substances over the surface, and ICs which play a role in ion regulation (Dubaissi & Papalopulu, 2011; Dubaissi et al., 2014; Haas et al., 2019). The development of the *Xenopus* ciliated epithelium at the external surface provides a useful tool for studying the development of this highly specialized and complex tissue, as well as providing a model for studying mucociliary diseases (Nenni et al., 2019; Walentek & Quigley, 2017).

There have been several recent reviews that address different aspects of ciliated epithelial development including centriole amplification and cilia polarity (Boutin & Kodjabachian, 2019; Brooks & Wallingford, 2014; Lewis & Stracker, 2020; Zhang & Mitchell, 2015a, 2015b). Here, we will focus on our current understanding of the transcriptional networks required for MCC specification and differentiation and the process of MCC radial intercalation in the *Xenopus* epidermis, which are required for the formation of a functional ciliated epithelium.

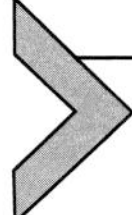

2. MCC specification and differentiation

2.1 Notch signaling controls the balance of ciliated and secretory cell fates

Notch signaling plays important roles in development, including cell fate determination and patterning (Henrique & Schweisguth, 2019). The single-pass transmembrane Notch receptor can bind to its membrane-anchored ligands including Delta, Jagged, Serrate, and Delta-like ligand (Dll) on neighboring cells. Following ligand-induced activation, the membrane-bound Notch receptor is cleaved, permitting the Notch intracellular domain (NICD) to translocate to the nucleus to regulate transcription of target genes involved in a variety of cellular processes, including proliferation, apoptosis, cell fate determination, metabolism function, and cytoskeletal regulators (Bray & Bernard, 2010). Importantly, Notch signaling has been shown to play an important role in the specification of cells in the *Xenopus* epidermis (Deblandre, Wettstein, Koyano-Nakagawa, & Kintner, 1999).

Early in *Xenopus* development, the Notch ligand Delta-1 is transiently expressed in a scattered pattern in the inner, sensorial layer of the epidermis. Peak Delta-1 expression occurs shortly before the onset of α-tubulin expression in MCCs at stage 11. Proper regulation of Delta-1 expression levels and patterns are critical during MCC specification, as cells expressing the highest levels of Delta-1 comprise the subset of cells that will adopt a ciliated cell fate. Importantly, newly fated MCCs activate Notch receptors on neighboring cells, resulting in lateral inhibition and repression of MCC differentiation (Fig. 1A) (Deblandre et al., 1999). The importance of Notch signaling in controlling these cell fate decisions is highlighted when Notch activity is altered. Activating Notch signaling by overexpression of the NICD prevents the progression of MCC differentiation, resulting in a complete loss of ciliated cells (Fig. 1B). Studies have also highlighted a role for the conserved transcription factor, Suppressor of Hairless Su(H), also known as CBF-1 or lag-1, downstream of Notch in this process. Notch signaling can be inhibited through expression of a dominant-negative, DNA-binding

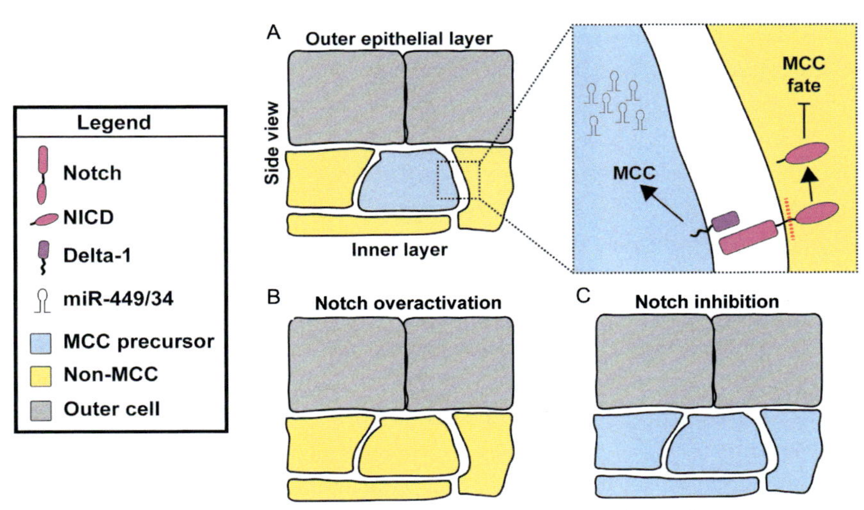

Fig. 1 Notch signaling inhibits MCC differentiation. (A) Delta-1 ligand binding results in cleavage of membrane-bound Notch and translocation of the NICD to the nucleus, which represses MCC specification. This results in a mosaic pattern of high Delta-1/low Notch-expressing cells that will adopt an MCC fate. In these MCC precursors, expression of miR-449 and miR-34 targets Delta-1 to promote MCC differentiation. (B) Overexpression of the NICD results in overactivation of Notch signaling and a complete loss of MCCs in *Xenopus* epidermis. (C) Inhibition of Notch signaling by expression of a dominant-negative downstream transcription factor Su(H) or altered expression of Delta-1 increases the number of MCCs in *Xenopus* skin.

mutant of Su(H) or a Delta-1 ligand that lacks an intracellular domain. Inhibition of Notch signaling through expression of either of these constructs leads to an increase in MCC differentiation (Fig. 1C) (Deblandre et al., 1999; Guseh et al., 2009; Quigley & Kintner, 2017). Similarly, decreasing Notch signaling by manipulating the levels of available Delta-1 ligand at the cell surface increases the number of MCCs in *Xenopus* skin (Fig. 1C) (Deblandre, Lai, & Kintner, 2001; Deblandre et al., 1999). These studies highlight that proper Notch signaling is a fundamental component of MCC specification. However, questions related to Notch signaling in the *Xenopus* skin remain. Although it is well known that Notch activation produced by Delta-induced lateral inhibition prevents MCC differentiation, the initial signal that leads to this asymmetric Notch/Delta signaling on neighboring cells is still not understood. This likely occurs through protein regulation via ubiquitination and/or epigenetic regulation, which would lead to increased expression of Delta-1 in a subset of cells in the sensorial layer (Luna-Escalante, Formosa-Jordan, & Ibanes, 2018; Sancho, Cremona, & Behrens, 2015). However, future studies are needed to directly test this idea.

Intriguingly, different Notch family members are required when generating ciliated epithelium in different organisms and tissues. In the *Xenopus* epidermis, Delta-1 is the predominant ligand that activates Notch signaling, whereas in the zebrafish pronephros jagged-2 is the Notch ligand that activates signaling in secretory cells to prevent MCC differentiation (Deblandre et al., 1999; Liu, Pathak, Kramer-Zucker, & Drummond, 2007; Ma & Jiang, 2007). Other Notch ligands, such as jagged-1 and -2, play predominant roles in MCC specification in the murine airway epithelium. Treatment with antibodies that target these ligands results in an irreversible conversion of Club cells into ciliated cells (Lafkas et al., 2015). Additionally, these ligands can also signal through other Notch receptors (Notch-2 and -3), further highlighting the complexity of these signaling networks (Mori et al., 2015; Pardo-Saganta et al., 2015). These findings underpin the importance of understanding mechanisms of Notch signaling in organ- and organismal-specific contexts.

Further adding to the complexity of Notch signaling is the presence of microRNAs (miRNAs), which post-transcriptionally modify Notch mRNA and influence downstream signaling. During multiciliogenesis in *Xenopus*, miR-449 and, to a lesser extent, miR-34, are significantly upregulated in MCCs (Marcet et al., 2011). Interestingly, these two miRNAs belong to the same superfamily that share a similar seed region that recognizes target sequences and are highly conserved in vertebrates. miR-449 can target either Delta-1 alone or both Delta-1 and Notch in *Xenopus* and human cells, respectively. Knockdown of miR-449 inhibits centriole multiplication and multiciliogenesis in MCCs without affecting the expression of MCC genes (e.g., α-tubulin and foxj1) or MCC intercalation (Marcet et al., 2011). In addition, defects in MCC differentiation due to loss of miR-449 expression are rescued by knockdown of Delta-1, suggesting that Delta-1 is a miR-449 target required for terminal differentiation following MCC specification (Marcet et al., 2011). It has been proposed that the role for miR-449 in MCC differentiation is also, in part, due to the ability of this miRNA to target genes involved in cell cycle regulation, which would permit cell cycle exit. However, loss of miR-449 is not sufficient to restart the cycle, indicating that once MCC differentiation has begun, the effect that miR-449 has on cell cycle regulation is minimal (Marcet et al., 2011). Since inhibition of Notch signaling is important for MCC specification and loss of miR-449 does not affect commitment to MCC fate, it is likely that Delta and Notch do not represent the only key targets involved in motile ciliogenesis. Additional studies have provided evidence that cellular defects resulting in loss of miR-449 function can also be rescued with the knockdown of Cp110, a centriolar protein that suppresses cilia assembly, and RRas, a small GTPase that inhibits apical actin network assembly (Chevalier et al., 2015; Song et al., 2014). It's likely that additional miR-449/34 targets are involved in the inhibition of motile cilia formation, such as histone deacetylase 6 (HDAC6), protein O-fucosyltransferase 1 (Pofut1), and CDK5 regulatory subunit-associated protein 2 (CDK5RAP2), which influence cilia formation, Notch ligand binding, and centriole replication, respectively (Barrera et al., 2010; Smith et al., 2018; Song et al., 2014; Tsao et al., 2009). Given the vast number of miRNAs that have been identified and the wide-array of target genes, it is likely that additional roles for miRNAs will be revealed in future studies.

2.2 Transcriptional regulation for MCC differentiation

2.2.1 Upstream regulation by geminin-like proteins

The most upstream regulation of MCC differentiation is controlled by two coiled-coil-containing paralogues, geminin C1 (GemC1/Gmnc) and multicilin (MCIDAS), both of which are part of the geminin family of proteins (Arbi, Pefani, Taraviras, & Lygerou, 2018; Balestrini, Cosentino, Errico, Garner, & Costanzo, 2010; Pefani et al., 2011; Vladar & Mitchell, 2016). Interestingly, both GemC1 and MCIDAS lack a DNA-binding domain and therefore must form complexes in order to induce transcription of MCC genes. Specifically, these transcriptional activators form complexes with transcription factors E2f4 or E2f5 and Dp1, to comprise EDG (E2f4/5-Dp1-GemC1) or EDM (E2f4/5-Dp1-MCIDAS) complexes. Within these complexes, the highly conserved C-terminal region of GemC1 and MCIDAS, called the TIRT domain, is required to bind to dimerized E2f4/5 and Dp1 (Ma, Quigley, Omran, & Kintner, 2014; Terre et al., 2016). The importance of the TIRT domain has been implicated in additional studies and specifically in pathologies associated with ciliary dysfunction. Of note, reduced generation of multiple cilia (RGMC), a rare mucociliary clearance disorder, is associated with mutations in the TIRT domain of MCIDAS including a glycine to aspartic acid mutation (G366D; c.1097G>A) or an arginine to histidine mutation (R381H; c.1142G>A) (Boon et al., 2014). The G366D mutation can be mimicked in *Xenopus* MCIDAS (G355D), resulting in disruption of EDM complex formation and downstream MCC-specific gene induction (Boon et al., 2014; Ma et al., 2014; Stubbs, Vladar, Axelrod, & Kintner, 2012; Terre et al., 2016). Similarly, a mutation in the TIRT domain of GemC1 (G313D) prevents the formation of the EDG complex (Terre et al., 2016). Identifying roles for the coiled-coil domains in GemC1 and MCIDAS is of interest, as these domains are available to facilitate interactions with other coiled-coil containing proteins. Recent work has demonstrated that the coiled-coiled domains of these two proteins can interact. However, the importance of this interaction in regulating downstream signaling, as well as the potential for an EDGM (E2f4/5-Dp1-GemC1-MCIDAS) complex, remains unknown (Terre et al., 2016).

Although there are many similarities between the EDG and EDM complexes and there is redundancy in their transcriptional targets, key differences in how these complexes induce the MCC fate have been identified. GemC1 appears to be at the top of the hierarchy of transcriptional activation of MCC fate, capable of transactivating the promoters of Foxj1, which is required for the function of motile cilia, and MCIDAS. However, studies have revealed a complex relationship between GemC1, Foxj1, and MCIDAS. Specifically, GemC1-induced Foxj1 expression can occur in the absence of MCIDAS, demonstrating that GemC1-induced MCC gene expression is independent of MCIDAS. Also, MCIDAS can induce Foxj1 and MCC gene expression, which then initiates a feedforward loop where the EDM complex can bind to the MCIDAS promoter, allowing for enhancement of MCC gene expression. However, MCIDAS does not have the ability to induce the expression of GemC1 (Fig. 2) (Arbi et al., 2016). Taken together, studies have elucidated differences in the regulation between the EDG and EDM complexes, but there is clear evidence that these complexes coordinate to provide some redundancy in order to enable sufficient induction of the MCC transcriptional profile.

GemC1, a core component of the EDG complex, is highly conserved in vertebrates. The protein was originally identified to be important for DNA replication, but later work revealed that GemC1 also functions downstream of Notch inhibition to induce the expression of key regulators of MCC formation (Balestrini et al., 2010; Terre et al., 2016). Importantly, in *Xenopus*, GemC1 expression is necessary and sufficient to trigger the MCC differentiation program (Zhou et al., 2015). In agreement with these findings, loss of GemC1 in mice results in an impairment of MCC development in the brain, respiratory system, and reproductive tract, as well as reduced growth,

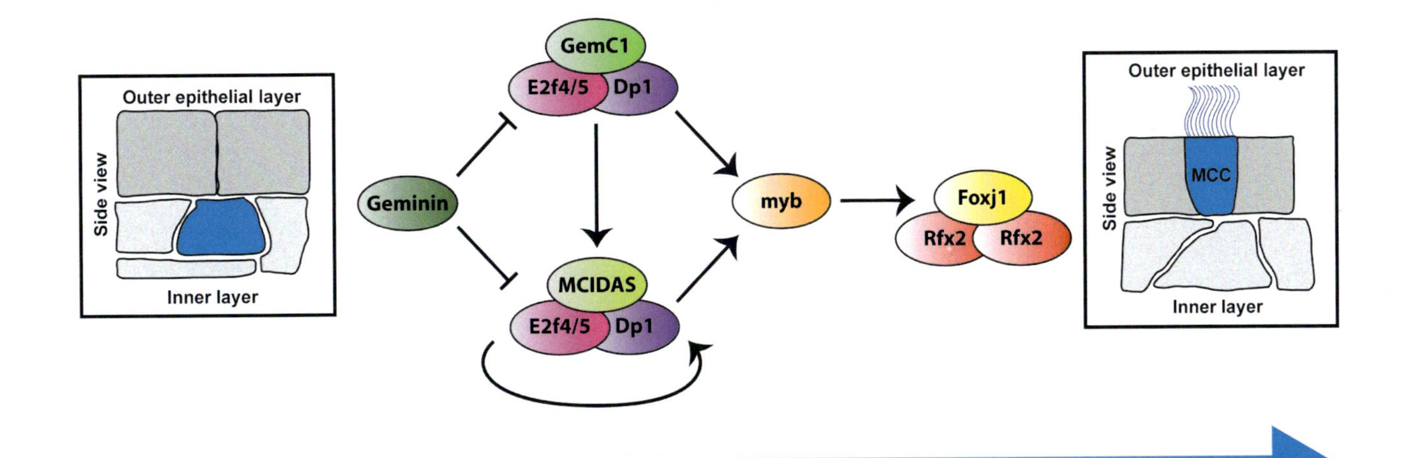

Fig. 2 The MCC transcriptional cascade in *Xenopus* skin. Prior to MCC specification, geminin inhibits GemC1 and MCIDAS. Once this inhibition in removed, GemC1, part of the EDG complex with E2f4/5 and Dp1, can induce the expression of MCIDAS. Then, MCIDAS, part of the EDM complex with E2f4/5 and Dp1, has feed-forward activity allowing for the induction of increased MCIDAS expression. Collectively, the EDG and EDM complexes activate myb, which then promotes MCC differentiation, centriole amplification, and ciliogenesis. The downstream transcription factor Foxj1, in collaboration with Rfx2, leads to the generation of motile cilia in MCCs.

bilateral hydrocephaly, and infertility. This is likely due, in part, to lack of induction of MCIDAS downstream of GemC1 (Terre et al., 2016; Zhou et al., 2015). Although it is known that GemC1 plays roles in DNA replication of cycling cells and development of post-mitotic MCCs, the interplay between these two opposing pathways remains unknown.

As previously mentioned, GemC1 induces the expression of MCIDAS, indicating that GemC1 functions upstream of MCIDAS to help amplify the MCC differentiation program. Loss of MCIDAS impairs centriole and subsequent cilia biogenesis, suggesting that MCIDAS function is essential for completion of MCC differentiation. Furthermore, MCIDAS has also been shown to be downstream of Notch signaling, as expression of MCIDAS promotes MCC differentiation, even in the presence of constitutively active Notch (NICD) (Stubbs et al., 2012). Taken together, it is likely that GemC1 and MCIDAS work together to coordinate MCC formation in two distinct steps: (1) First, GemC1 acts to specify MCC fate and (2) following MCC specification, MCIDAS drives MCC differentiation via expansive basal body production and multiciliation (Lu et al., 2019).

Both GemC1 and MCIDAS are paralogues that share the common ancestor geminin (Gmnn), another coiled-coil containing protein. However, unlike GemC1 and MCIDAS, Gmnn lacks the C-terminal TIRT domain, suggesting that it lacks the ability to form a complex similar to the EDM or EDG trimeric complexes (Arbi et al., 2018; Terre et al., 2016). Gmnn was originally identified to be important in regulating DNA replication and neural cell differentiation in two independent *Xenopus* screens. Additional studies revealed that Gmnn helps facilitate embryonic patterning, where it spatially restricts mesoderm, endoderm, and non-neural ectoderm to their proper embryonic locations (Kroll, Salic, Evans, & Kirschner, 1998; McGarry & Kirschner, 1998; Seo & Kroll, 2006). In line with this idea, misexpression of Gmnn in gastrula ectoderm disrupts tissue patterning, resulting in the prospective epidermis being converted into neural tissue (Kroll et al., 1998). Dynamic regulation of Gmnn expression levels also appears to be important for proper development. Gmnn is initially highly expressed in proliferating cells, but a decrease in expression is observed as the MCC differentiation program progresses. Ectopic overexpression of Gmnn represses many genes associated with cell commitment and differentiation and enhances the expression of genes that maintain pluripotency, further highlighting the importance of proper expression levels for MCC differentiation (Arbi et al., 2016; Lim, Hummert, Mills, & Kroll, 2011). Later studies revealed that Gmnn inhibits MCC development by forming heterodimers with GemC1 and MCIDAS via their coiled-coil domain, resulting in a decrease in their transactivation activities (Fig. 2) (Caillat et al., 2015, 2013; Ma et al., 2014). Taken together, a large body of work has identified complex and intricate networks of transcription factors and signaling cascades that work cooperatively with the EDG and EDM complexes to initiate MCC specification and differentiation.

2.2.2 Downstream activators of MCC differentiation

Once upstream signaling cascades have identified and initiated MCC specification in a subset of cells, several downstream activators of MCC differentiation are required to complete the process. One such activator is the E2f family of transcription factors, which serves as a master regulator of MCC differentiation as part of the EDG and EDM complexes (Ma et al., 2014; Terre et al., 2016). The function of different E2f family members has been highlighted in several model organisms. Prominent roles for E2f4 have been identified in mice, as lack of this protein results in decreased MCCs and increased secretory cells in the airway epithelium (Danielian et al., 2007). Alternatively, in zebrafish, E2f5 has a more prominent role in the development of MCCs (Chong, Zhang, Zhou, & Roy, 2018; Xie et al., 2020). These data suggest that the importance of specific E2f family members may be organism-dependent. In *Xenopus*, both E2f4 and E2f5 are required for MCIDAS-induced gene expression. Expression of dominant negative E2f4 or E2f5 lacking the C-terminal transactivation domain does not affect formation of the EDM complex or commitment to an MCC fate, but rather inhibits centriole assembly in MCCs and decreases the expression of genes associated with centriole biogenesis (Azimzadeh, Wong, Downhour, Sanchez Alvarado, & Marshall, 2012; Ma et al., 2014). The importance of EDM-induced transcriptional regulation in promoting MCC differentiation is highlighted in mouse embryonic fibroblasts (MEFs), where ectopic expression of MCIDAS and/or E2f4 is not sufficient to induce centriole assembly. However, co-expression of MCIDAS with an E2f4 mutant that had its transactivation domain replaced with a generic VP16 transcription activation domain synergized to induce basal body docking and motile cilia formation in MEFs (Kim, Ma, Shokhirev, Quigley, & Kintner, 2018). This suggests that supplementary transcriptional activation activity is required, in addition to the EDM complex, to promote MCC differentiation in non-epithelial cell types. Interestingly, E2f4 has also been shown to have non-canonical functions that are essential for generating a ciliated epithelium. E2f4 contains a nuclear export signal and begins to accumulate in the cytoplasm and form apical cytoplasmic organizing centers (e.g., deuterosomes) for the assembly and nucleation of centrioles following the induction of MCC genes. Both these nuclear and cytoplasmic roles of E2f4 are required for the formation of MCCs (Mori et al., 2017).

Downstream of the EDM complex is the transcription factor myb, whose overexpression can also drive the MCC differentiation program (Fig. 2) (Tan et al., 2013). In the developing mouse airway epithelium, myb is transiently expressed during development, where expression increases during centriole amplification and then decreases as centrioles dock at the apical surface. Within these cells, loss of myb expression prevents basal cells from fully differentiating into ciliated, secretory, and mucus-secreting cells, suggesting that myb is required to direct basal stem cells into an intermediate progenitor state. However, in *Xenopus* it is not known if myb plays a general role in promoting epithelial stem cell differentiation or if it is specific to directing cells exclusively into an MCC fate (Pan et al., 2014; Tan et al., 2013).

Foxj1 is a downstream target of myb and is required to activate the motile cilia pathway (Fig. 2). Loss of Foxj1 inhibits ciliogenesis in multiple organisms and tissues, including zebrafish Kupffer's vesicle, *Xenopus* gastrocoel roof plate, and the MCCs in *Xenopus* skin (Stubbs, Oishi, Izpisua Belmonte, & Kintner, 2008; Tan et al., 2013; Yu, Ng, Habacher, & Roy, 2008). Foxj1 is also required for the apical localization and docking of basal bodies in MCCs and the induction of proteins required for axoneme assembly, which is essential for proper ciliary function (You et al., 2004). This is in contrast to phenotypes observed with loss of MCIDAS, which prevents early stages of MCC differentiation and centriole amplification. When MCIDAS is overexpressed in cells lacking Foxj1, centriole amplification is induced, but ciliogenesis is severely disrupted, demonstrating the importance of Foxj1 in the induction of cilia outgrowth (Stubbs et al., 2012). More recent work has revealed that Foxj1 is not the only forkhead box protein that is important in regulating MCC formation. Foxn4 is an early target of the EDM complex and is upregulated prior to Foxj1. However, unlike the requirement of Foxj1 for the formation of motile cilia, Foxn4 is necessary for the correct timing of basal body docking and cilia extension (Campbell, Quigley, & Kintner, 2016).

Stabilization of Foxj1 at MCC promoters requires the transcription factor Rfx2, which is preferentially expressed in MCCs and targets many genes involved in ciliogenesis, cilia motility, and planar cell polarity.

Dimerization of Rfx2 acts as a scaffold to bring sites bound by Foxj1 into close proximity and induce coordinated cilia gene expression. This co-binding of Rfx2 and Foxj1 at the promoters of MCC genes leads to potent gene activation, whereas other combinations of transcription factors (e.g., Rfx2/E2f4 and Rfx2/myb) play more minor roles in this process (Chung et al., 2014, 2012; Quigley & Kintner, 2017). Loss of Rfx2 does not affect the expression of α-tubulin, but results in MCCs that contain only a few short axonemes extending from the surface. Therefore, Rfx2 is not required for the initial steps of MCC specification, but rather the formation of motile cilia assembly (Chung et al., 2012). In human and mouse ciliated epithelium, Rfx3 plays a similar role to Rfx2 and is required for sufficient expression of Foxj1 and the resultant generation of motile cilia, indicating a role for multiple Rfx family members in MCC maturation (Baas et al., 2006; Didon et al., 2013; El Zein et al., 2009). Interestingly, Rfx2 also induces genes involved in cell motility, which is essential for the process of radial intercalation, a key step in MCC maturation that occurs after MCC specification and differentiation.

3. Radial intercalation in the *Xenopus* epithelium

The process of radial intercalation is quite complex and includes activation of various signaling pathways, coordinated cellular movements, dynamic changes to cytoskeletal networks, and changes in cell-cell adhesion. While intercalation is an elaborate and coordinated event, we can divide the process into three stages: apical migration, apical insertion, and apical expansion (Fig. 3). During the initial stage of apical migration, intercalating cells located below the outer epithelial layer become polarized and extend dynamic protrusions that probe the surrounding environment, which allows these cells to begin wedging between the cell-cell junctions of the outer epithelial layer (Stubbs, Davidson, Keller, & Kintner, 2006). While cells are able to initiate wedging between outer cells at any point along the cell-cell junctions, intercalating cells insert into the outer epithelial layer almost exclusively at cell vertices where three or more outer cells meet (Stubbs et al., 2006). During the process of apical insertion, the intercalating cell establishes a small apical domain within the outer epithelium (Collins, Majekodunmi, & Mitchell, 2020; Sedzinski, Hannezo, Tu, Biro, & Wallingford, 2016) followed by a period of apical expansion, where the apical area of the cell grows until it reaches its final size (Sedzinski et al., 2016).

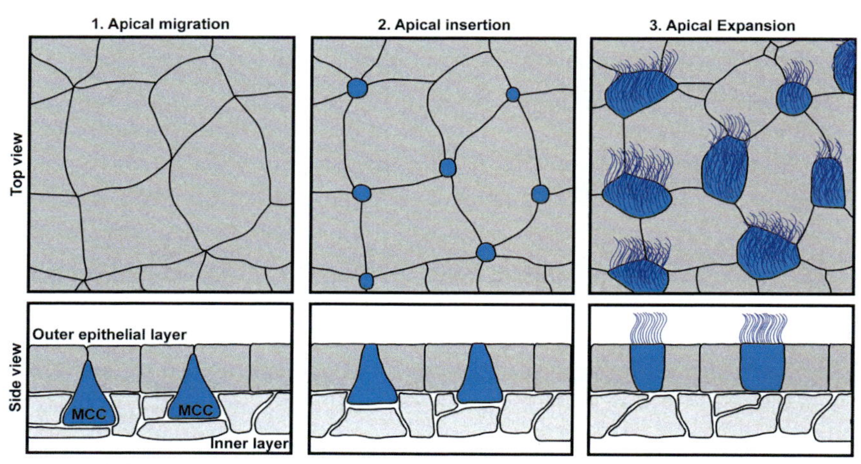

Fig. 3 Radial intercalation in *Xenopus* skin. Top and side views of the three stages of radial intercalation: (1) apical migration (left), differentiated MCCs in the inner layer migrate toward the outer epithelial layer and begin wedging between outer cells; (2) apical insertion (middle), MCCs establish a small apical domain in the outer epithelial layer; (3) apical expansion (right), MCCs expand their apical area and fully incorporate into the outer epithelium.

In addition to influencing MCC specification, perturbations in Notch signaling also indirectly influence intercalation. Inhibiting Notch activity in the inner ectodermal layer dramatically increases the number of MCCs and ICs that differentiate within the inner layer (Stubbs et al., 2006). One could predict that this would lead to additional MCCs and ICs intercalating into the outer epithelium, but surprisingly, there is only a slight increase in the number of MCCs that intercalate. As a result, MCCs incorporated into the outer layer remain stereotypically spaced, with many MCCs remaining trapped below the outer layer (Stubbs et al., 2006). In contrast, there is a robust increase in the number of ICs that intercalate into the outer epithelium, with many ICs intercalating adjacent to one another. One possible explanation for this observation is the size discrepancy between the two cell types. MCCs are much larger in area and may have packing limitations when it comes to their ability to penetrate available outer cell vertices, while ICs are columnar in shape, permitting them to squeeze into physically constrained spaces. While additional studies may provide further insight into how sites of MCC and IC intercalation are defined, these data clearly demonstrate that manipulating MCC and IC specification within the inner layer can influence the population of cells that are able to intercalate into the outer layer. Many other signaling pathways and cellular components that influence radial intercalation have been identified and are described in detail below.

3.1 The role of cytoskeletal networks during radial intercalation

3.1.1 Actin cytoskeleton

The actin cytoskeleton influences many aspects of cellular behavior, including motility (Mitchison & Cramer, 1996), adhesion (Cavey & Lecuit, 2009; Zhang et al., 2005), and generation/transmission of cellular forces (Kovar & Pollard, 2004; Mogilner & Oster, 2003), all of which are involved in radial intercalation. Recent genomic analysis of the cilia-associated transcription factor Rfx2 identified many cytoskeletal proteins as part of the Rfx2 transcriptome, and Rfx2 morphant MCCs exhibit intercalation defects, suggesting a prominent role for cytoskeletal elements in this process (Chung et al., 2014). Indeed, several studies have highlighted essential roles for the actin cytoskeleton in various stages of intercalation.

One of the earliest steps of MCC radial intercalation is the massive amplification of centrioles adjacent to the nucleus, which then migrate apically to become positioned between the nucleus and apical edge of the cell (Klos Dehring et al., 2013). Apical migration and docking of centrioles, which become the basal bodies that template the 100+ cilia in mature MCCs, is required for proper intercalation and MCC function. Work in other systems has demonstrated that the apical migration of this cluster of centrioles relies on dynamic changes in the actin cytoskeleton (Boisvieux-Ulrich, Laine, & Sandoz, 1990), and disruption of actin dynamics in *Xenopus* MCCs affects apical migration of centrioles and basal body docking (Ioannou, Santama, & Skourides, 2013; Kulkarni, Griffin, Date, Liem, & Khokha, 2018). Specifically, the highly conserved ATPase nucleotide binding protein 1 (Nubp1) has been implicated in centriole duplication (Christodoulou, Lederer, Surrey, Vernos, & Santama, 2006) and was demonstrated to play a role in MCC intercalation. Depletion of Nubp1 disrupts the subapical actin network (which contributes to apical migration of centrioles) and the apical actin network (required for docking and patterning of basal bodies), and ultimately leads to defects in intercalation (Ioannou et al., 2013). These findings suggest that multiple pools of actin contribute to early events required for proper MCC intercalation.

Once MCCs emerge in the outer epithelium, cells expand their apical domain circumferentially to reach their final apical size. This process undoubtedly involves force generation and transmission of cellular forces between the intercalating cell and the neighboring outer cells. However, until recently, the mechanical cues regulating this process were not well understood. One possible model of apical expansion could be that as the intercalating cell emerges into the outer epithelium, the surrounding outer cells apply tension by pulling on the apical membrane of the intercalating cell to drive apical expansion. However, when this idea was tested, the data did not support a neighbor pulling model (Sedzinski et al., 2016). Another possible model of apical expansion involves cell autonomous pushing forces generated within the intercalating MCC. Theoretical modeling of this hypothesis predicts that a rounded cell shape would emerge if MCC pushing forces drive apical expansion. Indeed, when the shapes of intercalating MCCs are tracked by time-lapse imaging, cell shapes closely mirror those predicted by MCC-generated pushing forces (Sedzinski et al., 2016). This is not to say that pushing forces generated within the MCC are the only forces involved in apical expansion, as pushing forces must be met with some resistance. Emerging MCCs are mechanically linked to the surrounding outer cells via cell-cell adhesion complexes (such as adherens junctions), which interact with intracellular pools of actin within the outer epithelial cells, and it is known that there is tension at cell-cell junctions within a layer of epithelial cells (Charras & Yap, 2018; Harris, Daeden, & Charras, 2014). Laser microdissection of perpendicular junctions surrounding emerging MCCs revealed that junctional tension of the outer epithelial cells is low during the bulk of apical expansion, but increases near the end of the process (Sedzinski et al., 2016). Thus, the current model of force dynamics during apical emergence suggests that low junctional tension of surrounding outer cells and 2D forces generated within an MCC drives apical expansion to an area beyond the critical point of collapse. After reaching the critical stage, junctional tension in the neighboring outer cells increases to help shape the mature MCC (Fig. 4A). One possible source of force generation that may provide pushing forces required for apical expansion is the actin cytoskeleton. In agreement with this idea, there is a robust positive correlation between apical actin concentration and the expanding apical domain (Sedzinski et al., 2016). In addition, the Rfx2 target and regulator of actin polymerization, formin-1, is also enriched at the apical domain of expanding MCCs, and disruption of formin-1 expression or function results in a defect in apical expansion (Fig. 4B) (Sedzinski et al., 2016). While these

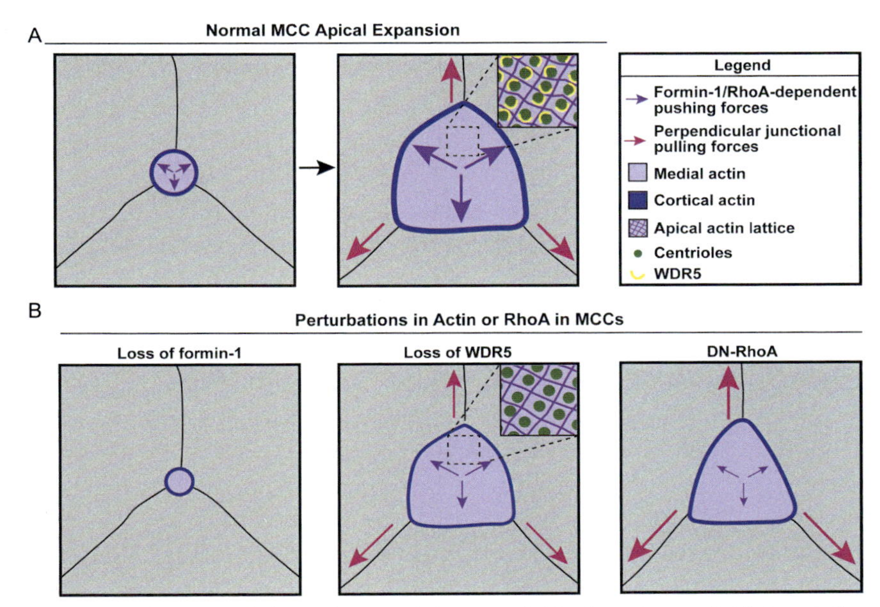

Fig. 4 Contributions of the actin cytoskeleton and RhoA during radial intercalation. (A) During normal MCC apical expansion, RhoA activity and formin-dependent actin polymerization generate initial MCC pushing forces. After reaching a critical apical area, junctional tension in the neighboring outer cells increases to help shape the mature MCC, and WDR5 stabilizes apical actin to facilitate proper expansion. (B) Perturbations in actin and RhoA disrupt MCC expansion. Loss of formin-1 diminishes MCC pushing forces, resulting in apical domain collapse (left), loss of WDR5 leads to increased apical actin turnover and decreased actin stability, resulting in decreased expansion and a smaller apical domain (middle), and expression of a DN-RhoA in MCCs results to a smaller apical area due to decreased MCC pushing forces and greater contribution of perpendicular junctional pulling forces.

data suggest a role for formin-mediated actin dynamics during apical expansion, there are many other regulators of actin dynamics and organization that likely also play a role in this process.

The chromatin modifier WDR5 was recently identified as a regulator of actin dynamics during radial intercalation. WDR5 is a core component of the human histone H3-Lys4-methyltransferase (H3K4MT) complexes involved in chromatin modification and regulation of transcription (Trievel & Shilatifard, 2009; Wysocka et al., 2005). WDR5 was first identified as a potential regulator of MCC function through genetic analyses of congenital heart disease and heterotaxy (Zaidi et al., 2013), which revealed associated respiratory issues, a consequence of MCC dysfunction in the airway epithelia (Kennedy et al., 2007; Swisher et al., 2011). Consistent with the respiratory issues found in human genetic analyses, WDR5 depletion in *Xenopus* embryos disrupts fluid flow over embryos, indicative of ciliary dysfunction. Interestingly, studies using truncated WDR5 mutants revealed an H3K4MT-independent function for the protein in MCCs (Kulkarni et al., 2018). WDR5-depleted embryos exhibit a significant decrease in MCC apical area and loss of actin accumulation and stability, suggesting a potential for decreased pushing forces required for apical expansion (Fig. 4B) (Kulkarni et al., 2018). Localization analyses revealed that WDR5 co-localizes with the cluster of centrioles in MCCs as they apically migrate prior to apical insertion and are later positioned between basal bodies and the surrounding actin lattice at the apical surface in fully intercalated MCCs (Fig. 4A). Thus, WDR5 has been proposed to act as a bridge, providing a physical link between basal bodies and actin, but binding assays to confirm physical interactions of these proteins have not yet been performed. In summary, recent work has highlighted that actin stabilization and actin-dependent force generation are critical for proper expansion of intercalating MCCs. While these studies provide insight into the essential roles of the actin cytoskeleton during radial intercalation, additional analyses are needed to provide further mechanistic insights into actin dynamics and cell mechanics during this complex process.

3.1.2 Microtubule cytoskeleton

Microtubules (MTs) comprise the structural component of cilia and are well characterized for their roles in cell polarity and migration (Bouchet & Akhmanova, 2017; Etienne-Manneville, 2013). MTs are inherently polarized structures defined by a dynamic fast growing (+) end and a more stable slow growing (−) end (Hendershott & Vale, 2014; Jiang et al., 2014; Mitchison & Kirschner, 1984). In most migrating cells, the centriole-derived MT organizing center (MTOC) positioned near the nucleus creates a polarized array of MTs with the (+) ends oriented toward the leading edge. This polarization has been proposed to be important for trafficking of vesicles, organelles, and signaling molecules toward the leading edge and establishes directional migration and persistence (Etienne-Manneville, 2013). MCC intercalation also requires directional migration and multiple polarity cues. As an intercalating cell is incorporated into the outer epithelium, proper apicobasal polarity and establishment of new cell-cell adhesions is essential to keep the epithelial layer intact.

Apical localization and function of the Par polarity components Par3, Par6, and atypical protein kinase C (aPKC) are also essential for MCC intercalation. The Par complex influences many aspects of cell polarity and has previously been implicated in centriole positioning between the nucleus and leading edge during cell migration (Gomes, Jani, & Gundersen, 2005; Schmoranzer et al., 2009). Apical localization of centrioles in MCCs is also mediated by the Par complex and is essential for proper intercalation (Fig. 5A). Par components are enriched at the apical (leading) edge of intercalating MCCs, and defects in Par function (via expression of DN-Par3 or a

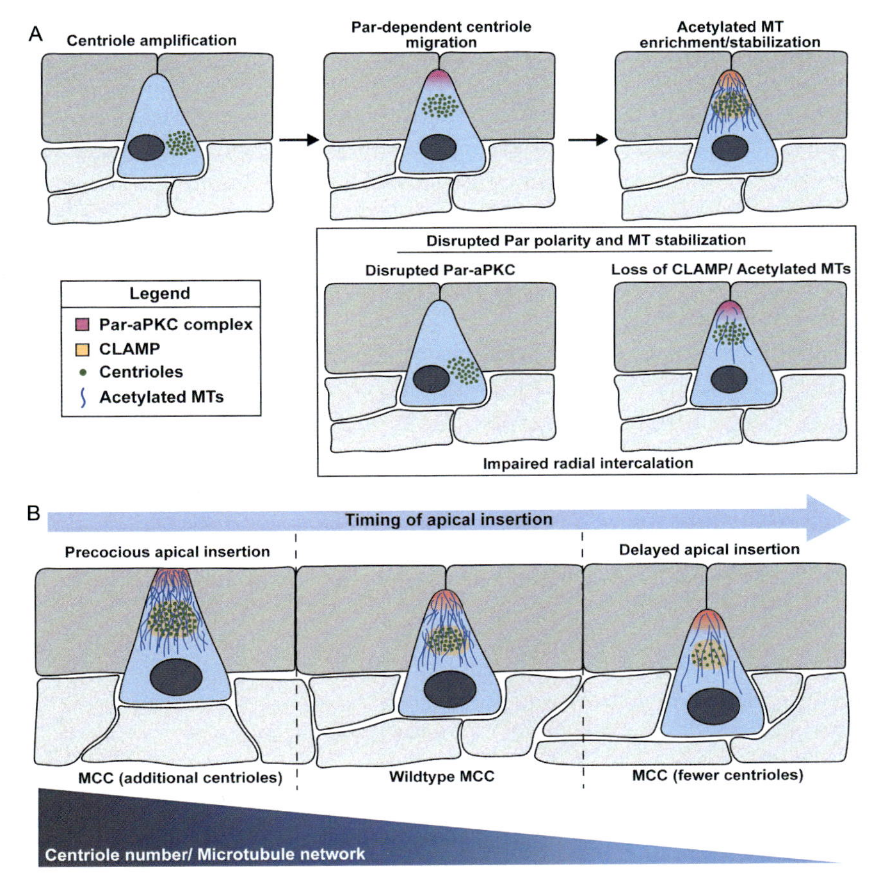

Fig. 5 Par- and MT-dependent regulation of radial intercalation. (A) The Par-aPKC complex mediates apical positioning of centrioles and accumulation of the MT-binding protein CLAMP, which leads to apical enrichment of stabilized MTs in the axis of apical migration. Disruptions in Par-aPKC function or MT stability results in intercalation defects (boxed inset). (B) The number of centrioles in an intercalating MCC influences MT accumulation and timing of apical insertion. MCCs with extra centrioles exhibit increased MT density and undergo precocious apical insertion, whereas MCCs with fewer centrioles have a reduced MT network and exhibit delayed apical insertion.

kinase-dead aPKC) disrupt the apical localization of centrioles and results in defects in intercalation (Werner et al., 2014). Wildtype MCCs exhibit a robust apical MT network, which is significantly decreased in Par-defective cells, suggesting MTs are crucial during the radial intercalation process. In line with this hypothesis, pharmacological inhibition of MT polymerization significantly impairs intercalation (Werner, Del Castillo, Ventrella, Brotslaw, & Mitchell, 2018; Werner et al., 2014). The apical accumulation of MTs in MCCs is dependent upon the MT-stabilizing protein CLAMP/Spef1, which directly binds to MTs (Chan, Fowler, Choo, & Kalitsis, 2005; Dougherty et al., 2005). In *Xenopus*, CLAMP morphant MCCs exhibit a significant reduction in MT accumulation and defects in intercalation (Werner et al., 2014). Of note, the same defects in intercalation are also observed in ICs, further highlighting the importance of MTs in multiple intercalating cell types. Taken together, these data support a model in which the Par complex defines the orientation of apical migration during intercalation, and apical enrichment of CLAMP promotes an axis of enriched MTs that helps drive migration and intercalation of MCCs into the outer epithelial layer.

More recent work has indicated that the number of centrioles in an intercalating cell also influences MT accumulation and intercalation. MCCs contain approximately 150 centrioles which maintain an extensive MT network, whereas ICs contain two centrioles and have a much less robust network. This has provided an ideal system to examine how the number of centrioles and MTs in an intercalating cell facilitates radial intercalation. Recent single cell RNASeq data has demonstrated that MCCs and ICs differentiate at the same developmental stage, several stages prior to the onset of intercalation (Briggs et al., 2018; Nieuwkoop & Faber, 1994). However, MCCs initiate and complete radial intercalation prior to ICs, with a clear disparity in the timing of apical insertion into the outer epithelial epithelium (Collins et al., 2020). These data suggest that the number of centrioles and MTs in an intercalating cell may influence the timing of apical insertion (Fig. 5B). Indeed, increasing the number of centrioles in MCCs or ICs increases the MT network and results in precocious intercalation, while decreasing the number of centrioles in MCCs decreases the MT network and delays apical insertion (Collins et al., 2020). While MTs appear to show some polarity, with (+) ends enriched at the apical edge of intercalating cells, ectopic expression of MTs with the (−) end targeted to the apical membrane also results in precocious intercalation, indicating that an increase in stabilized MTs can accelerate apical insertion independent of orientation (Collins et al., 2020).

There is clearly a role for Par polarity cues in establishing an enrichment of stabilized MTs in intercalating cells, but planar cell polarity (PCP) cues are required to coordinate cell behavior across the 2D plane of the epithelial layer (Zallen, 2007). Core components of the PCP pathway, Vangl2 and Prickle3 (Pk3, the predominant prickle family member in *Xenopus* skin), localize to the apical surface of intercalating MCCs, and inhibition of Vangl2 or Pk3 expression or function significantly impair MCC intercalation (Ossipova et al., 2015). Disheveled (Dvl), a key component of both the canonical Wnt and PCP signaling pathways, also influences MCC intercalation. Expression of *Xenopus* Dvl constructs Xdd1 (which acts as a dominant negative) and Xdd2 (a C-terminus deletion mutant) (Sokol, 1996) disrupt Dvl signaling and impair MCC intercalation, supporting a role for Dvl signaling in this process (Ossipova et al., 2015). However, other reports have not observed the same intercalation defects with Dvl mutants, suggesting more work on this question is warranted (Mitchell et al., 2009; Park, Mitchell, Abitua, Kintner, & Wallingford, 2008). Interestingly, the MT-based kinesin KIF13B was identified as a Dvl interactor, and KIF13B morphant MCCs also exhibit defects in intercalation (Ossipova et al., 2015). These findings suggest that MT-based trafficking and PCP signaling may work together to influence intercalation, and it is possible that kinesin-based trafficking may be involved in establishing the apical localization of core PCP components.

The exact mechanisms of how accumulation of MTs influence apical insertion are not entirely understood. MT-based trafficking has been shown to be important for intercalation of MCCs, and depletion of the MT-based kinesin KIF13B impairs MCC migration during intercalation (Ossipova et al., 2015). If MT-based trafficking was the only mechanism that influenced apical insertion, one would predict proper directionality of the MTs to be an essential condition of MT enrichment. However, accumulation of MTs accelerates apical insertion independent of MT directionality, arguing against MT-based trafficking as the sole mechanism responsible for apical insertion (Collins et al., 2020). It is possible that accumulation of MTs may provide structural stability and mechanical resilience to an intercalating cell emerging among mechanically linked outer cells. Additional studies need to be done in order to dissect the numerous functions of the MT networks that could be contributing to radial intercalation.

3.1.3 Coordination of actin and MT dynamics

While the roles of the actin cytoskeleton and MT cytoskeleton are often discussed separately, these cytoskeletal networks almost certainly work in concert to mediate proper radial intercalation. During early steps of radial intercalation, both actin and MTs mediate apical migration of centrioles in MCCs and defects in apical localization of centrioles impairs basal body docking (Ioannou et al., 2013; Ossipova et al., 2015; Werner et al., 2014). Upon apical localization of centrioles, there is an enrichment of stable, acetylated MTs and actin cables that link from the cluster of centrioles to the cell cortex (Ioannou et al., 2013; Werner et al., 2014). Accumulation of MTs, which is dependent on the number of centrioles in the intercalating cell and stabilization via polarity cues and MT-associated proteins, promotes apical insertion into the outer epithelium (Collins et al., 2020; Werner et al., 2014). Apical expansion requires accumulation and stabilization of the actin cytoskeleton at the apical domain, which is thought to generate pushing forces within the intercalating cell and drive growth of the apical surface (Kulkarni et al., 2018; Sedzinski et al., 2016). While it is possible that MTs contribute to apical expansion, work thus far has only highlighted the role for the actin cytoskeleton in this process.

Furthermore, the role of these cytoskeletal networks in the context of radial intercalation during development has been of particular focus here, but it must also be noted that highly specialized cells, such as MCCs, have very elaborate cytoskeletal networks that are essential for proper function in mature cells. A number of studies have demonstrated both actin and MTs to be essential for proper basal body spacing, polarization, and metachronal synchrony, all of which are required for proper ciliary beating and function (Kulkarni et al., 2018; Mahuzier et al., 2018; Werner et al., 2011). Interestingly, a similar role has also been established for the focal adhesion-associated proteins FAK, paxillin, and vinculin. These proteins also appear to form “ciliary adhesion” complexes at basal bodies in MCCs, facilitating linkage between basal bodies and the actin cytoskeleton (Antoniades, Stylianou, & Skourides, 2014).

In addition, proteins involved in actin and MT regulation during intercalation have also been shown to influence mature function of MCCs. For example, WDR5, which influences actin organization and accumulation during apical expansion, also affects the planar distribution of basal bodies across the apical cell membrane and coordinated orientation of basal bodies and rootlets (Kulkarni et al., 2018). Not surprisingly, these defects in basal body patterning lead to defects in ciliary-induced flow. In addition, the MT-associated protein CLAMP, which is critical for enriching MTs at the apical edge of intercalating MCCs (Werner et al., 2014), is also required for establishing both cell and cilia polarity in mature MCCs

(Kim, Zhang, et al., 2018). Furthermore, basal bodies at the apical domain of mature MCCs are linked to both MTs and actin, which promotes location coordination of cilia orientation and metachronal cilia beating, respectively (Werner et al., 2011).

Finally, most studies investigating actin and MTs in intercalation have focused on the role of these cytoskeletal networks in MCCs. However, other cell types, including ICs and SSCs, also undergo radial intercalation and do not have the same enrichment of apical actin and MTs that is observed in MCCs. Pharmacological inhibition of MT polymerization also results in intercalation defects in ICs, indicating that MTs contribute to proper intercalation of multiple cell types (Werner et al., 2014). The actin cytoskeleton also undoubtedly plays a role in IC and SSC intercalation. However, the apical actin network is much less robust in non-ciliated cells, leading to questions of how force generation and apical expansion may occur in other cell types.

3.2 Role of small GTPases during radial intercalation

Small GTPases regulate signaling involved in numerous cellular processes, such as cytoskeletal dynamics, migration, and intracellular trafficking. Here, we highlight roles for Rho GTPases and Rab GTPases, known for their roles in regulating cytoskeletal networks and trafficking, respectively, during MCC intercalation.

3.2.1 RhoA

RhoA is a master regulator of the actin cytoskeleton and influences actin dynamics by regulating the activity of actin nucleating proteins (Hall, 1998). Given that changes in actin dynamics play a key role in mediating radial intercalation, it is not surprising that RhoA is also essential for the intercalation process. As detailed previously, the elaborate actin network at the apical domain of MCCs is essential for basal body docking, planar polarization, and apical expansion, and RhoA plays a role in all of these processes.

Similar to the accumulation of actin at the apical MCC surface during expansion, active RhoA also accumulates throughout the process (Sedzinski, Hannezo, Tu, Biro, & Wallingford, 2017). Interestingly, RhoA activity seems to influence the localization of the actin nucleator formin-1 (and subsequent actin accumulation) at the MCC apical surface. In control MCCs, formin-1 is found evenly distributed throughout the apical domain, whereas expression of dominant negative (DN)-RhoA results in accumulation of formin-1 and actin in the central region of the apical domain and coincides with an overall decrease in actin assembly and increased actin turnover (Sedzinski et al., 2017). Expression of DN-RhoA in MCCs also slows the rate of apical expansion and results in a smaller final apical area (Sedzinski et al., 2017). It has been proposed that 2D pushing forces generated within MCCs drive apical expansion, and these data suggest that RhoA is also important for this force generation. In line with this idea, MCCs with impaired RhoA function exhibit iterative pulses of contraction and constriction during apical expansion, suggesting that MCCs with impaired RhoA function are not able to generate sustained pushing forces required to overpower the resisting forces of the surrounding outer cells (Sedzinski et al., 2017). Furthermore, MCCs normally have a rounded cell shape during apical emergence, indicative of MCC-generated pushing forces as the predominant force during that phase of expansion. However, MCCs expressing DN-RhoA exhibit a more polygonal shape, which suggests a decrease in MCC pushing forces and increased influence of junctional tension of neighboring outer cells (Fig. 4B) (Sedzinski et al., 2017). Taken together, these data suggest that active RhoA regulates actin accumulation and formin-1 activity, both of which contribute to 2D pushing forces that drive expansion of the MCC apical domain. It is possible that RhoA regulates the activity of other actin nucleators (in addition to formin-1) to influence apical expansion, but additional studies are needed for a more complete understanding of how RhoA signaling within MCCs influences pushing forces and expansion.

RhoA activity in outer epithelial cells also influences radial intercalation. Expression of constitutively active (CA)-RhoA in outer cells surrounding an intercalating MCC results in slower apical expansion of the MCC and a smaller apical size (Sedzinski et al., 2016). In contrast, decreasing RhoA activity in outer cells results in accelerated MCC apical expansion and a larger final apical area. Other work has demonstrated that transient flares of RhoA activity are important for epithelial barrier integrity when tight junctions become locally compromised due to changes in cell shape and tissue tension (Stephenson et al., 2019). Given that MCC intercalation requires dynamic changes in epithelial cell shape, tension, and establishment of new cell-cell junctions, it is possible that localized RhoA flares facilitate proper intercalation and incorporation into the outer epithelial layer. However, a role for RhoA flares has not yet been directly tested in this context. Nevertheless, our current understanding of RhoA function during intercalation indicates that precise regulation of RhoA activity in both MCCs and outer cells is essential force generation and balance required during radial intercalation.

Other proteins that have demonstrated roles in radial intercalation have also been shown to influence RhoA localization and/or activity during intercalation. For example, embryos depleted of the nucleotide binding protein Nubp1 (which influences actin accumulation at the apical domain) also display disrupted RhoA localization (Ioannou et al., 2013). RhoA activity also coordinates other processes during radial intercalation and maturation of MCCs, including apical migration and docking of basal bodies, as well as planar polarization (Pan, You, Huang, & Brody, 2007; Park et al., 2008). It is likely that RhoA additionally contributes to early stages of intercalation (apical migration/apical insertion) that also require cellular motility and dynamic changes to the actin cytoskeleton, but studies addressing RhoA regulation during these stages have not yet been performed leaving gaps in our current understanding of RhoA signaling throughout the entire intercalation process.

3.2.2 Other Rho GTPases (Rac/Cdc42)

Also included in the Rho family of GTPases are Rac1 and Cdc42, which elicit their own distinct effects on cytoskeletal dynamics during various biological processes (Hall, 1998). These GTPases almost certainly play significant roles in various aspects of radial intercalation. However, the roles of both Rac1 and Cdc42 during intercalation remain largely unexplored at this time and are ripe for discovery. There are significant reasons to believe that Cdc42 activity is essential for apical insertion and expansion, given the connection of Cdc42 to the Par polarity complex (Etienne-Manneville, 2004). In its active GTP-bound form, Cdc42 is able to physically interact with Par6, which promotes activation of aPKC and leads to the establishment of various polarity cues (Etienne-Manneville, 2004; Garrard et al., 2003). Of note, Cdc42-dependent regulation of the Par-aPKC polarity complex is involved in the orientation of centrioles and MTs in migrating cells and helps to determine migratory polarity and persistence (Etienne-Manneville, 2004; Ridley, 2015). Therefore, it is attractive to hypothesize that Cdc42 helps establish the polarized enrichment of the Par complex and MT accumulation at the apical edge of the intercalating cells as they migrate toward the outer epithelium (Werner et al., 2014). However, the role of Cdc42 has not yet been investigated in this context and additional studies are needed to test this hypothesis.

As previously mentioned, the apical actin network in MCCs is highly specialized and complex (Ioannou et al., 2013; Kulkarni et al., 2018; Sedzinski et al., 2016; Werner et al., 2011), which suggests that many regulators of actin dynamics are required for establishing the robust actin network, and Rac1 is a top candidate for future studies. Of note, Rac1 activity has been shown to be important for maturation of MCCs. The unconventional bipartite guanine nucleotide exchange factor (GEF) ELMO-DOCK1 activates Rac1 and contributes to proper basal body docking and spacing during MCC maturation (Epting et al., 2015). Thus, Rac1 activity is known to be important for MCC function, but potential roles for the GTPase during intercalation have not yet been explored.

3.2.3 Rab GTPases

Rab GTPases are the largest family of small GTPases and primarily function to regulate intracellular and vesicle-mediated transport (Zhen & Stenmark, 2015). Specificity and directionality of membrane trafficking pathways are achieved through activation of distinct Rab GTPases. One such Rab GTPase, Rab11, is involved in establishing apicobasal polarity in many systems and facilitates cell-cell adhesion via recycling of E-cadherin (Desclozeaux et al., 2008; Ossipova et al., 2014; Roeth, Sawyer, Wilner, & Peifer, 2009). This led to the hypothesis that Rab11 may be involved in radial intercalation in *Xenopus*, as intercalation requires establishment of new cell-cell junctions and polarity cues to maintain a functional epithelial layer. Rab11 is enriched in MCCs and disruption of its function (through expression of DN-Rab11 or Rab11 morpholino) impairs the ability of MCCs to intercalate into the outer epithelial layer (Kim, Lake, Haremaki, Weinstein, & Sokol, 2012). Interestingly, the MCCs that remain trapped under the outer epithelial layer and fail to intercalate are still able to establish ZO-1 and aPKC-positive domains, suggesting that they have traditional apical properties. However, in these cells, the localization of the cluster of centrioles and aPKC-positive domains are randomly oriented relative to the outer cells (compared to control cells with centrioles and aPKC oriented toward the outer cells). The presence of aPKC-positive domains (a hallmark of apical polarity) suggests that Rab11-dependent trafficking may not be required for establishing apicobasal polarity in MCCs, but instead contributes to polarity required for orienting migration and incorporation into the outer epithelial layer (as these cells are not able to properly orient within the tissue) (Kim et al., 2012). In addition, Rab11 morphant MCCs that are able to intercalate exhibit a smaller apical area, suggesting there may be a defect in apical expansion as well. However, how Rab11 may contribute to apical expansion remains unknown at this time. Interestingly, ICs expressing DN-Rab11 appear to intercalate at levels similar to control ICs, suggesting that Rab11 function might be MCC-specific during intercalation (Kim et al., 2012). As there are at least 60 unique Rab GTPases, future studies investigating the roles of other Rab family members or perhaps cell-type specific roles for distinct Rab family members, would greatly add to our current understanding of specific trafficking pathways required during radial intercalation.

3.3 Interactions with the surrounding microenvironment (cell-ECM/cell-cell interactions)

3.3.1 Cell-ECM interactions

The extracellular matrix (ECM) provides important mechanical and signaling cues to surrounding cells in the tissue via interactions with cellular adhesion receptors (Tsang, Cheung, Chan, & Cheah, 2010). Many distinct extracellular glycoproteins contribute to ECM composition, including fibronectin, collagen, and laminin (Van Agtmael & Bruckner-Tuderman, 2010). While integrins are the best-characterized cell-ECM adhesion receptors, other adhesion receptors also interact with ECM proteins. The laminin receptor dystroglycan (Dg) is abundantly expressed in epithelial cells and has been shown to influence radial intercalation in *Xenopus* skin. Interestingly, Dg expression is influenced by Rfx2 and Notch activity, two transcription factors that have well-defined roles in MCC specification and radial intercalation (Chung et al., 2014; Sirour et al., 2011). Dg is highly expressed in the inner (sensorial) layer of the *Xenopus* epidermis, but is not expressed in MCCs (Sirour et al., 2011). However, loss of Dg expression disrupts MCC intercalation, suggesting that the adhesion receptor influences intercalation in a non-cell-autonomous manner (Sirour et al., 2011). Closer examination revealed depletion of Dg in *Xenopus* embryos results in loss of laminin, disruption of fibronectin organization, and defects in E-cadherin localization at cell-cell contacts (Sirour et al., 2011). Changes in the ECM microenvironment and cell adhesion molecules leads to many MCCs being trapped under the outer epithelial layer and MCCs that are able to intercalate exhibit a smaller apical domain and irregular spacing (Sirour et al., 2011). The contributions of Dg-dependent activation of adhesion signaling cascades and Dg-dependent ECM deposition during this process remain unclear. However, these results highlight the ability of cellular interactions with the surrounding microenvironment to influence cell adhesion and cellular behaviors required for intercalation.

Relatively little is known about how other ECM components or cellular ECM adhesion receptors may influence intercalation. As previously mentioned, integrins are the best-characterized family of ECM receptors and integrin function and fibronectin assembly are required for tissue morphogenesis and cell movements in *Xenopus* during developmental processes prior to radial intercalation in the skin (Marsden & DeSimone, 2001). It is possible that integrin function and other ECM components contribute to MCC intercalation, but additional studies need to be performed in order to identify specific roles for integrins and other adhesion receptors.

3.3.2 Cell-cell interactions

Radial intercalation requires dynamic changes in cell-cell adhesion as intercalating cells are incorporated into the outer epithelial layer. Therefore, cell-cell adhesion complexes, including cadherin-based adherens junctions, tight junctions, and desmosomes, must be dynamically regulated. Yet mechanisms regulating cell-cell adhesion and the contributions of various adhesion complexes remains largely unknown. As previously mentioned, Rab11-dependent trafficking has been shown to influence radial intercalation of MCCs (Kim et al., 2012). Previous work has demonstrated that Rab11-dependent trafficking of E-cadherin contributes to cell polarity during epithelial cell morphogenesis and stability of adherens junctions in other systems (Desclozeaux et al., 2008; Roeth et al., 2009). Disruptions in cell-ECM interactions resulting from loss of Dg expression also disrupts E-cadherin localization at cell contacts and appears to result in defects in cell-cell adhesion (Sirour et al., 2011).

Recent work highlighted the role of the desmosomal protein desmoplakin (Dsp) in radial intercalation. Within desmosomes, Dsp interacts with desmosomal cadherins and intermediate filaments, thus linking the desmosome to the cytoskeleton (Bornslaeger, Corcoran, Stappenbeck, & Green, 1996). During intercalation, there appears to be a decrease in Dsp at outer epithelial cell vertices at the site of intercalation just prior to apical insertion, and once the intercalating cell apically emerges into the outer layer, Dsp becomes enriched at the junctions (Bharathan & Dickinson, 2019). Loss of Dsp expression impairs intercalation of MCCs, although the exact mechanism is not entirely understood. Dsp morphant MCCs that intercalate generally have a smaller apical area, suggesting impaired apical expansion. In line with this idea, loss of Dsp results in perturbations of the actin cytoskeleton, as well as the MT and intermediate filament networks (Bharathan & Dickinson, 2019). Furthermore, expression of a Dsp truncation mutant that is not able to associate with cytoskeletal filaments significantly impairs intercalation, suggesting that interactions with cytoskeletal filaments are important during intercalation. However, loss of Dsp is also associated with decreased E-cadherin at cell-cell junctions (Bharathan & Dickinson, 2019), which may signal disruption of overall cell-cell adhesion and contribute to intercalation deficits. Taken together, current studies highlight a potential role for cadherins during the process of intercalation, but additional studies are needed to identify the molecular mechanisms of cadherin-mediated adhesion during intercalation. In the future, it will be interesting to tease apart how Dsp and other proteins involved in cell adhesion coordinate dynamic changes in cell-cell adhesion and association with various cytoskeletal elements throughout radial intercalation.

4. Conclusions

The *Xenopus* ciliated epithelium is a powerful model system that permits the study of specification and maintenance of highly specialized cells, such as MCCs, and fundamental biological processes, such as radial intercalation. Recent advances in CRISPR/Cas9 genome editing and single cell RNA sequencing has greatly enhanced our ability to dissect complex signaling pathways and mechanisms required for MCC specification and function, which could have implications in human diseases characterized by ciliary dysfunction.

The ability to undertake complex, global analyses has also highlighted the interconnectedness of signaling involved in MCC specification and function. Notably, the transcription factor Rfx2, which works in concert with Foxj1 to facilitate transcription of MCC genes, also has transcriptional targets, such as Dg and the ECM protein slit 2, which are important for MCC intercalation into the outer epithelium (Chung et al., 2014; Medioni, Astier, Zmojdzian, Jagla, & Semeriva, 2008; Sirour et al., 2011; Wright et al., 2012). This finding highlights the possibility that MCC transcriptional cascades may have underappreciated roles in other key processes that are required for the formation of a functional ciliated epithelium and may be an area of particular interest moving forward.

While many advances in understanding MCC specification and function have been made in recent years, the highly complex and specialized properties of MCCs leave significant questions unanswered. Future studies targeted at understanding the complexities and interconnectedness of MCC transcriptional cascades and critical cellular processes will further contribute to our understanding of how a ciliated epithelium is established and maintained, addressing fundamental questions in cell and developmental biology.

Acknowledgments

Work in the Mitchell lab is supported by the following grants: NIH/NIGMS R01GM089970 (BJM), NIH/NIGMS R01GM113922 (BJM), and T32AR060710 (RV). We thank the Mitchell lab members for critical reading of this manuscript.

References

Angerilli, A., Smialowski, P., & Rupp, R. A. (2018). The Xenopus animal cap transcriptome: Building a mucociliary epithelium. *Nucleic Acids Research*, *46*, 8772–8787.

Antoniades, I., Stylianou, P., & Skourides, P. A. (2014). Making the connection: Ciliary adhesion complexes anchor basal bodies to the actin cytoskeleton. *Developmental Cell*, *28*, 70–80.

Arbi, M., Pefani, D. E., Kyrousi, C., Lalioti, M. E., Kalogeropoulou, A., Papanastasiou, A. D., et al. (2016). GemC1 controls multiciliogenesis in the airway epithelium. *EMBO Reports*, *17*, 400–413.

Arbi, M., Pefani, D. E., Taraviras, S., & Lygerou, Z. (2018). Controlling centriole numbers: Geminin family members as master regulators of centriole amplification and multiciliogenesis. *Chromosoma*, *127*, 151–174.

Azimzadeh, J., Wong, M. L., Downhour, D. M., Sanchez Alvarado, A., & Marshall, W. F. (2012). Centrosome loss in the evolution of planarians. *Science*, *335*, 461–463.

Baas, D., Meiniel, A., Benadiba, C., Bonnafe, E., Meiniel, O., Reith, W., et al. (2006). A deficiency in RFX3 causes hydrocephalus associated with abnormal differentiation of ependymal cells. *The European Journal of Neuroscience*, *24*, 1020–1030.

Balestrini, A., Cosentino, C., Errico, A., Garner, E., & Costanzo, V. (2010). GEMC1 is a TopBP1-interacting protein required for chromosomal DNA replication. *Nature Cell Biology*, *12*, 484–491.

Barrera, J. A., Kao, L. R., Hammer, R. E., Seemann, J., Fuchs, J. L., & Megraw, T. L. (2010). CDK5RAP2 regulates centriole engagement and cohesion in mice. *Developmental Cell*, *18*, 913–926.

Bharathan, N. K., & Dickinson, A. J. G. (2019). Desmoplakin is required for epidermal integrity and morphogenesis in the *Xenopus laevis* embryo. *Developmental Biology*, *450*, 115–131.

Boisvieux-Ulrich, E., Laine, M. C., & Sandoz, D. (1990). Cytochalasin D inhibits basal body migration and ciliary elongation in quail oviduct epithelium. *Cell and Tissue Research*, *259*, 443–454.

Boon, M., Wallmeier, J., Ma, L., Loges, N. T., Jaspers, M., Olbrich, H., et al. (2014). MCIDAS mutations result in a mucociliary clearance disorder with reduced generation of multiple motile cilia. *Nature Communications*, *5*, 4418.

Bornslaeger, E. A., Corcoran, C. M., Stappenbeck, T. S., & Green, K. J. (1996). Breaking the connection: Displacement of the desmosomal plaque protein desmoplakin from cell-cell interfaces disrupts anchorage of intermediate filament bundles and alters intercellular junction assembly. *The Journal of Cell Biology*, *134*, 985–1001.
Bouchet, B. P., & Akhmanova, A. (2017). Microtubules in 3D cell motility. *Journal of Cell Science*, *130*, 39–50.
Boutin, C., & Kodjabachian, L. (2019). Biology of multiciliated cells. *Current Opinion in Genetics & Development*, *56*, 1–7.
Bray, S., & Bernard, F. (2010). Notch targets and their regulation. *Current Topics in Developmental Biology*, *92*, 253–275.
Briggs, J. A., Weinreb, C., Wagner, D. E., Megason, S., Peshkin, L., Kirschner, M. W., et al. (2018). The dynamics of gene expression in vertebrate embryogenesis at single-cell resolution. *Science*, *360*, eaar5780.
Brooks, E. R., & Wallingford, J. B. (2014). Multiciliated cells. *Current Biology*, *24*, R973–R982.
Caillat, C., Fish, A., Pefani, D. E., Taraviras, S., Lygerou, Z., & Perrakis, A. (2015). The structure of the GemC1 coiled coil and its interaction with the Geminin family of coiled-coil proteins. *Acta Crystallographica. Section D, Biological Crystallography*, *71*, 2278–2286.
Caillat, C., Pefani, D. E., Gillespie, P. J., Taraviras, S., Blow, J. J., Lygerou, Z., et al. (2013). The Geminin and Idas coiled coils preferentially form a heterodimer that inhibits Geminin function in DNA replication licensing. *The Journal of Biological Chemistry*, *288*, 31624–31634.
Campbell, E. P., Quigley, I. K., & Kintner, C. (2016). Foxn4 promotes gene expression required for the formation of multiple motile cilia. *Development*, *143*, 4654–4664.
Cavey, M., & Lecuit, T. (2009). Molecular bases of cell-cell junctions stability and dynamics. *Cold Spring Harbor Perspectives in Biology*, *1*, a002998.
Chan, S. W., Fowler, K. J., Choo, K. H., & Kalitsis, P. (2005). Spef1, a conserved novel testis protein found in mouse sperm flagella. *Gene*, *353*, 189–199.
Charras, G., & Yap, A. S. (2018). Tensile forces and mechanotransduction at cell-cell junctions. *Current Biology*, *28*, R445–R457.
Chevalier, B., Adamiok, A., Mercey, O., Revinski, D. R., Zaragosi, L. E., Pasini, A., et al. (2015). miR-34/449 control apical actin network formation during multiciliogenesis through small GTPase pathways. *Nature Communications*, *6*, 8386.
Choksi, S. P., Lauter, G., Swoboda, P., & Roy, S. (2014). Switching on cilia: Transcriptional networks regulating ciliogenesis. *Development*, *141*, 1427–1441.
Chong, Y. L., Zhang, Y., Zhou, F., & Roy, S. (2018). Distinct requirements of E2f4 versus E2f5 activity for multiciliated cell development in the zebrafish embryo. *Developmental Biology*, *443*, 165–172.
Christodoulou, A., Lederer, C. W., Surrey, T., Vernos, I., & Santama, N. (2006). Motor protein KIFC5A interacts with Nubp1 and Nubp2, and is implicated in the regulation of centrosome duplication. *Journal of Cell Science*, *119*, 2035–2047.
Chung, M. I., Kwon, T., Tu, F., Brooks, E. R., Gupta, R., Meyer, M., et al. (2014). Coordinated genomic control of ciliogenesis and cell movement by RFX2. *eLife*, *3*, e01439.
Chung, M. I., Peyrot, S. M., LeBoeuf, S., Park, T. J., McGary, K. L., Marcotte, E. M., et al. (2012). RFX2 is broadly required for ciliogenesis during vertebrate development. *Developmental Biology*, *363*, 155–165.
Collins, C., Majekodunmi, A., & Mitchell, B. (2020). Centriole number and the accumulation of microtubules modulate the timing of apical insertion during radial intercalation. *Current Biology*, *30*, 1958–1964.e3.

Danielian, P. S., Bender Kim, C. F., Caron, A. M., Vasile, E., Bronson, R. T., & Lees, J. A. (2007). E2f4 is required for normal development of the airway epithelium. *Developmental Biology, 305*, 564–576.
Deblandre, G. A., Lai, E. C., & Kintner, C. (2001). Xenopus neuralized is a ubiquitin ligase that interacts with XDelta1 and regulates Notch signaling. *Developmental Cell, 1*, 795–806.
Deblandre, G. A., Wettstein, D. A., Koyano-Nakagawa, N., & Kintner, C. (1999). A two-step mechanism generates the spacing pattern of the ciliated cells in the skin of Xenopus embryos. *Development, 126*, 4715–4728.
Desclozeaux, M., Venturato, J., Wylie, F. G., Kay, J. G., Joseph, S. R., Le, H. T., et al. (2008). Active Rab11 and functional recycling endosome are required for E-cadherin trafficking and lumen formation during epithelial morphogenesis. *American Journal of Physiology. Cell Physiology, 295*, C545–C556.
Didon, L., Zwick, R. K., Chao, I. W., Walters, M. S., Wang, R., Hackett, N. R., et al. (2013). RFX3 modulation of FOXJ1 regulation of cilia genes in the human airway epithelium. *Respiratory Research, 14*, 70.
Dougherty, G. W., Adler, H. J., Rzadzinska, A., Gimona, M., Tomita, Y., Lattig, M. C., et al. (2005). CLAMP, a novel microtubule-associated protein with EB-type calponin homology. *Cell Motility and the Cytoskeleton, 62*, 141–156.
Drysdale, T. A., & Elinson, R. P. (1992). Cell-migration and induction in the development of the surface ectodermal pattern of the Xenopus-Laevis tadpole. *Development, Growth & Differentiation, 34*, 51–59.
Dubaissi, E., & Papalopulu, N. (2011). Embryonic frog epidermis: A model for the study of cell-cell interactions in the development of mucociliary disease. *Disease Models & Mechanisms, 4*, 179–192.
Dubaissi, E., Rousseau, K., Lea, R., Soto, X., Nardeosingh, S., Schweickert, A., et al. (2014). A secretory cell type develops alongside multiciliated cells, ionocytes and goblet cells, and provides a protective, anti-infective function in the frog embryonic mucociliary epidermis. *Development, 141*, 1514–1525.
El Zein, L., Ait-Lounis, A., Morle, L., Thomas, J., Chhin, B., Spassky, N., et al. (2009). RFX3 governs growth and beating efficiency of motile cilia in mouse and controls the expression of genes involved in human ciliopathies. *Journal of Cell Science, 122*, 3180–3189.
Epting, D., Slanchev, K., Boehlke, C., Hoff, S., Loges, N. T., Yasunaga, T., et al. (2015). The Rac1 regulator ELMO controls basal body migration and docking in multiciliated cells through interaction with Ezrin. *Development, 142*, 174–184.
Etienne-Manneville, S. (2004). Cdc42—The centre of polarity. *Journal of Cell Science, 117*, 1291–1300.
Etienne-Manneville, S. (2013). Microtubules in cell migration. *Annual Review of Cell and Developmental Biology, 29*, 471–499.
Garrard, S. M., Capaldo, C. T., Gao, L., Rosen, M. K., Macara, I. G., & Tomchick, D. R. (2003). Structure of Cdc42 in a complex with the GTPase-binding domain of the cell polarity protein, Par6. *The EMBO Journal, 22*, 1125–1133.
Gomes, E. R., Jani, S., & Gundersen, G. G. (2005). Nuclear movement regulated by Cdc42, MRCK, myosin, and actin flow establishes MTOC polarization in migrating cells. *Cell, 121*, 451–463.
Guseh, J. S., Bores, S. A., Stanger, B. Z., Zhou, Q., Anderson, W. J., Melton, D. A., et al. (2009). Notch signaling promotes airway mucous metaplasia and inhibits alveolar development. *Development, 136*, 1751–1759.
Haas, M., Gomez Vazquez, J. L., Sun, D. L, Tran, H. T., Brislinger, M., Tasca, A., et al. (2019). DeltaN-Tp63 mediates Wnt/beta-catenin-induced inhibition of differentiation in basal stem cells of mucociliary epithelia. *Cell Reports, 28*, 3338–3352.e3336.

Hall, A. (1998). Rho GTPases and the actin cytoskeleton. *Science*, *279*, 509–514.

Harris, A. R., Daeden, A., & Charras, G. T. (2014). Formation of adherens junctions leads to the emergence of a tissue-level tension in epithelial monolayers. *Journal of Cell Science*, *127*, 2507–2517.

Hendershott, M. C., & Vale, R. D. (2014). Regulation of microtubule minus-end dynamics by CAMSAPs and Patronin. *Proceedings of the National Academy of Sciences of the United States of America*, *111*, 5860–5865.

Henrique, D., & Schweisguth, F. (2019). Mechanisms of Notch signaling: A simple logic deployed in time and space. *Development*, *146*, dev172148.

Hildebrandt, F., Benzing, T., & Katsanis, N. (2011). Ciliopathies. *The New England Journal of Medicine*, *364*, 1533–1543.

Ioannou, A., Santama, N., & Skourides, P. A. (2013). *Xenopus laevis* nucleotide binding protein 1 (xNubp1) is important for convergent extension movements and controls ciliogenesis via regulation of the actin cytoskeleton. *Developmental Biology*, *380*, 243–258.

Jiang, K., Hua, S., Mohan, R., Grigoriev, I., Yau, K. W., Liu, Q., et al. (2014). Microtubule minus-end stabilization by polymerization-driven CAMSAP deposition. *Developmental Cell*, *28*, 295–309.

Kennedy, M. P., Omran, H., Leigh, M. W., Dell, S., Morgan, L., Molina, P. L., et al. (2007). Congenital heart disease and other heterotaxic defects in a large cohort of patients with primary ciliary dyskinesia. *Circulation*, *115*, 2814–2821.

Kim, K., Lake, B. B., Haremaki, T., Weinstein, D. C., & Sokol, S. Y. (2012). Rab11 regulates planar polarity and migratory behavior of multiciliated cells in Xenopus embryonic epidermis. *Developmental Dynamics: An Official Publication of the American Association of Anatomists*, *241*, 1385–1395.

Kim, S., Ma, L., Shokhirev, M. N., Quigley, I., & Kintner, C. (2018). Multicilin and activated E2f4 induce multiciliated cell differentiation in primary fibroblasts. *Scientific Reports*, *8*, 12369.

Kim, S. K., Zhang, S., Werner, M. E., Brotslaw, E. J., Mitchell, J. W., Altabbaa, M. M., et al. (2018). CLAMP/Spef1 regulates planar cell polarity signaling and asymmetric microtubule accumulation in the Xenopus ciliated epithelia. *The Journal of Cell Biology*, *217*, 1633–1641.

Klos Dehring, D. A., Vladar, E. K., Werner, M. E., Mitchell, J. W., Hwang, P., & Mitchell, B. J. (2013). Deuterosome-mediated centriole biogenesis. *Developmental Cell*, *27*, 103–112.

Kovar, D. R., & Pollard, T. D. (2004). Insertional assembly of actin filament barbed ends in association with formins produces piconewton forces. *Proceedings of the National Academy of Sciences of the United States of America*, *101*, 14725–14730.

Kroll, K. L., Salic, A. N., Evans, L. M., & Kirschner, M. W. (1998). Geminin, a neuralizing molecule that demarcates the future neural plate at the onset of gastrulation. *Development*, *125*, 3247–3258.

Kulkarni, S. S., Griffin, J. N., Date, P. P., Liem, K. F., Jr., & Khokha, M. K. (2018). WDR5 stabilizes actin architecture to promote multiciliated cell formation. *Developmental Cell*, *46*, 595–610.e593.

Lafkas, D., Shelton, A., Chiu, C., de Leon Boenig, G., Chen, Y., Stawicki, S. S., et al. (2015). Therapeutic antibodies reveal Notch control of transdifferentiation in the adult lung. *Nature*, *528*, 127–131.

Lewis, M., & Stracker, T. H. (2020). Transcriptional regulation of multiciliated cell differentiation. *Seminars in Cell & Developmental Biology*, S1084–9521-9. https://doi.org/10.1016/j.semcdb.2020.04.007. published online ahead of print, 2020 Apr 30.

Lim, J. W., Hummert, P., Mills, J. C., & Kroll, K. L. (2011). Geminin cooperates with Polycomb to restrain multi-lineage commitment in the early embryo. *Development*, *138*, 33–44.

Liu, Y., Pathak, N., Kramer-Zucker, A., & Drummond, I. A. (2007). Notch signaling controls the differentiation of transporting epithelia and multiciliated cells in the zebrafish pronephros. *Development*, *134*, 1111–1122.
Lu, H., Anujan, P., Zhou, F., Zhang, Y., Chong, Y. L., Bingle, C. D., et al. (2019). Mcidas mutant mice reveal a two-step process for the specification and differentiation of multiciliated cells in mammals. *Development*, *146*, dev172643.
Luna-Escalante, J. C., Formosa-Jordan, P., & Ibanes, M. (2018). Redundancy and cooperation in Notch intercellular signaling. *Development*, *145*, dev154807.
Ma, M., & Jiang, Y. J. (2007). Jagged2a-notch signaling mediates cell fate choice in the zebrafish pronephric duct. *PLoS Genetics*, *3*, e18.
Ma, L., Quigley, I., Omran, H., & Kintner, C. (2014). Multicilin drives centriole biogenesis via E2f proteins. *Genes & Development*, *28*, 1461–1471.
Mahuzier, A., Shihavuddin, A., Fournier, C., Lansade, P., Faucourt, M., Menezes, N., et al. (2018). Ependymal cilia beating induces an actin network to protect centrioles against shear stress. *Nature Communications*, *9*, 2279.
Marcet, B., Chevalier, B., Luxardi, G., Coraux, C., Zaragosi, L. E., Cibois, M., et al. (2011). Control of vertebrate multiciliogenesis by miR-449 through direct repression of the Delta/Notch pathway. *Nature Cell Biology*, *13*, 693–699.
Marsden, M., & DeSimone, D. W. (2001). Regulation of cell polarity, radial intercalation and epiboly in Xenopus: Novel roles for integrin and fibronectin. *Development*, *128*, 3635–3647.
McGarry, T. J., & Kirschner, M. W. (1998). Geminin, an inhibitor of DNA replication, is degraded during mitosis. *Cell*, *93*, 1043–1053.
Medioni, C., Astier, M., Zmojdzian, M., Jagla, K., & Semeriva, M. (2008). Genetic control of cell morphogenesis during *Drosophila melanogaster* cardiac tube formation. *The Journal of Cell Biology*, *182*, 249–261.
Mitchell, B., Stubbs, J. L., Huisman, F., Taborek, P., Yu, C., & Kintner, C. (2009). The PCP pathway instructs the planar orientation of ciliated cells in the Xenopus larval skin. *Current Biology*, *19*, 924–929.
Mitchison, T. J., & Cramer, L. P. (1996). Actin-based cell motility and cell locomotion. *Cell*, *84*, 371–379.
Mitchison, T., & Kirschner, M. (1984). Dynamic instability of microtubule growth. *Nature*, *312*, 237–242.
Mogilner, A., & Oster, G. (2003). Force generation by actin polymerization II: The elastic ratchet and tethered filaments. *Biophysical Journal*, *84*, 1591–1605.
Mori, M., Hazan, R., Danielian, P. S., Mahoney, J. E., Li, H., Lu, J., et al. (2017). Cytoplasmic E2f4 forms organizing centres for initiation of centriole amplification during multiciliogenesis. *Nature Communications*, *8*, 15857.
Mori, M., Mahoney, J. E., Stupnikov, M. R., Paez-Cortez, J. R., Szymaniak, A. D., Varelas, X., et al. (2015). Notch3-Jagged signaling controls the pool of undifferentiated airway progenitors. *Development*, *142*, 258–267.
Nenni, M. J., Fisher, M. E., James-Zorn, C., Pells, T. J., Ponferrada, V., Chu, S., et al. (2019). Xenbase: Facilitating the use of Xenopus to model human disease. *Frontiers in Physiology*, *10*, 154.
Nieuwkoop, P. D., & Faber, J. (1994). *Normal table of Xenopus laevis (Daudin)*. New York, NY: Garland Publishing Inc.
Ossipova, O., Chu, C. W., Fillatre, J., Brott, B. K., Itoh, K., & Sokol, S. Y. (2015). The involvement of PCP proteins in radial cell intercalations during Xenopus embryonic development. *Developmental Biology*, *408*, 316–327.
Ossipova, O., Kim, K., Lake, B. B., Itoh, K., Ioannou, A., & Sokol, S. Y. (2014). Role of Rab11 in planar cell polarity and apical constriction during vertebrate neural tube closure. *Nature Communications*, *5*, 3734.

Pan, J. H., Adair-Kirk, T. L., Patel, A. C., Huang, T., Yozamp, N. S., Xu, J., et al. (2014). Myb permits multilineage airway epithelial cell differentiation. *Stem Cells*, *32*, 3245–3256.

Pan, J., You, Y., Huang, T., & Brody, S. L. (2007). RhoA-mediated apical actin enrichment is required for ciliogenesis and promoted by Foxj1. *Journal of Cell Science*, *120*, 1868–1876.

Pardo-Saganta, A., Tata, P. R., Law, B. M., Saez, B., Chow, R. D., Prabhu, M., et al. (2015). Parent stem cells can serve as niches for their daughter cells. *Nature*, *523*, 597–601.

Park, T. J., Mitchell, B. J., Abitua, P. B., Kintner, C., & Wallingford, J. B. (2008). Dishevelled controls apical docking and planar polarization of basal bodies in ciliated epithelial cells. *Nature Genetics*, *40*, 871–879.

Pefani, D. E., Dimaki, M., Spella, M., Karantzelis, N., Mitsiki, E., Kyrousi, C., et al. (2011). Idas, a novel phylogenetically conserved geminin-related protein, binds to geminin and is required for cell cycle progression. *The Journal of Biological Chemistry*, *286*, 23234–23246.

Quigley, I. K., & Kintner, C. (2017). Rfx2 stabilizes Foxj1 binding at chromatin loops to enable multiciliated cell gene expression. *PLoS Genetics*, *13*, e1006538.

Quigley, I. K., Stubbs, J. L., & Kintner, C. (2011). Specification of ion transport cells in the Xenopus larval skin. *Development*, *138*, 705–714.

Reiter, J. F., & Leroux, M. R. (2017). Genes and molecular pathways underpinning ciliopathies. *Nature Reviews. Molecular Cell Biology*, *18*, 533–547.

Ridley, A. J. (2015). Rho GTPase signalling in cell migration. *Current Opinion in Cell Biology*, *36*, 103–112.

Roeth, J. F., Sawyer, J. K., Wilner, D. A., & Peifer, M. (2009). Rab11 helps maintain apical crumbs and adherens junctions in the Drosophila embryonic ectoderm. *PLoS One*, *4*, e7634.

Sancho, R., Cremona, C. A., & Behrens, A. (2015). Stem cell and progenitor fate in the mammalian intestine: Notch and lateral inhibition in homeostasis and disease. *EMBO Reports*, *16*, 571–581.

Schmoranzer, J., Fawcett, J. P., Segura, M., Tan, S., Vallee, R. B., Pawson, T., et al. (2009). Par3 and dynein associate to regulate local microtubule dynamics and centrosome orientation during migration. *Current Biology*, *19*, 1065–1074.

Sedzinski, J., Hannezo, E., Tu, F., Biro, M., & Wallingford, J. B. (2016). Emergence of an apical epithelial cell surface in vivo. *Developmental Cell*, *36*, 24–35.

Sedzinski, J., Hannezo, E., Tu, F., Biro, M., & Wallingford, J. B. (2017). RhoA regulates actin network dynamics during apical surface emergence in multiciliated epithelial cells. *Journal of Cell Science*, *130*, 420–428.

Seo, S., & Kroll, K. L. (2006). Geminin's double life: Chromatin connections that regulate transcription at the transition from proliferation to differentiation. *Cell Cycle*, *5*, 374–379.

Sirour, C., Hidalgo, M., Bello, V., Buisson, N., Darribere, T., & Moreau, N. (2011). Dystroglycan is involved in skin morphogenesis downstream of the Notch signaling pathway. *Molecular Biology of the Cell*, *22*, 2957–2969.

Smith, Q., Macklin, B., Chan, X. Y., Jones, H., Trempel, M., Yoder, M. C., et al. (2018). Differential HDAC6 activity modulates ciliogenesis and subsequent mechanosensing of endothelial cells derived from pluripotent stem cells. *Cell Reports*, *24*, 1930.

Sokol, S. Y. (1996). Analysis of dishevelled signalling pathways during Xenopus development. *Current Biology*, *6*, 1456–1467.

Song, R., Walentek, P., Sponer, N., Klimke, A., Lee, J. S., Dixon, G., et al. (2014). miR-34/449 miRNAs are required for motile ciliogenesis by repressing cp110. *Nature*, *510*, 115–120.

Stephenson, R. E., Higashi, T., Erofeev, I. S., Arnold, T. R., Leda, M., Goryachev, A. B., et al. (2019). Rho flares repair local tight junction leaks. *Developmental Cell*, *48*, 445–459.e445.

Stubbs, J. L., Davidson, L., Keller, R., & Kintner, C. (2006). Radial intercalation of ciliated cells during Xenopus skin development. *Development*, *133*, 2507–2515.

Stubbs, J. L., Oishi, I., Izpisua Belmonte, J. C., & Kintner, C. (2008). The forkhead protein Foxj1 specifies node-like cilia in Xenopus and zebrafish embryos. *Nature Genetics*, *40*, 1454–1460.

Stubbs, J. L., Vladar, E. K., Axelrod, J. D., & Kintner, C. (2012). Multicilin promotes centriole assembly and ciliogenesis during multiciliate cell differentiation. *Nature Cell Biology*, *14*, 140–147.

Swisher, M., Jonas, R., Tian, X., Lee, E. S., Lo, C. W., & Leatherbury, L. (2011). Increased postoperative and respiratory complications in patients with congenital heart disease associated with heterotaxy. *The Journal of Thoracic and Cardiovascular Surgery*, *141*, 637–644. 644 e631–633.

Tan, F. E., Vladar, E. K., Ma, L., Fuentealba, L. C., Hoh, R., Espinoza, F. H., et al. (2013). Myb promotes centriole amplification and later steps of the multiciliogenesis program. *Development*, *140*, 4277–4286.

Terre, B., Piergiovanni, G., Segura-Bayona, S., Gil-Gomez, G., Youssef, S. A., Attolini, C. S., et al. (2016). GEMC1 is a critical regulator of multiciliated cell differentiation. *The EMBO Journal*, *35*, 942–960.

Trievel, R. C., & Shilatifard, A. (2009). WDR5, a complexed protein. *Nature Structural & Molecular Biology*, *16*, 678–680.

Tsang, K. Y., Cheung, M. C., Chan, D., & Cheah, K. S. (2010). The developmental roles of the extracellular matrix: Beyond structure to regulation. *Cell and Tissue Research*, *339*, 93–110.

Tsao, P. N., Vasconcelos, M., Izvolsky, K. I., Qian, J., Lu, J., & Cardoso, W. V. (2009). Notch signaling controls the balance of ciliated and secretory cell fates in developing airways. *Development*, *136*, 2297–2307.

Van Agtmael, T., & Bruckner-Tuderman, L. (2010). Basement membranes and human disease. *Cell and Tissue Research*, *339*, 167–188.

Vladar, E. K., & Mitchell, B. J. (2016). It's a family act: The geminin triplets take center stage in motile ciliogenesis. *The EMBO Journal*, *35*, 904–906.

Walentek, P., & Quigley, I. K. (2017). What we can learn from a tadpole about ciliopathies and airway diseases: Using systems biology in Xenopus to study cilia and mucociliary epithelia. *Genesis*, *55*.

Werner, M., Del Castillo, U., Ventrella, R., Brotslaw, E., & Mitchell, B. (2018). The small molecule AMBMP disrupts microtubule growth, ciliogenesis, cell polarity, and cell migration. *Cytoskeleton (Hoboken)*, *75*, 450–457.

Werner, M. E., Hwang, P., Huisman, F., Taborek, P., Yu, C. C., & Mitchell, B. J. (2011). Actin and microtubules drive differential aspects of planar cell polarity in multiciliated cells. *The Journal of Cell Biology*, *195*, 19–26.

Werner, M. E., & Mitchell, B. J. (2013). Using Xenopus skin to study cilia development and function. *Methods in Enzymology*, *525*, 191–217.

Werner, M. E., Mitchell, J. W., Putzbach, W., Bacon, E., Kim, S. K., & Mitchell, B. J. (2014). Radial intercalation is regulated by the Par complex and the microtubule-stabilizing protein CLAMP/Spef1. *The Journal of Cell Biology*, *206*, 367–376.

Wright, K. M., Lyon, K. A., Leung, H., Leahy, D. J., Ma, L., & Ginty, D. D. (2012). Dystroglycan organizes axon guidance cue localization and axonal pathfinding. *Neuron*, *76*, 931–944.

Wysocka, J., Swigut, T., Milne, T. A., Dou, Y., Zhang, X., Burlingame, A. L., et al. (2005). WDR5 associates with histone H3 methylated at K4 and is essential for H3 K4 methylation and vertebrate development. *Cell*, *121*, 859–872.

Xie, H., Kang, Y., Wang, S., Zheng, P., Chen, Z., Roy, S., et al. (2020). E2f5 is a versatile transcriptional activator required for spermatogenesis and multiciliated cell differentiation in zebrafish. *PLoS Genetics*, *16*, e1008655.
You, Y., Huang, T., Richer, E. J., Schmidt, J. E., Zabner, J., Borok, Z., et al. (2004). Role of f-box factor foxj1 in differentiation of ciliated airway epithelial cells. *American Journal of Physiology. Lung Cellular and Molecular Physiology*, *286*, L650–L657.
Yu, X., Ng, C. P., Habacher, H., & Roy, S. (2008). Foxj1 transcription factors are master regulators of the motile ciliogenic program. *Nature Genetics*, *40*, 1445–1453.
Zaidi, S., Choi, M., Wakimoto, H., Ma, L., Jiang, J., Overton, J. D., et al. (2013). De novo mutations in histone-modifying genes in congenital heart disease. *Nature*, *498*, 220–223.
Zallen, J. A. (2007). Planar polarity and tissue morphogenesis. *Cell*, *129*, 1051–1063.
Zhang, J., Betson, M., Erasmus, J., Zeikos, K., Bailly, M., Cramer, L. P., et al. (2005). Actin at cell-cell junctions is composed of two dynamic and functional populations. *Journal of Cell Science*, *118*, 5549–5562.
Zhang, S., & Mitchell, B. J. (2015a). Basal bodies in Xenopus. *Cilia*, *5*, 2.
Zhang, S., & Mitchell, B. J. (2015b). Centriole biogenesis and function in multiciliated cells. *Methods in Cell Biology*, *129*, 103–127.
Zhen, Y., & Stenmark, H. (2015). Cellular functions of Rab GTPases at a glance. *Journal of Cell Science*, *128*, 3171–3176.
Zhou, F., Narasimhan, V., Shboul, M., Chong, Y. L., Reversade, B., & Roy, S. (2015). Gmnc is a master regulator of the multiciliated cell differentiation program. *Current Biology*, *25*, 3267–3273.

CHAPTER TWO

Xenopus neural tube closure: A vertebrate model linking planar cell polarity to actomyosin contractions

Miho Matsuda* and Sergei Y. Sokol*
Department of Cell, Developmental and Regenerative Biology, Icahn School of Medicine at Mount Sinai, New York, NY, United States
*Corresponding authors: e-mail address: miho.matsuda@mssm.edu; sergei.sokol@mssm.edu

Contents

Abstract

Planar cell polarity (PCP) refers to the coordinated polarization of cells within the plane of a tissue. PCP is a controlled by a group of conserved proteins organized in a specific signaling pathway known as the PCP pathway. A hallmark of PCP signaling is the asymmetric localization of "core" PCP protein complexes at the cell cortex, although endogenous PCP cues needed to establish this asymmetry remain unknown. While the PCP pathway was originally discovered as a mechanism directing the planar organization of *Drosophila* epithelial tissues, subsequent studies in *Xenopus* and other vertebrates demonstrated a critical role for this pathway in the regulation of actomyosin-dependent morphogenetic processes, such as neural tube closure. Large size and external

Current Topics in Developmental Biology, Volume 145
ISSN 0070-2153
https://doi.org/10.1016/bs.ctdb.2021.04.001

development of amphibian embryos allows live cell imaging, placing *Xenopus* among the best models of vertebrate neurulation at the molecular, cellular and organismal level. This review describes cross-talk between core PCP proteins and actomyosin contractility that ultimately leads to tissue-scale movement during neural tube closure.

1. Introduction

Planar cell polarity (PCP) is a common property of cells within the plane of a tissue. The core PCP pathway was originally discovered in forward mutagenesis screens for regulators of cuticle patterning in *Drosophila melanogaster* (Adler, 2012; Goodrich & Strutt, 2011). These screens identified "core PCP" components, including the transmembrane proteins Flamingo (Fmi/Celsr), Frizzled (Fz/Fzd) and Van Gogh (Vang/Vangl, also known as Strabismus (Stbm)), and the cytoplasmic components, including Disheveled (Dsh/Dvl), Prickle (Pk) and Diego (Dgo, corresponding to the vertebrate proteins Diversin and Inversin (Inv)). The first abbreviated name refers to the fly homolog, whereas the second name refers to the vertebrate homolog. These core proteins are evolutionally highly conserved and organized in a specific signaling pathway. Although multiple mechanisms establish PCP in various tissues (Butler & Wallingford, 2017; Goodrich & Strutt, 2011; Zallen, 2007), this review focuses only on the pathway mediated by core PCP proteins.

A key feature of the PCP pathway is the establishment of mutually exclusive localization of two protein complexes, Fmi-Fz-Dsh and Fmi-Vang-Pk, to opposite sides of the cell. The polarized distribution of core PCP proteins was first described in *Drosophila melanogaster* (Axelrod, 2001; Strutt, 2001; Usui et al., 1999), and later confirmed in various vertebrate tissues (reviewed by Butler & Wallingford, 2017). This segregation of core PCP complexes to opposite sides of cells is believed to be critical for PCP signaling.

The core PCP pathway modulates epithelial patterning in *Drosophila* by coordinating patterns of bristles and cuticular hairs (Adler, 2012; Devenport, 2016). The core PCP pathway also regulates many morphogenetic events in vertebrate embryos, such as neural tube (NT) closure (Davey & Moens, 2017; Gray, Roszko, & Solnica-Krezel, 2011). NT closure is one of the major PCP-dependent morphogenetic processes that require highly organized cell movements and cell shape changes driven by the actomyosin cytoskeleton. Importantly, the disruption of the actomyosin system also results in NT closure defects, suggesting that the PCP pathway might influence NT closure via actomyosin reorganization. However, how PCP is established in the NT and how PCP proteins modulate actomyosin contractility for NT closure remains to be understood. Due to the prominent polarization of *Xenopus* neuroepithelial cells, frog embryos are among the best models for vertebrate PCP.

Several features make *Xenopus* embryos superior to other experimental models for NT closure. First, one female lays many eggs, providing a large amount of biochemical material. Protein extracts can be obtained from total embryos or a specific embryonic tissue after microdissection. Second, gene and protein expression can be targeted to specific embryonic regions, for instance the neural plate, facilitating the study of neuroectodermal PCP. Third, *Xenopus* embryos develop externally and the movements of individual cells or groups of cells can be imaged live in the closing NT without any additional manipulation or dissection. Lastly, unlike zebrafish embryos, *Xenopus* embryos undergo neural plate folding similar to mammals, in which the neural epithelium with apical-basal polarity folds at hinge points and fuses at the dorsal midline. Thus, studies of NT closure in *Xenopus* may advance our understanding of the etiology of NT defects in humans. With the emphasis on the *Xenopus* model, this review discusses possible connections between PCP proteins and actomyosin networks mediating vertebrate NT closure.

2. Mechanisms leading to PCP

2.1 Assembly of core PCP complexes

The core PCP protein complexes predominantly localize to the apical cell-cell contact region (Devenport & Fuchs, 2008; Mahaffey, Grego-Bessa, Liem Jr., & Anderson, 2013). In *Drosophila* pupal wings, the PCP complexes initially orient toward the wing margin. Cell rearrangements realign the PCP complexes along the proximodistal axis (Aigouy et al., 2010; Strutt & Strutt, 2009). Fmi, Fz and Dsh enrich at the distal end of each cell, and Fmi, Vang and Pk accumulate at the proximal end (Fig. 1A) (Strutt & Strutt, 2009). The mechanisms controlling the complementary distribution of core PCP complexes are not completely clear. Fmi is an atypical cadherin with a seven-pass transmembrane domain. Similar to classical cadherins, Fmi forms the homophilic complex *in trans* between two adjacent cells (Shimada, Usui, Yanagawa, Takeichi, & Uemura, 2001; Usui et al., 1999), or *in cis* within the same cell (Stahley, Basta, Sharan, & Devenport, 2021). Both *trans* and *cis* interactions are required for the polarized distribution of the core PCP complexes. Fmi also forms heterophilic complex with

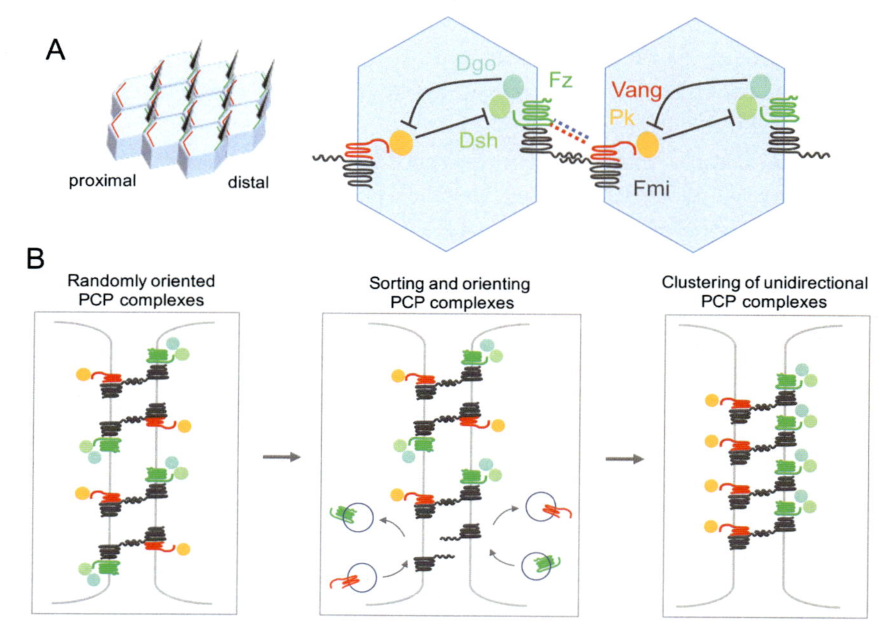

Fig. 1 The asymmetric distribution of the "core" PCP protein complexes. (A) The asymmetric distribution of core PCP protein complexes in *Drosophila* epithelia. The Vang-Pk complex (red) and the Fz-Dsh complex (green) localize to the proximal and the distal ends of individual cells, respectively. Trichome formation at the distal cell ends is indicated. Fmi forms homophilic complexes between adjacent cells. The Fmi-Vang complex preferably interacts with the Fmi-Fz complex on the opposing cell cortex. The cytoplasmic components of each core PCP complex are thought to negatively regulate the formation of the complementary PCP complex. (B) The sorting and clustering of the core complexes. The PCP complexes initially orient in random directions. Protein trafficking, ubiquitin-mediated proteolysis and endocytosis contribute to the sorting and clustering of the core PCP complexes.

Fz or Vang within the same cell. The Fmi-Fz complex on one cell is hypothesized to preferentially interact with the Fmi-Vang complex on the neighboring cell (Chen et al., 2008). This leads to the complementary localization of the Fmi-Fz and Fmi-Vang complexes across the plasma membrane, resulting in PCP (Fig. 1A).

Cytoplasmic PCP components including Pk, Dsh and Dgo, are essential for the stabilization of PCP clusters at each end of the cell (Yang & Mlodzik, 2015). Pk associates with Vang, whereas Dsh forms a complex with Fz. However, whether these physical interactions are constitutive or regulated by PCP signaling is poorly understood. Nevertheless, the formation of core PCP complexes is reinforced by ubiquitin-mediated protein degradation and protein trafficking (Narimatsu et al., 2009; Strutt, Searle, Thomas-Macarthur, Brookfield, & Strutt, 2013). Live imaging of fluorescent PCP proteins revealed directional transport of Fz- and Dsh-containing vesicles to and from the cell cortex, indicating roles of micrtotubules and vesicular trafficking in PCP (Matis, Russler-Germain, Hu, Tomlin, & Axelrod, 2014; Olofsson, Sharp, Matis, Cho, & Axelrod, 2014; Shimada, Yonemura, Ohkura, Strutt, & Uemura, 2006). These processes amplify initial differences, leading to the complementary distribution of core PCP clusters to opposite cell sides (Fig. 1B).

2.2 PCP cues

PCP "cues" have been proposed to instruct the orientation of PCP complexes within a tissue. Cell contact-based communication via core PCP complexes is one way to instruct PCP across the tissue. Consistent with this idea, PCP components are required for maintaining each other's localization (Adler, 2012; Goodrich & Strutt, 2011). The Fmi-Fz-Dsh complex and the Fmi-Vang-Pk complex form mutually exclusive clusters on opposite edges of the cell and associate across the plasma membrane through the extracellular domain of Fmi (Chen et al., 2008). Therefore, once the complementary pattern of PCP complexes is established in one cell, for instance near the Wnt source (Fig. 2A), the same orientation can be conferred to the adjacent cell (Fig. 2B). Long-range PCP could be established by the reiteration of this process (Fig. 2C). However, this cell contact-based model still needs to be directly demonstrated in vivo.

Another way to instruct PCP is by diffusion of a secreted growth factor, such as a Wnt protein that can modulate Frizzled receptor activity. In *Drosophila*, ectopic expression of Wingless (Wg) or Wnt4 reoriented the polarity of hairs in the surrounding epithelium in the wing (Lim, Norga, Chen, & Choi, 2005; Wu, Roman, Carvajal-Gonzalez, & Mlodzik, 2013). In vertebrates, dominant interfering forms of Wnt5a or Wnt11 and the knockdown of Wnt11b disrupted the polarized distribution of Vangl2 in the *Xenopus* neural plate cells (Chu & Sokol, 2016; Ossipova, Kim, & Sokol, 2015). Moreover, ectopic Wnt5a and Wnt11 induced polarized distribution of Prickle3 and Vangl2 in *Xenopus* ectoderm (Chu & Sokol, 2016). Consistent with Wnt ligand requirement for PCP, Wnt5a genetically interacted with Vangl2 to orient hair cells in the mouse cochlea (Qian et al., 2007). Wnt5a and the receptor tyrosine kinase Ror2 cooperated to polarize Vangl2 in mouse limb cells and promote limb elongation (Gao et al., 2011).

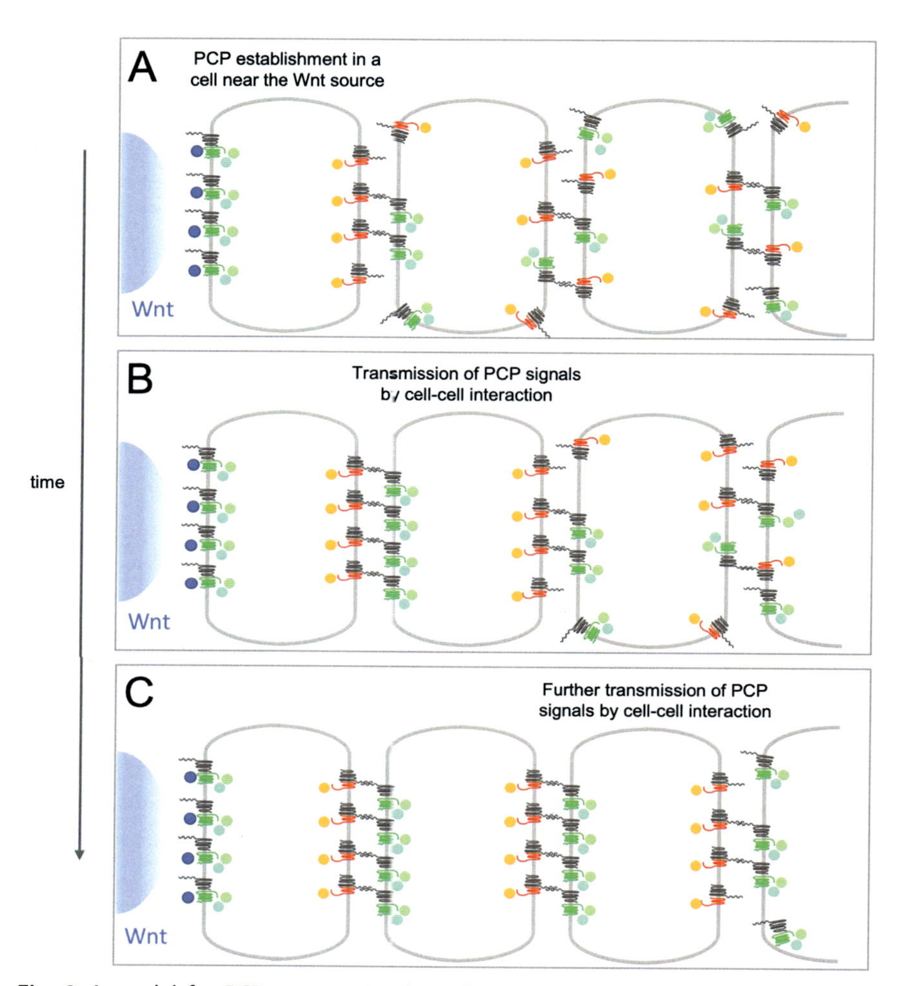

Fig. 2 A model for PCP propagation by cell-cell contact. (A) Wnt ligands bind and activate Fz receptors at the cell surface proximal to the source. Due to their mutually exclusive subcellular localization, the Fmi-Vang-Pk complexes accumulate at the opposite side of the cell. (B) The Fmi-Vang-Pk complexes associate with the Fmi-Fz-Dsh complexes across the cell membrane of the neighboring cell, reconstructing the original polarized distribution of the PCP complexes. (C) Global PCP can be established after this process is repeated multiple times.

Although Wnt proteins can directly polarize individual cells in the tissue, the diffusion range of Wnt proteins may be limited to a few cell diameters (Chaudhary et al., 2019). Moreover, membrane-tethered Wg is able to functionally replace endogenous Wg in *Drosophila* (Alexandre, Baena-Lopez, & Vincent, 2014). Moreover, in the mouse intestinal crypt, Wnt3 was proposed to travel from its source in a cell-bound manner through cell division rather than by diffusion (Farin et al., 2016). These observations challenge the idea that Wnt proteins function as long-range PCP cues.

Despite the evidence that Wnt ligands are required for PCP both in flies and vertebrates (Chu & Sokol, 2016; Minegishi et al., 2017; Wu et al., 2013), recent studies suggest that Wnt proteins are dispensable, at least in the *Drosophila* wing (Ewen-Campen, Comyn, Vogt, & Perrimon, 2020; Yu et al., 2020). Similarly, interfering with Wnt secretion did not inhibit planar polarization of Dvl1/2 or Fzd3/6 in the mouse cochlea (Najarro et al., 2020). These observations suggest the existence of additional PCP cues. As alternative long-range cues for PCP, mechanical forces can arise during morphogenesis, for instance, during cell divisions or cell shape changes within the tissue or in neighboring tissues. Gastrulation movements may produce mechanical strains in the embryonic ectoderm triggering PCP along the anteroposterior (AP) axis (Chien, Keller, Kintner, & Shook, 2015). In *Drosophila* embryos, contraction of the wing hinge epithelium creates anisotropic tension in the wing blade cells, leading to cell elongation, oriented cell divisions, and rearrangements along the proximal-distal axis (Aigouy et al., 2010). In mice, depletion of Wdr1, an F-actin binding protein that enhances cofilin/destrin-mediated F-actin disassembly, caused loss of PCP in the epidermis (Luxenburg et al., 2015). The proposed model is that Wdr1 and cofilin reorganize the actomyosin network and generate cortical tension that acts as a global cue orienting cells in the developing epidermis. These reports suggest that mechanical forces driven by the actomyosin cytoskeleton can instruct PCP, but the direct demonstration of these forces in vivo is still pending.

3. PCP and actomyosin contractility

Once the polarization of core PCP complexes is achieved, the downstream effectors would be expected to influence various cellular targets. It is becoming increasingly clear that the actomyosin system is a major cellular target of the PCP pathway. Oriented PCP puncta composed of core protein complexes commonly mark the location of actin-rich structures, such as insect trichomes or mouse stereocilia (Adler, 2012; Dabdoub & Kelley, 2005). This suggests that PCP complexes provide positional information that initiates actin reorganization in close proximity. In addition, PCP components regulate actomyosin-dependent cell behavior. The known connections between the core PCP components and the actomyosin cytoskeleton are described below.

3.1 Rho GTPases and Myosin II in actomyosin contractility and force generation

Rho small GTPases are highly conserved signaling proteins that control reorganization of the actomyosin cytoskeleton (Jaffe & Hall, 2005). Rho GTPase cycles between a GTP-bound form and a GDP-bound form. This cycle is regulated by guanine nucleotide exchange factors (GEFs) which activate Rho by replacing GDP with GTP, and GTPase activating proteins (GAPs) which inactivate Rho by stimulating GTP hydrolysis (Fig. 3A) (Heasman & Ridley, 2008). Rho-GTP binds downstream effectors that directly modulate the actomyosin cytoskeleton (Sit & Manser, 2011), such as Rho-associated kinase (ROCK). Rho-GTP binding to ROCK releases the autoinhibition of ROCK. ROCK phosphorylates a group of proteins that regulate the actomyosin cytoskeleton (Fig. 3A) (Amin et al., 2013). Studies in *Drosophila* have implicated Rho and ROCK in PCP signaling (Strutt, Weber, & Mlodzik, 1997; Winter et al., 2001).

Filamentous actin (F-actin) and non-muscle myosin II (NMII) form actomyosin networks. A single NMII filament consists of two myosin heavy chains (MYHs), two myosin essential light chains (MELCs) and two myosin regulatory light chains (MRLCs) (Fig. 3B). MYHs assemble tail-to-tail to form a bipolar NMII filament, which oligomerizes to form NMII clusters (Fig. 3B). The head domain of NMII filaments binds anti-parallel F-actin and moves along F-actin toward the barbed (+) end via its motor activity (Fig. 3C). This sliding movement between F-actin and NMII filaments generates the contractile force (Murrell, Oakes, Lenz, & Gardel, 2015). Thus, the formation of NMII bipolar filaments is critical for actomyosin contractility.

Phosphorylation of MRLC by ROCK is required to initiate NMII bipolar filament formation. This phosphorylation changes the structural conformation of the MYH tail domain, allowing the tail-to-tail assembly (Fig. 3B) (Shutova & Svitkina, 2018). ROCK also phosphorylates myosin phosphatase targeting subunit (MYPT1) to inactivate myosin light chain phosphatase (MLCP) and indirectly increases MRLC phosphorylation (Fig. 3A) (Grassie, Moffat, Walsh, & MacDonald, 2011). These studies establish the mechanism of actomyosin regulation by ROCK.

3.2 Linking PCP signaling to the Rho-ROCK-Myosin II pathway

Whereas loss-of-function studies indicate that core PCP components are necessary for coordinated morphogenetic events such as NT closure, the

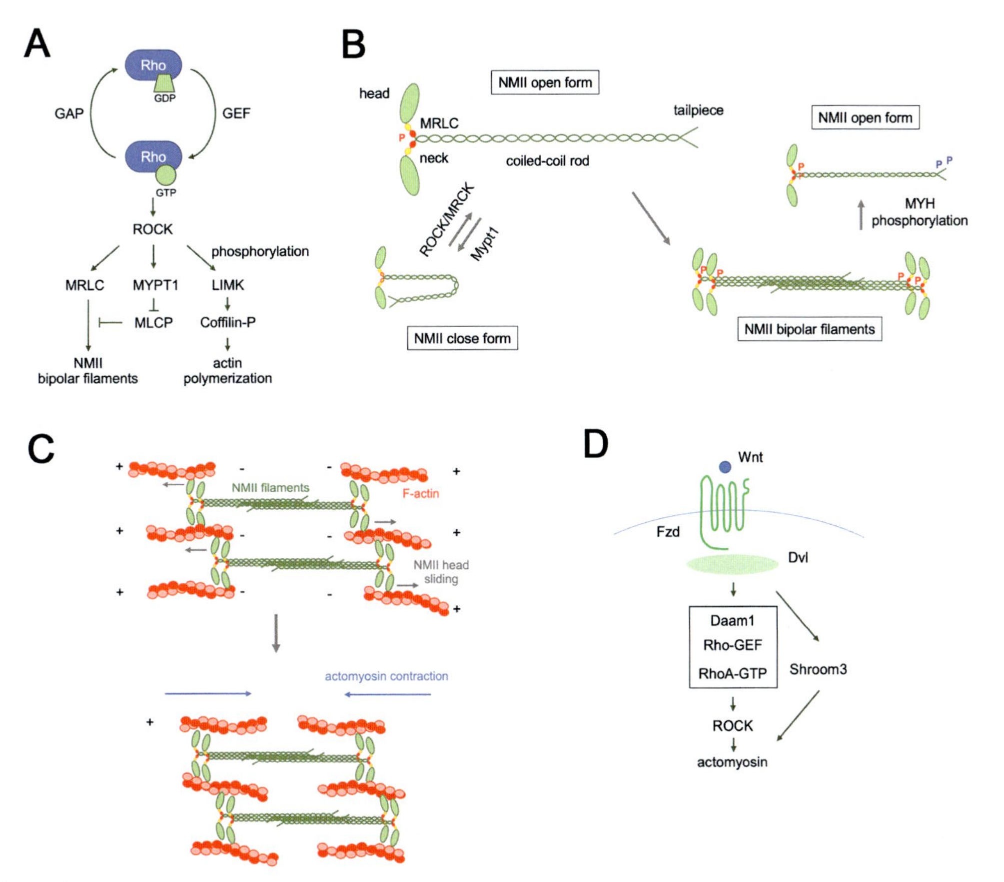

Fig. 3 See figure legend on next page.

underlying mechanisms remain unclear. In one model, Wnt binding to Fzd leads to the formation of the Dvl-containing signalosome at the cell cortex (Bilic et al., 2007). Whereas Dvl promotes β-catenin stabilization in canonical Wnt signaling (Logan & Nusse, 2004), Dvl can also activate the Rho-ROCK-MyoII branch to modulate the actomyosin cytoskeleton (Fig. 3D). In this model, Dvl recruits Daam1, a formin family protein which promotes actin polymerization and has been linked to RhoA activation (Habas, Kato, & He, 2001; Liu et al., 2008; Sato et al., 2006). Taken together, published evidence indicates that Wnt/Fzd signaling activates ROCK, linking PCP signaling to actomyosin contractility. This model is at odds with the data in *Drosophila* that Wnt gain-of-function phenotype mimics Fz loss-of-function (Wu et al., 2013).

Contrary to the above model, other reports provide evidence of colocalization of MRLC with Pk at the anterior cell faces in *Ciona* notochord and *Xenopus* neural plate (Butler & Wallingford, 2018; Newman-Smith, Kourakis, Reeves, Veeman, & Smith, 2015). These findings suggest that actomyosin contractions might be activated by the Vangl-Pk rather than Fzd-Dvl complexes, but the direct evidence is missing. This question can be addressed by live imaging studies at high resolution.

Shroom3 is another candidate molecule connecting PCP proteins to the Rho-ROCK-MyoII system. Originally identified as a factor required for NT closure in mice (Hildebrand & Soriano, 1999), Shroom3 binds F-actin (Bolinger, Zasadil, Rizaldy, & Hildebrand, 2010) and recruits ROCK to cell-cell contact sites to locally increase MRLC phosphorylation and actomyosin contractile forces (Fig. 3D) (Haigo, Hildebrand, Harland, & Wallingford, 2003; Hildebrand, 2005; Nishimura & Takeichi, 2008; Simoes

Fig. 3 Cross-talk between the PCP pathway and the Rho-ROCK-Myosin II pathway. (A) An overview of RhoA-ROCK-Myosin II pathway. Rho-GEF and Rho-GAP mediate GTP-GDP exchange of RhoA. Rho-GTP binding activates ROCK, leading to the phosphorylation of MRLC, MYPT1 and LIMK. This promotes the formation of non-muscle myosin II (NMII) bipolar filaments and actin polymerization. (B) The regulation of NMII bipolar filament formation. A single NMII filament consists of two myosin heavy chains (MYH, green), two myosin essential light chains (MELC, yellow), and two myosin regulatory light chains (MRLC, red). MRLC phosphorylation (marked by P in red) changes the structural conformation of NMII and promotes NMII bipolar filament formation. MYH phosphorylation (marked by P in blue) suppresses NMII bipolar filament formation. (C) The generation of actomyosin contractile forces. The head domain of NMII bipolar filaments bind F-actin and "walks" along F-actin toward the barbed end. (D) Non-canonical Wnt-NMII pathway. Wnt ligand binds Fzd receptor, recruiting Dvl to the cell cortex. Dvl recruits ROCK via its direct binding to Daam1 or Shroom3.

Sde, Mainieri, & Zallen, 2014). Mouse embryos that are mutant for both Shroom3 and Vangl2 or Wnt5a exhibit increased frequency of NT defects (NTDs), demonstrating genetic interactions (McGreevy, Vijayraghavan, Davidson, & Hildebrand, 2015). Shroom3 associates with Dvl2 and preferentially localizes to AP cell boundaries in the neural plate. This localization was abolished in *vangl2*$^{Lp/Lp}$ mutant mice. Notably, planar polarization of Dvl2 and ROCK was not changed in *shroom3*$^{gt/gt}$ mutant mice (McGreevy et al., 2015). These observations suggest that the PCP pathway might activate the Rho-ROCK-MyoII system by controlling Shroom3 localization.

4. The PCP pathway in neural tube closure

Interference with core PCP proteins causes shortened body axis and NT defects (Sokol, 1996; Topczewski et al., 2001; Wallingford & Harland, 2002; Wallingford et al., 2000; Ybot-Gonzalez et al., 2007). Whereas defective NT closure is a common phenotypic consequence of core PCP gene mutations in mouse and human embryos (Greene & Copp, 2005; Nikolopoulou, Galea, Rolo, Greene, & Copp, 2017; Wilde, Petersen, & Niswander, 2014), the cellular and molecular mechanisms behind these phenotypes remain to be clarified.

4.1 Cell behaviors and PCP protein localization in the neural plate

Neurulation is manifested by various behaviors of neuroepithelial cells, depending on their location and the developmental stage (Butler & Wallingford, 2017; Nishimura, Honda, & Takeichi, 2012). Some cells undergo oriented cell divisions along the AP axis (Gong, Mo, & Fraser, 2004). Other cells intercalate or elongate mediolaterally (Elul, Koehl, & Keller, 1997) (Fig. 4A and B). The cells also undergo apical constriction (Fig. 4C) (Suzuki, Morita, & Ueno, 2012). While all these behaviors are likely to play a role in neurulation, their frequencies in the neural plate and their relative contribution to NT closure remain unknown. For example, whereas the shrinking of the A-P cell junctions (Fig. 4B) is conserved during NT closure, it is less clear whether this process reflects anisotropic apical constriction or regulates protrusion-dependent cell intercalation during neural convergent extension (Huebner & Wallingford, 2019).

During NT closure, various cell behaviors are accompanied with progressive polarization of core PCP components. For instance, the Vangl2-Pk3 complex predominantly distributes at the anterior end of individual

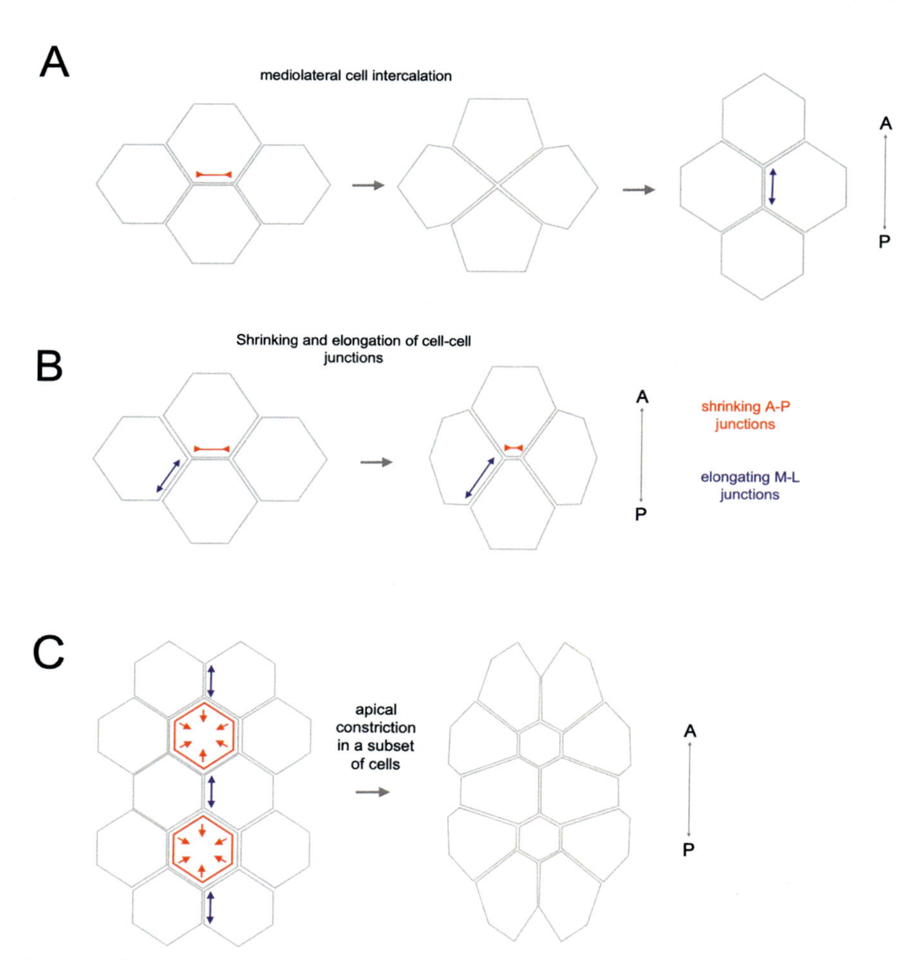

Fig. 4 Cell behaviors that contribute to neural plate folding. (A) Neighbor cell exchange. Shrinking of cell junctions in the direction perpendicular to the AP axis initiates mediolateral cell intercalation, leads to the formation of a vertex and the extension of the junctions perpendicular to the ML axis. (B) The same process of shrinking some junctions and extension of other junctions in the absence of cell neighbor exchange triggers cell shape change (elongation) during neural plate folding. (C) Apical constriction observed in a subset of cells along the AP axis may cause the neuroepithelium to bend (as indicated).

neuroepithelial cells in *Xenopus* embryos (Ossipova, Chuykin, Chu, & Sokol, 2015). This accumulation of Vangl2 at the anterior side of each cell is observed only in the posterior part of the neural plate where NT folding is first initiated. In the anterior region where the neuroepithelium is still flat, Vangl2 does not show polarized distribution (Ossipova, Chuykin, et al., 2015). The NT closes from posterior to anterior direction in *Xenopus* embryos, suggesting that PCP is initially established in the posterior neural plate and later transmitted to the anterior neural plate by a diffusing morphogen or via cell-contact-based communication (Fig. 2). Alternatively, the temporal progression of polarization of PCP complexes could be programmed to correspond to NT closure at different AP positions.

Besides Vangl2 and Pk2/3 segregation to the *anterior* of each neuroepithelial cell, other proteins become asymmetrically localized in each cell along the mediolateral (ML) axis of the neural plate. Rab11, the exocyst component Sec15 and Diversin, a homolog of the *Drosophila* core PCP protein Diego, have been found concentrated at the medial border of each cell in the lateral neural plate, generating bilaterally symmetric PCP (Ossipova et al., 2014). Thus, the neural plate exhibits *two different axes of PCP*, that likely correspond to different cell behaviors. For example, the anterior localization of Vangl2 may be relevant to cell elongation and AP axis extension, whereas the medial enrichment of Rab11 and Diversin could accompany the convergence along the ML axis. Nevertheless, the two PCP axes are coupled, because the depletion of Vangl2 abolishes the polarization of Rab11 along the ML axis (Ossipova et al., 2014). Of note, two orthogonal PCP axes, Vangl-dependent and Fat/Ds-dependent, have been recently described in planarians (Vu et al., 2019).

4.2 Cross-talk between PCP protein localization and actomyosin contractility

The polarized distribution of core PCP proteins in neuroepithelium correlates with the activation of the Rho-ROCK-MyoII system in the closing NT. Live imaging revealed progressive accumulation of fluorescent protein-tagged Pk2 and Vangl2 at shrinking junctions but depletion from elongating junctions (Butler & Wallingford, 2018). The shrinking junctions were enriched with GFP-MRLC, although whether there was a causal connection remained unclear (Butler & Wallingford, 2018). Phosphorylated MRLC (pMRLC) also showed accumulation at the A-P cell faces in the chick (Nishimura & Takeichi, 2008) and mouse NT (Grego-Bessa, Hildebrand, & Anderson, 2015). This pMRLC accumulation was inhibited by Y-27632, a pharmacological ROCK inhibitor, or PDZ-RhoGEF knockdown (Nishimura et al., 2012; Nishimura & Takeichi, 2008). These observations suggest a causal relationship between PCP signaling and actomyosin contractility.

The neural plate is characterized by actomyosin cables that extend across multiple cells along their AP boundaries, which are proposed to promote NT closure. The formation of these supracellular actomyosin cables in the chick neural plate required Celsr1 (Nishimura et al., 2012). Both Celsr1 and Dvl2 preferentially localized to A-P cell boundaries and were necessary for F-actin and pMRLC accumulation (Nishimura et al., 2012). PDZ-RhoGEF and Daam1 are candidate proteins that modulate the actomyosin system downstream of Celsr1 and Dvl2 (Nishimura et al., 2012). Daam1 physically interacts with PDZ-RhoGEF in a Dvl2-dependent manner and enhances PDZ-RhoGEF activity and actomyosin contractions at the posterior cell faces. Supporting this model, interference with the function of Daam1 or GEF-H1 leads to failure in NT closure (Itoh, Ossipova, & Sokol, 2014; Nakaya et al., 2020). Although Dvl2 associates with several RhoGEFs (Nishimura et al., 2012; Tanegashima, Zhao, & Dawid, 2008; Tsuji et al., 2010), it remains unclear which GEF is involved in PCP. The proposed model is that Dvl2 recruited by Fzd and Celsr leads to anisotropic activation of actomyosin complexes at the same cell junctions (Fig. 5A). This model assumes that the Fzd-Dvl side of the cell is contractile, whereas the colocalization of MRLC and Pk in ascidian and *Xenopus* embryos suggests that the Vangl-Pk-enriched cortex has enhanced contractility (Fig. 5B) (Butler & Wallingford, 2018; Newman-Smith et al., 2015).

Consistent with the view that Wnt and/or PCP signaling locally activate the Rho-ROCK-MyoII system, Wnt5a has been reported to cause actomyosin accumulation in *Xenopus* ectoderm (Choi & Sokol, 2009).

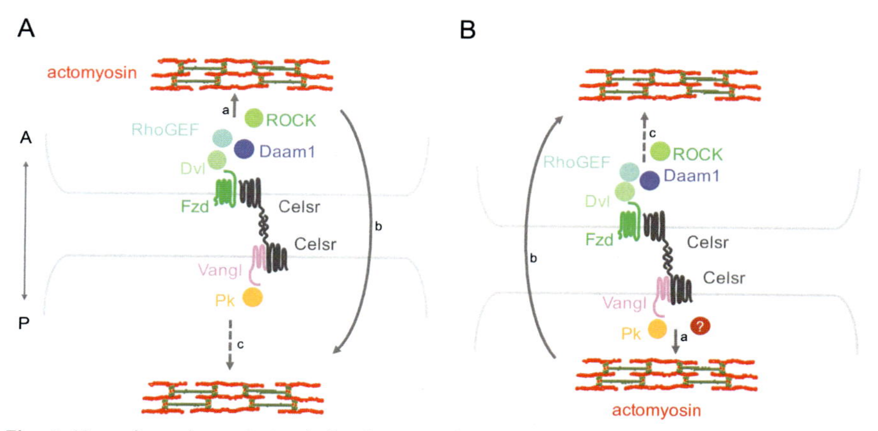

Fig. 5 How does the polarized distribution of PCP complexes lead to actomyosin contractility during neural plate folding? (A) Fzd-mediated recruitment of Dvl activates RhoA and Daam1, leading to actomyosin contractions at the posterior cell edge. (B) The Vangl-Prickle complex promotes Myosin II phosphorylation at the anterior cell edge. In both models, the increased actomyosin contractility on one cell side is transmitted to the other side via PCP protein interactions and/or cell-cell adhesion, possibly coupled with mechanotransduction (large arrows). A, anterior; P, posterior.

Furthermore, dominant-negative forms of Wnt and Vangl2 knockdown disrupted the anterior polarized distribution of Vangl2 as well as MRLC phosphorylation in the closing NT (Chu & Sokol, 2016; Ossipova, Kim, et al., 2015). Importantly, Vangl2 regulates actomyosin complexes and promotes apical constriction in a non-cell autonomous manner (Galea et al., 2021). These findings suggest that polarized core PCP complexes are involved in the activation of the Rho-ROCK-MyoII system, leading to cell junction remodeling and supracellular actomyosin cable formation during NT closure.

Conversely, the activity of ROCK and Myosin II is necessary for the accumulation of polarized core PCP complexes. Overexpression of the dominant-negative form of ROCK or the constitutively active form of Mypt1 disrupted the anterior enrichment of Vangl2 in *Xenopus* neuroepithelial cells (Ossipova, Kim, et al., 2015). Similarly, inhibition of Myosin II activity by blebbistatin interfered with planar polarization of Pk in ascidian notochord cells (Newman-Smith et al., 2015). This requirement of the Rho-ROCK-MyoII module for PCP reflects the regulatory feedback between actomyosin contractility and PCP signaling.

5. Conclusions

The core PCP pathway coordinates cell polarity in *Drosophila* epithelia and has critical roles in vertebrate morphogenetic events, including NT closure. Genetic and biochemical studies in multiple models over three decades identified main molecular players and revealed feedback regulation between core PCP proteins and actomyosin contractile complexes that drive neurulation. While PCP signaling is necessary for actomyosin contractility during NT closure, Myosin II functions to control PCP protein localization.

Due to the ease of experimental manipulation and monitoring, the *Xenopus* neural plate has become one of the best models for vertebrate PCP. This model allows research at various levels, from whole tissue dynamics and individual cell behaviors to molecular signaling pathways at subcellular resolution. The polarized segregation of core PCP complexes is initially observed at the posterior neural plate, but later becomes apparent in the anterior neuroectoderm. This dynamic change could be intrinsically coupled with the progression of neurulation or instructed by a diffusible signal or cell-cell communication. Future studies of biochemical and mechanical signals are needed to elucidate how the PCP pathway coordinates actomyosin-dependent cell behaviors during vertebrate NT closure.

Acknowledgments

We thank Ilya Chuykin, Poulomi Ray and Jakub Harnos for comments on the manuscript and stimulating discussions. We apologize to those colleagues whose work has not been discussed in due to space limitations. The work in the Sokol laboratory has been supported by the NIH grants GM122492, HD092990, DE027665 and NS100759.

References

Adler, P. N. (2012). The frizzled/stan pathway and planar cell polarity in the Drosophila wing. *Current Topics in Developmental Biology*, *101*, 1–31.

Aigouy, B., Farhadifar, R., Staple, D. B., Sagner, A., Roper, J. C., Julicher, F., et al. (2010). Cell flow reorients the axis of planar polarity in the wing epithelium of Drosophila. *Cell*, *142*, 773–786.

Alexandre, C., Baena-Lopez, A., & Vincent, J. P. (2014). Patterning and growth control by membrane-tethered wingless. *Nature*, *505*, 180–185.

Amin, E., Dubey, B. N., Zhang, S. C., Gremer, L., Dvorsky, R., Moll, J. M., et al. (2013). Rho-kinase: Regulation, (dys)function, and inhibition. *Biological Chemistry*, *394*, 1399–1410.

Axelrod, J. D. (2001). Unipolar membrane association of dishevelled mediates frizzled planar cell polarity signaling. *Genes & Development*, *15*, 1182–1187.

Bilic, J., Huang, Y. L., Davidson, G., Zimmermann, T., Cruciat, C. M., Bienz, M., et al. (2007). Wnt induces LRP6 signalosomes and promotes dishevelled-dependent LRP6 phosphorylation. *Science*, *316*, 1619–1622.

Bolinger, C., Zasadil, L., Rizaldy, R., & Hildebrand, J. D. (2010). Specific isoforms of drosophila shroom define spatial requirements for the induction of apical constriction. *Developmental Dynamics*, *239*, 2078–2093.

Butler, M. T., & Wallingford, J. B. (2017). Planar cell polarity in development and disease. *Nature Reviews. Molecular Cell Biology*, *18*, 375–388.

Butler, M. T., & Wallingford, J. B. (2018). Spatial and temporal analysis of PCP protein dynamics during neural tube closure. *eLife*, *7*, e36456.

Chaudhary, V., Hingole, S., Frei, J., Port, F., Strutt, D., & Boutros, M. (2019). Robust Wnt signaling is maintained by a Wg protein gradient and Fz2 receptor activity in the developing Drosophila wing. *Development*, *146*, dev174789.

Chen, W. S., Antic, D., Matis, M., Logan, C. Y., Povelones, M., Anderson, G. A., et al. (2008). Asymmetric homotypic interactions of the atypical cadherin flamingo mediate intercellular polarity signaling. *Cell*, *133*, 1093–1105.

Chien, Y. H., Keller, R., Kintner, C., & Shook, D. R. (2015). Mechanical strain determines the axis of planar polarity in ciliated epithelia. *Current Biology*, *25*, 2774–2784.

Choi, S. C., & Sokol, S. Y. (2009). The involvement of lethal giant larvae and Wnt signaling in bottle cell formation in Xenopus embryos. *Developmental Biology*, *336*, 68–75.

Chu, C. W., & Sokol, S. Y. (2016). Wnt proteins can direct planar cell polarity in vertebrate ectoderm. *eLife*, *5*, e16463.

Dabdoub, A., & Kelley, M. W. (2005). Planar cell polarity and a potential role for a Wnt morphogen gradient in stereociliary bundle orientation in the mammalian inner ear. *Journal of Neurobiology*, *64*, 446–457.

Davey, C. F., & Moens, C. B. (2017). Planar cell polarity in moving cells: Think globally, act locally. *Development*, *144*, 187–200.

Devenport, D. (2016). Tissue morphodynamics: Translating planar polarity cues into polarized cell behaviors. *Seminars in Cell & Developmental Biology*, *55*, 99–110.

Devenport, D., & Fuchs, E. (2008). Planar polarization in embryonic epidermis orchestrates global asymmetric morphogenesis of hair follicles. *Nature Cell Biology*, *10*, 1257–1268.

Elul, T., Koehl, M. A., & Keller, R. (1997). Cellular mechanism underlying neural convergent extension in Xenopus laevis embryos. *Developmental Biology*, *191*, 243–258.
Ewen-Campen, B., Comyn, T., Vogt, E., & Perrimon, N. (2020). No evidence that Wnt ligands are required for planar cell polarity in Drosophila. *Cell Reports*, *32*, 108121.
Farin, H. F., Jordens, I., Mosa, M. H., Basak, O., Korving, J., Tauriello, D. V., et al. (2016). Visualization of a short-range Wnt gradient in the intestinal stem-cell niche. *Nature*, *530*, 340–343.
Galea, G. L., Maniou, E., Edwards, T. J., Marshall, A. R., Ampartzidis, I., Greene, N. D. E., et al. (2021). Cell non-autonomy amplifies disruption of neurulation by mosaic Vangl2 deletion in mice. *Nature Communications*, *12*, 1159.
Gao, B., Song, H., Bishop, K., Elliot, G., Garrett, L., English, M. A., et al. (2011). Wnt signaling gradients establish planar cell polarity by inducing Vangl2 phosphorylation through Ror2. *Developmental Cell*, *20*, 163–176.
Gong, Y., Mo, C., & Fraser, S. E. (2004). Planar cell polarity signalling controls cell division orientation during zebrafish gastrulation. *Nature*, *430*, 689–693.
Goodrich, L. V., & Strutt, D. (2011). Principles of planar polarity in animal development. *Development*, *138*, 1877–1892.
Grassie, M. E., Moffat, L. D., Walsh, M. P., & MacDonald, J. A. (2011). The myosin phosphatase targeting protein (MYPT) family: A regulated mechanism for achieving substrate specificity of the catalytic subunit of protein phosphatase type 1delta. *Archives of Biochemistry and Biophysics*, *510*, 147–159.
Gray, R. S., Roszko, I., & Solnica-Krezel, L. (2011). Planar cell polarity: Coordinating morphogenetic cell behaviors with embryonic polarity. *Developmental Cell*, *21*, 120–133.
Greene, N. D., & Copp, A. J. (2005). Mouse models of neural tube defects: Investigating preventive mechanisms. *American Journal of Medical Genetics. Part C, Seminars in Medical Genetics*, *135C*, 31–41.
Grego-Bessa, J., Hildebrand, J., & Anderson, K. V. (2015). Morphogenesis of the mouse neural plate depends on distinct roles of cofilin 1 in apical and basal epithelial domains. *Development*, *142*, 1305–1314.
Habas, R., Kato, Y., & He, X. (2001). Wnt/frizzled activation of rho regulates vertebrate gastrulation and requires a novel Formin homology protein Daam1. *Cell*, *107*, 843–854.
Haigo, S. L., Hildebrand, J. D., Harland, R. M., & Wallingford, J. B. (2003). Shroom induces apical constriction and is required for hingepoint formation during neural tube closure. *Current Biology*, *13*, 2125–2137.
Heasman, S. J., & Ridley, A. J. (2008). Mammalian rho GTPases: New insights into their functions from in vivo studies. *Nature Reviews. Molecular Cell Biology*, *9*, 690–701.
Hildebrand, J. D. (2005). Shroom regulates epithelial cell shape via the apical positioning of an actomyosin network. *Journal of Cell Science*, *118*, 5191–5203.
Hildebrand, J. D., & Soriano, P. (1999). Shroom, a PDZ domain-containing actin-binding protein, is required for neural tube morphogenesis in mice. *Cell*, *99*, 485–497.
Huebner, R. J., & Wallingford, J. B. (2019). Coming to consensus: A unifying model emerges for convergent extension. *Developmental Cell*, *48*, 126.
Itoh, K., Ossipova, O., & Sokol, S. Y. (2014). GEF-H1 functions in apical constriction and cell intercalations and is essential for vertebrate neural tube closure. *Journal of Cell Science*, *127*, 2542–2553.
Jaffe, A. B., & Hall, A. (2005). Rho GTPases: Biochemistry and biology. *Annual Review of Cell and Developmental Biology*, *21*, 247–269.
Lim, J., Norga, K. K., Chen, Z., & Choi, K. W. (2005). Control of planar cell polarity by interaction of DWnt4 and four-jointed. *Genesis*, *42*, 150–161.
Liu, W., Sato, A., Khadka, D., Bharti, R., Diaz, H., Runnels, L. W., et al. (2008). Mechanism of activation of the Formin protein Daam1. *Proceedings of the National Academy of Sciences of the United States of America*, *105*, 210–215.

Logan, C. Y., & Nusse, R. (2004). The Wnt signaling pathway in development and disease. *Annual Review of Cell and Developmental Biology*, *20*, 781–810.
Luxenburg, C., Heller, E., Pasolli, H. A., Chai, S., Nikolova, M., Stokes, N., et al. (2015). Wdr1-mediated cell shape dynamics and cortical tension are essential for epidermal planar cell polarity. *Nature Cell Biology*, *17*, 592–604.
Mahaffey, J. P., Grego-Bessa, J., Liem, K. F., Jr., & Anderson, K. V. (2013). Cofilin and Vangl2 cooperate in the initiation of planar cell polarity in the mouse embryo. *Development*, *140*, 1262–1271.
Matis, M., Russler-Germain, D. A., Hu, Q., Tomlin, C. J., & Axelrod, J. D. (2014). Microtubules provide directional information for core PCP function. *eLife*, *3*, e02893.
McGreevy, E. M., Vijayraghavan, D., Davidson, L. A., & Hildebrand, J. D. (2015). Shroom3 functions downstream of planar cell polarity to regulate myosin II distribution and cellular organization during neural tube closure. *Biology Open*, *4*, 186–196.
Minegishi, K., Hashimoto, M., Ajima, R., Takaoka, K., Shinohara, K., Ikawa, Y., et al. (2017). A Wnt5 activity asymmetry and intercellular signaling via PCP proteins polarize node cells for left-right symmetry breaking. *Developmental Cell*, *40*, 439–452 e434.
Murrell, M., Oakes, P. W., Lenz, M., & Gardel, M. L. (2015). Forcing cells into shape: The mechanics of actomyosin contractility. *Nature Reviews. Molecular Cell Biology*, *16*, 486–498.
Najarro, E. H., Huang, J., Jacobo, A., Quiruz, L. A., Grillet, N., & Cheng, A. G. (2020). Dual regulation of planar polarization by secreted Wnts and Vangl2 in the developing mouse cochlea. *Development*, *147*, dev191981.
Nakaya, M. A., Gudmundsson, K. O., Komiya, Y., Keller, J. R., Habas, R., Yamaguchi, T. P., et al. (2020). Placental defects lead to embryonic lethality in mice lacking the Formin and PCP proteins Daam1 and Daam2. *PLoS One*, *15*, e0232025.
Narimatsu, M., Bose, R., Pye, M., Zhang, L., Miller, B., Ching, P., et al. (2009). Regulation of planar cell polarity by Smurf ubiquitin ligases. *Cell*, *137*, 295–307.
Newman-Smith, E., Kourakis, M. J., Reeves, W., Veeman, M., & Smith, W. C. (2015). Reciprocal and dynamic polarization of planar cell polarity core components and myosin. *eLife*, *4*, e05361.
Nikolopoulou, E., Galea, G. L., Rolo, A., Greene, N. D., & Copp, A. J. (2017). Neural tube closure: Cellular, molecular and biomechanical mechanisms. *Development*, *144*, 552–566.
Nishimura, T., Honda, H., & Takeichi, M. (2012). Planar cell polarity links axes of spatial dynamics in neural-tube closure. *Cell*, *149*, 1084–1097.
Nishimura, T., & Takeichi, M. (2008). Shroom3-mediated recruitment of Rho kinases to the apical cell junctions regulates epithelial and neuroepithelial planar remodeling. *Development*, *135*, 1493–1502.
Olofsson, J., Sharp, K. A., Matis, M., Cho, B., & Axelrod, J. D. (2014). Prickle/spiny-legs isoforms control the polarity of the apical microtubule network in planar cell polarity. *Development*, *141*, 2866–2874.
Ossipova, O., Chuykin, I., Chu, C. W., & Sokol, S. Y. (2015). Vangl2 cooperates with Rab11 and myosin V to regulate apical constriction during vertebrate gastrulation. *Development*, *142*, 99–107.
Ossipova, O., Kim, K., Lake, B. B., Itoh, K., Ioannou, A., & Sokol, S. Y. (2014). Role of Rab11 in planar cell polarity and apical constriction during vertebrate neural tube closure. *Nature Communications*, *5*, 3734.
Ossipova, O., Kim, K., & Sokol, S. Y. (2015). Planar polarization of Vangl2 in the vertebrate neural plate is controlled by Wnt and myosin II signaling. *Biology Open*, *4*, 722–730.
Qian, D., Jones, C., Rzadzinska, A., Mark, S., Zhang, X., Steel, K. P., et al. (2007). Wnt5a functions in planar cell polarity regulation in mice. *Developmental Biology*, *306*, 121–133.
Sato, A., Khadka, D. K., Liu, W., Bharti, R., Runnels, L. W., Dawid, I. B., et al. (2006). Profilin is an effector for Daam1 in non-canonical Wnt signaling and is required for vertebrate gastrulation. *Development*, *133*, 4219–4231.

Shimada, Y., Usui, T., Yanagawa, S., Takeichi, M., & Uemura, T. (2001). Asymmetric colocalization of flamingo, a seven-pass transmembrane cadherin, and dishevelled in planar cell polarization. *Current Biology*, *11*, 859–863.

Shimada, Y., Yonemura, S., Ohkura, H., Strutt, D., & Uemura, T. (2006). Polarized transport of frizzled along the planar microtubule arrays in Drosophila wing epithelium. *Developmental Cell*, *10*, 209–222.

Shutova, M. S., & Svitkina, T. M. (2018). Common and specific functions of nonmuscle myosin II paralogs in cells. *Biochemistry (Mosc)*, *83*, 1459–1468.

Simoes Sde, M., Mainieri, A., & Zallen, J. A. (2014). Rho GTPase and Shroom direct planar polarized actomyosin contractility during convergent extension. *The Journal of Cell Biology*, *204*, 575–589.

Sit, S. T., & Manser, E. (2011). Rho GTPases and their role in organizing the actin cytoskeleton. *Journal of Cell Science*, *124*, 679–683.

Sokol, S. Y. (1996). Analysis of dishevelled signalling pathways during Xenopus development. *Current Biology*, *6*, 1456–1467.

Stahley, S. N., Basta, L. P., Sharan, R., & Devenport, D. (2021). Celsr1 adhesive interactions mediate the asymmetric organization of planar polarity complexes. *eLife*, *10*, e62097.

Strutt, D. I. (2001). Asymmetric localization of frizzled and the establishment of cell polarity in the Drosophila wing. *Molecular Cell*, *7*, 367–375.

Strutt, H., Searle, E., Thomas-Macarthur, V., Brookfield, R., & Strutt, D. (2013). A Cul-3-BTB ubiquitylation pathway regulates junctional levels and asymmetry of core planar polarity proteins. *Development*, *140*, 1693–1702.

Strutt, H., & Strutt, D. (2009). Asymmetric localisation of planar polarity proteins: Mechanisms and consequences. *Seminars in Cell & Developmental Biology*, *20*, 957–963.

Strutt, D. I., Weber, U., & Mlodzik, M. (1997). The role of RhoA in tissue polarity and frizzled signalling. *Nature*, *387*, 292–295.

Suzuki, M., Morita, H., & Ueno, N. (2012). Molecular mechanisms of cell shape changes that contribute to vertebrate neural tube closure. *Development, Growth & Differentiation*, *54*, 266–276.

Tanegashima, K., Zhao, H., & Dawid, I. B. (2008). WGEF activates Rho in the Wnt-PCP pathway and controls convergent extension in Xenopus gastrulation. *The EMBO Journal*, *27*, 606–617.

Topczewski, J., Sepich, D. S., Myers, D. C., Walker, C., Amores, A., Lele, Z., et al. (2001). The zebrafish glypican knypek controls cell polarity during gastrulation movements of convergent extension. *Developmental Cell*, *1*, 251–264.

Tsuji, T., Ohta, Y., Kanno, Y., Hirose, K., Ohashi, K., & Mizuno, K. (2010). Involvement of p114-RhoGEF and Lfc in Wnt-3a- and dishevelled-induced RhoA activation and neurite retraction in N1E-115 mouse neuroblastoma cells. *Molecular Biology of the Cell*, *21*, 3590–3600.

Usui, T., Shima, Y., Shimada, Y., Hirano, S., Burgess, R. W., Schwarz, T. L., et al. (1999). Flamingo, a seven-pass transmembrane cadherin, regulates planar cell polarity under the control of frizzled. *Cell*, *98*, 585–595.

Vu, H. T., Mansour, S., Kucken, M., Blasse, C., Basquin, C., Azimzadeh, J., et al. (2019). Dynamic polarization of the multiciliated planarian epidermis between body plan landmarks. *Developmental Cell*, *51*, 526–542 e526.

Wallingford, J. B., & Harland, R. M. (2002). Neural tube closure requires dishevelled-dependent convergent extension of the midline. *Development*, *129*, 5815–5825.

Wallingford, J. B., Rowning, B. A., Vogeli, K. M., Rothbacher, U., Fraser, S. E., & Harland, R. M. (2000). Dishevelled controls cell polarity during Xenopus gastrulation. *Nature*, *405*, 81–85.

Wilde, J. J., Petersen, J. R., & Niswander, L. (2014). Genetic, epigenetic, and environmental contributions to neural tube closure. *Annual Review of Genetics*, *48*, 583–611.

Winter, C. G., Wang, B., Ballew, A., Royou, A., Karess, R., Axelrod, J. D., et al. (2001). Drosophila rho-associated kinase (Drok) links frizzled-mediated planar cell polarity signaling to the actin cytoskeleton. *Cell, 105*, 81–91.

Wu, J., Roman, A. C., Carvajal-Gonzalez, J. M., & Mlodzik, M. (2013). Wg and Wnt4 provide long-range directional input to planar cell polarity orientation in Drosophila. *Nature Cell Biology, 15*, 1045–1055.

Yang, Y., & Mlodzik, M. (2015). Wnt-frizzled/planar cell polarity signaling: Cellular orientation by facing the wind (Wnt). *Annual Review of Cell and Developmental Biology, 31*, 623–646.

Ybot-Gonzalez, P., Savery, D., Gerrelli, D., Signore, M., Mitchell, C. E., Faux, C. H., et al. (2007). Convergent extension, planar-cell-polarity signalling and initiation of mouse neural tube closure. *Development, 134*, 789–799.

Yu, J. J. S., Maugarny-Cales, A., Pelletier, S., Alexandre, C., Bellaiche, Y., Vincent, J. P., et al. (2020). Frizzled-dependent planar cell polarity without secreted Wnt ligands. *Developmental Cell, 54*, 583–592 e585.

Zallen, J. A. (2007). Planar polarity and tissue morphogenesis. *Cell, 129*, 1051–1063.

CHAPTER THREE

Modeling endoderm development and disease in *Xenopus*

Nicole A. Edwards[a,*] and Aaron M. Zorn[a,b,*]
[a]Division of Developmental Biology, Center for Stem Cell and Organoid Medicine, Perinatal Institute, Cincinnati Children's Hospital Medical Center, Cincinnati, OH, United States
[b]Department of Pediatrics, University of Cincinnati College of Medicine, Cincinnati, OH, United States
*Corresponding authors: e-mail address: nicole.edwards@cchmc.org; aaron.zorn@cchmc.org

Contents

Abstract

The endoderm is the innermost germ layer that forms the linings of the respiratory and gastrointestinal tracts, and their associated organs, during embryonic development. *Xenopus* embryology experiments have provided fundamental insights into how the endoderm develops in vertebrates, including the critical role of TGFβ-signaling in endoderm induction,elucidating the gene regulatory networks controlling germ layer development and the key molecular mechanisms regulating endoderm patterning and morphogenesis. With new genetic, genomic, and imaging approaches, *Xenopus* is now routinely used to model human disease, discover mechanisms underlying endoderm organogenesis, and inform differentiation protocols for pluripotent stem cell differentiation and regenerative medicine applications. In this chapter, we review historical and

Current Topics in Developmental Biology, Volume 145
ISSN 0070-2153
https://doi.org/10.1016/bs.ctdb.2021.01.001

current discoveries of endoderm development in *Xenopus*, then provide examples of modeling human disease and congenital defects of endoderm-derived organs using *Xenopus*.

1. Introduction

The endoderm is the innermost germ layer of the vertebrate embryo, which gives rise to the epithelial lining of the respiratory and gastrointestinal (GI) systems as well as the parenchyma of associated organs such as the liver and pancreas. The fundamental principles of embryonic development are highly conserved between humans and animal models, and studies in the amphibian *Xenopus* have been instrumental to our understanding of endoderm organogenesis. Experiments in *Xenopus* have shed light on the genomic regulatory networks controlling initial endoderm specification, the importance of endoderm and mesoderm interactions in gut tube patterning, and have revealed some of the key molecular mechanisms controlling organ induction and morphogenesis. Studies in *Xenopus* are increasingly providing insight into how disruptions in these processes can result in congenital diseases of endoderm-derived organs in humans. Indeed, it is estimated that approximately 83% of human disease-associated genes have *Xenopus* orthologs (Hellsten et al., 2010; Session et al., 2016). Additionally, paradigms learned from *Xenopus* endoderm development continue to inform strategies to differentiate human pluripotent stem cells (hPSCs) into different endoderm organ lineages for regenerative medicine applications. In this chapter, we review our current understanding of endoderm development in *Xenopus* and highlight how *Xenopus* can model human defects and diseases of endoderm organs, with an emphasis on recent work integrating new genomic, imaging, and genome editing technologies.

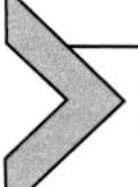

2. Experimental advantages of *Xenopus* to study endoderm development

Xenopus have all of the same endoderm-derived organs as humans, including a functional respiratory system, and have a long history of being used to study organogenesis. Two *Xenopus* species are used in research—the larger allotetraploid *Xenopus laevis*, which is better for embryology, and the somewhat smaller *Xenopus tropicalis*, which is diploid and increasingly being used for genetic studies. Both species have well annotated genomes (Hellsten et al., 2010; Session et al., 2016) supported by Xenbase (www.xenbase.org), a publicly available comprehensive *Xenopus* research knowledgebase containing curated literature, genomic, gene expression and phenotypic datasets (Karimi et al., 2018). Xenbase also facilitates human disease modeling by integrating functional data from *Xenopus* research with human disease-associated genes (Nenni et al., 2019). CRISPR-Cas9 gene editing is now routinely used in both *Xenopus laevis* and *Xenopus tropicalis* to generate F0 tadpole mutants and mutant *Xenopus* lines to effectively model endoderm-related developmental defects and investigate human disease gene function (Bhattacharya, Marfo, Li, Lane, & Khokha, 2015; Blitz, Biesinger, Xie, & Cho, 2013; Guo et al., 2014) (Table 1). Combined with its unique experimental advantages, these resources make *Xenopus* a powerful model for investigating the molecular basis of human endoderm development and disease (Hwang, Marquez, & Khokha, 2019).

Large, abundant, and externally developing *Xenopus* embryos enable experimental approaches that are difficult or impossible in mammalian embryos. For example, the ability to obtain thousands of synchronously developing sibling embryos has enabled detailed genomic, epigenetic, and proteomic studies of gastrulation and early endoderm development (Charney, Paraiso, Blitz, & Cho, 2017; Mukherjee et al., 2020; Sun et al., 2015; Wühr et al., 2014). Micro-injection of antisense morpholinos oligos (MOs) or CRISPR/Cas9 gene editing components are the typical methods to knockdown or mutate gene products for loss-of-function experiments, whereas injection of expression plasmids or synthetic mRNAs is commonly used in gain-of-function experiments. A well-characterized fate map, where presumptive endoderm cells and the organs that they develop into can be identified in early cleavage stage embryos (Chalmers & Slack, 2000; Dale & Slack, 1987; Moody, 1987a, 1987b), allows researchers to target microinjection of experimental reagents specifically to the endoderm, foregut or hindgut, and even right or left hand sides of the primitive gut (Davis et al., 2017; McLin, Rankin, & Zorn, 2007; Muller, Prather, & Nascone-Yoder, 2003; Stevens et al., 2017; Womble et al., 2018) (Fig. 1A). Temporal control of transcription factors can be achieved by injection of mRNA encoding hormone-inducible GR:fusion proteins that can be activated by the addition of dexamethasone to the media to bypass effects on gastrulation or early endoderm patterning (Kolm & Sive, 1995). As *Xenopus* embryos develop in a simple saline solution, treatment with small molecule antagonists, agonists, or chemical compounds is possible at any time point, permitting large-scale pharmacological screens and manipulation of signaling pathways at different stages of embryogenesis. These approaches have been used to model drug and

Table 1 Congenital defects and diseases of endoderm-derived organs that have been modeled using Xenopus.

Organ system	Congenital defect/ disease	Gene/treatment
Mouth and pharynx	Hypoplastic mouth	*ctnnd1* Alharatani et al. (2020)
	Oral clefts, craniofacial defects	*raldh2* Kennedy and Dickinson (2012) *ism1* Lansdon et al. (2018) *sf3b4* Devotta, Juraver-Geslin, Gonzalez, Hong, and Saint-Jeannet (2016) *smchd1* Gordon et al. (2017) *cadherin11, adam13* Abbruzzese, Becker, Kashef, and Alfandari (2016)
Trachea, lungs and esophagus	Complete tracheoesophageal clefts	*gli2/3, rab11a* Nasr et al. (2019) *isl1* Kim et al. (2019)
	Lung agenesis	*osr1/2* Rankin, Gallas, Neto, Gómez-Skarmeta, and Zorn (2012) *tbx5, wnt2/2b* Steimle et al. (2018)
Pancreas	Neonatal diabetes	*rfx6* Pearl, Jarikji, and Horb (2011) *ngn3* Jensen, Rosenberg, Hecksher-Sørensen, and Serup (2007), Wang et al. (2006) *pcdb1* Simaite et al. (2014) *pax6* Nakayama et al. (2015) *abcc8* Proks et al. (2006) *slc2a2* Michau et al. (2013) *kcnj11* Girard et al. (2006), Vedovato et al. (2016) *wfs1* Osman et al. (2003)
	Annular pancreas, pancreatic hypoplasia	*tm4sf3* Jarikji et al. (2009) *pdx1* Guo et al. (2014) *ptf1a* Lei et al. (2012)
Liver	Abnormal lobation/ left-right asymmetry	*pitx2c* Womble, Amin, and Nascone-Yoder (2018)
	Drug and chemical induced liver injury	Paracetamol Saide, Sherwood, and Wheeler (2019) Copper Carotenuto et al. (2020) Benzo[*a*]pyrene Regnault et al. (2014)
Stomach and intestines	Intestinal atresia and malrotation	*jnk* Dush and Nascone-Yoder (2013) *foxf1* Tseng, Shah, and Jamrich (2004) *vangl2* Dush and Nascone-Yoder (2019)
	Colon cancer	*smad7* Pittman et al. (2009) *apc* Van Nieuwenhuysen et al. (2015)

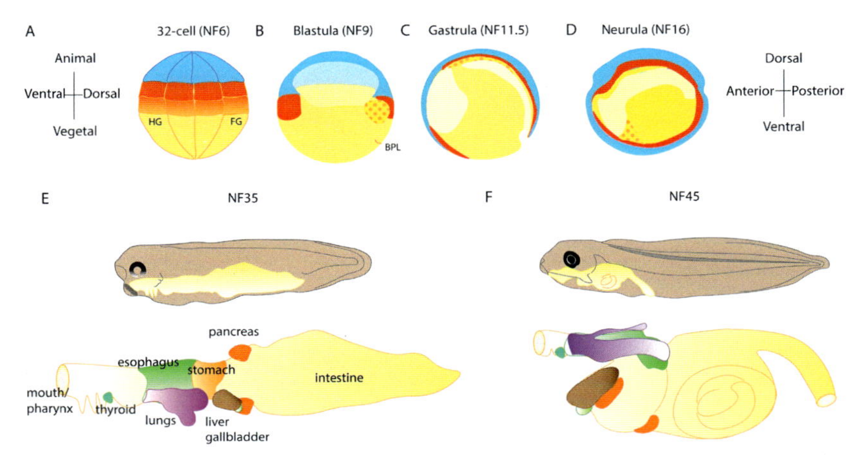

Fig. 1 Developmental fate map of the *Xenopus* endoderm. (A) At the 32-cell (NF6) stage, the germ layers are patterned in an animal to vegetal configuration (blue: ectoderm, red: mesoderm, yellow: endoderm) with the dorsal vegetal cells fated to become foregut and the ventral vegetal cells hindgut endoderm. (B) Gastrulation initiates at the blastopore lip (BPL) and future anterior foregut endoderm (AFG) cells originate at the organizer (yellow spots). (C) Gastrulation proceeds with the future AFG incrementally internalizing and migrating along the archenteron roof. (D) After gastrulation is complete, the AFG cells are located in the ventral endoderm mass. (E) At the tailbud stage (NF35) the gut tube is an elongated structure with developing organ buds. (F) By NF45, the gut has shortened, coiled, and organs have differentiated.

environmental toxicity-induced damage in *Xenopus* livers and to determine when key signaling pathways are necessary for endoderm organogenesis (Table 1, Fig. 2B).

Transgenics are also used for tissue specific expression in *Xenopus* and a large collection of versatile gateway compatible transgenic vectors (Love et al., 2011) are available in the community to study gene function, assay DNA cis-regulatory enhancer elements, or to generate fluorescent reporter lines (recently reviewed in Horb et al., 2019). A number of transgenic constructs driving expression in different endoderm lineages have been established such as *hhex* (gastrula anterior endoderm), *elastase* (exocrine pancreas), *pdx1* (endocrine pancreas), *transthyretin* (liver), *ifabp* (intestine), and *nkx2-5* (foregut and heart) (Beck & Slack, 1999; McLin et al., 2007; Sparrow et al., 2000). Temporal control can be added to tissue specific transgenes through the use of combinatorial doxycycline-inducible rtTA transgenic constructs to precisely regulate the spatiotemporal expression of wild type or dominant mutant constructs (Love et al., 2011; Nasr et al., 2019;

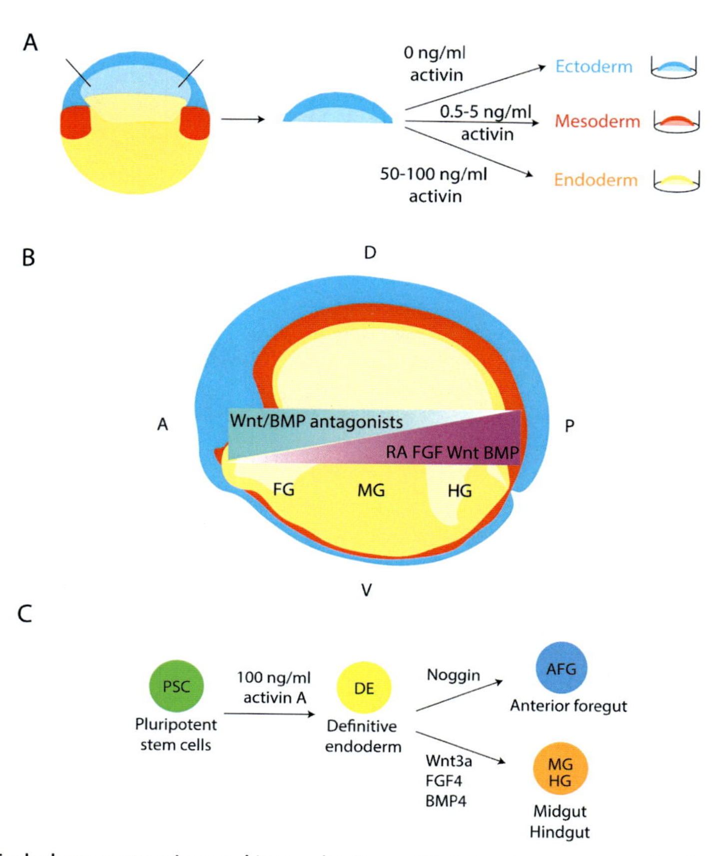

Fig. 2 Endoderm patterning and its applications. (A) Animal cap assay: animal cap cells of the *Xenopus* blastula can be dissected from the embryo and treated with varying levels of activin to direct differentiation into the three germ layers; high levels of activin promote endoderm differentiation. (B) The primitive gut tube is patterned during gastrula and neurula stages by signals between the mesoderm and endoderm. High levels of RA/FGF/Wnt/BMP from the posterior mesoderm promote mid and hindgut fates, while secreted Wnt/BMP antagonists in the anterior embryo keep these signals low to promote foregut fates. (C) Pluripotent stem cells can be directed into definitive endoderm with high levels of activin, then foregut, midgut, or hindgut fates using similar signals discovered in the *Xenopus* embryo.

Sterner et al., 2019). It is worth noting that mammalian enhancers and promoters frequently work in *Xenopus* (Beck & Slack, 1999; Juárez-Morales, Martinez-De Luna, Zuber, Roberts, & Lewis, 2017), emphasizing the highly conserved developmental programs. Many of these and other established transgenic lines are available from *Xenopus* resource centers

(National *Xenopus* Resource, http://www.mbl.edu/*Xenopus*, RRID:SCR_013731; European *Xenopus* Resource Center, http://www.port.ac.uk/research/exrc, RRID:SCR_007164) (Horb et al., 2019).

Another strength of the robust *Xenopus* embryo is relatively easy microsurgery and organ explant culture, a feature that has been particularly valuable in investigating the paracrine signaling between the endoderm and mesoderm during cardiopulmonary and gut development (Foley & Mercola, 2005; Hempel & Kühl, 2016; Horb & Slack, 2001; Nascone & Mercola, 1995; Rankin et al., 2016; Steimle et al., 2018). The fact that early *Xenopus* embryos can absorb oxygen from the media and have yolk protein in each cell means that they can survive severe defects, which in mammals results in early embryonic lethality and no embryos to analyze due to the loss of the heart or placenta. Combining all these different experimental advantages of *Xenopus* enables complex molecular epistasis experiments, testing how multiple genetic components interact in ways that are not really possible with mouse genetics.

3. Endoderm formation

In all vertebrate species the three primary germ layers endoderm, mesoderm, and ectoderm are specified during gastrulation. Fate maps of *Xenopus* blastula show that prior to gastrulation the embryo is patterned along the animal-vegetal (top-bottom) axis with the topmost animal cells giving rise to ectoderm, mesoderm from a ring of equator cells and the yolk-rich vegetal cells becoming endoderm (Fig. 1A). This mass of endoderm tissue in *Xenopus* is equivalent to the epithelial sheet of definitive endoderm in mammals, which emerges from the primitive streak and gives rise to the gut tube of the embryo, in contrast to the extra embryonic endoderm surrounding the epiblast prior to gastrulation which contributes primarily to the yolk sac (Zorn & Wells, 2009). While the morphology of the early endoderm tissue and the mechanics of gastrulation differ between mammals and amphibians, most of the molecular mechanisms underlying endoderm specification, patterning, and organogenesis are remarkably conserved between vertebrate species (Zorn & Wells, 2009).

Experiments in *Xenopus* from the 1990s were instrumental in defining how embryonic cells develop into the three germ layers (ectoderm, mesoderm, and endoderm). Particularly informative was the use the use of blastula "animal cap" cells. Normally fated to give rise to ectoderm, it was discovered that multipotent animal cap explants can be instructed to differentiate into mesoderm and endoderm tissues by culturing with Activin/TGFβ growth factors (Asashima et al., 1990; Okabayashi & Asashima, 2003; Smith & Green, 1990) (Fig. 2A). This original "organoid" system was used to define a conserved TGFβ/Nodal-dependent pathway promoting *in vivo* mesendoderm formation in all vertebrate embryos with high Nodal activity promoting endoderm and lower levels promoting mesoderm (Zorn & Wells, 2009) (Fig. 2A). This is now incorporated in routine protocols to differentiate hPSCs into definitive endoderm (D'Amour et al., 2005; Kubo et al., 2004; Loh et al., 2014) (Fig. 2C). Animal caps can in fact be directed toward many endoderm organ lineages including pharynx, intestine, pancreas, and liver using different iterations of activin and retinoic acid treatment (Ariizumi, Michiue, & Asashima, 2017). Another critical advance was the discovery of the transcription factor Sox17 as the first marker of the endodermal lineage in early vertebrate embryos (Hudson, Clements, Friday, Stott, & Woodland, 1997; Sinner, Rankin, Lee, & Zorn, 2004; Zorn et al., 1999). This opened the door to molecular analysis of the gene regulatory networks (GRN) controlling endoderm development, which is ongoing to this day (Charney et al., 2017).

In *Xenopus,* endoderm formation starts with maternally deposited *vegT* mRNA, which encodes a T-box transcription factor necessary to induce zygotic expression of Nodal ligands in vegetal cells of the blastula that are fated to become endoderm (Xanthos, Kofron, Wylie, & Heasman, 2001). Nodal expression is amplified in the dorsal vegetal cells by maternal Wnt/β-catenin, which helps specify and patterns the nascent endoderm and promotes the dorsal-anterior endoderm, a critical component of the Spemann-Mangold organizer (Afouda et al., 2020; Crease, Dyson, & Gurdon, 1998). The activation of Nodal ligand expression by maternal transcription factors is a *Xenopus*-specific feature of endoderm development. In mammals, Nodal expression in the anterior primitive streak is controlled by zygotic Wnt signaling (Zorn & Wells, 2009). However, the molecular mechanisms by which Nodal signaling regulates endoderm development appears to be conserved in all vertebrates.

Nodal induces both endoderm and mesoderm in a dose-dependent manner. Activity is highest in the vegetal cells, which secrete Nodal ligands and hence autocrine signaling between vegetal cells induces the expression of endoderm-promoting transcription factors (Smith & Green, 1990; Zorn & Wells, 2009). These Nodal ligands also diffuse to the overlaying equatorial cells, where moderate levels of Nodal promote a mesoderm GRN (Fig. 3). In vegetal cells, Sox17, Foxa, Gata4-6, and Mix-like transcription factors

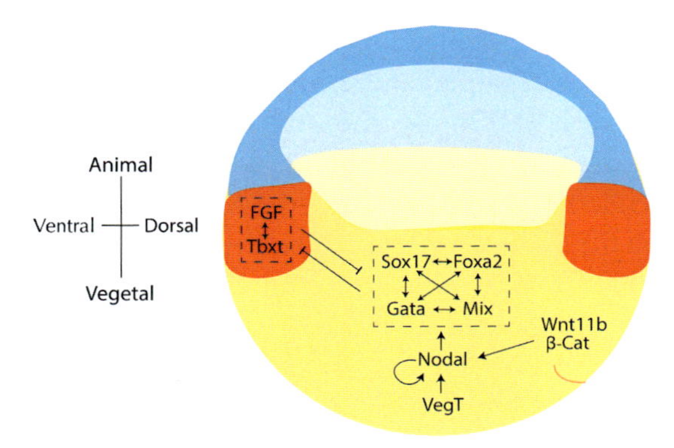

Fig. 3 Gene regulatory networks (GRN) governing mesendoderm formation. Maternal VegT and dorsal Wnt/β-catenin signaling promote high levels of Nodal in the vegetal cells of the pre-gastrula embryo. High Nodal signaling promotes the expression of endoderm GRN transcription factors—Sox17, Gata, Foxa2, and Mix, which stabilize the endoderm fate while reciprocally inhibiting the mesoderm GRN (FGF, Tbxt).

interact in an endoderm GRN to promote endoderm fate and repress mesoderm gene expression, while a mutually antagonistic Fgf-Tbxt (brachyury) GRN commits cells to the mesoderm lineage and represses endoderm identity (Charney et al., 2017; Sinner et al., 2006). Cutting edge genomics techniques used in *Xenopus* gastrula have recently uncovered key principles of the complex vertebrate endoderm GRN, including how dynamic epigenetic modifications control chromatin accessibility, how endoderm super enhancers coordinate lineage-specific transcription, and how combinatorial transcription factors interact with cell signaling effectors such as Smad and β-catenin to orchestrate the endoderm transcriptional program (Afouda et al., 2020; Charney et al., 2017; Gentsch, Spruce, Owens, & Smith, 2019; Mukherjee et al., 2020; Paraiso et al., 2019).

Work in *Xenopus* has also informed our understanding of the morphogenetic processes of gastrulation that internalizes the endoderm germ layer and elongates the anterior-posterior (A-P) axis. Gastrulation is initiated at the future dorsal side of the embryo by Wnt/β-catenin signaling. Internalization of the mesendoderm proceeds through cell involution, elongation, and migration along the blastocoel roof, forming and lining the archenteron, or future gut cavity (Fig. 1B and C). One of the main biomechanical mechanisms is convergent extension and elongation of the axial mesendoderm regulated by the non-canonical Wnt/planar cell polarity pathway and FGF signaling (reviewed in Butler & Wallingford, 2017; Keller & Sutherland, 2020; Winklbauer, 2020). Studies on the biomechanics of *Xenopus* mesendoderm tissue are ongoing and

Xenopus embryology experiments recently discovered that endoderm cell ingression is a major cellular mechanism driving vegetal rotation in gastrulation (Wen & Winklbauer, 2017). By the end of gastrulation, the endoderm becomes completely internalized. At this point the leading edge of the dorsal-anterior endoderm has migrated to the future ventral foregut, whereas the endoderm on the dorsal roof of the archenteron is a single layer of cells and ventral endodermal mass is several layers thick (Fig. 1D) (Popov et al., 2017).

3.1 Patterning the primitive gut tube

Initial patterning of the endoderm into broad dorsal-anterior and ventral-posterior domains is coincident with germ layer induction in the blastula with maternal Wnt/β-catenin specifying dorsal-anterior endoderm fate in the organizer. During gastrula and neural stages, zygotic Wnt, BMP, FGF and retinoic acid (RA) signals from the ventral-posterior mesoderm elaborate this further and pattern the endoderm along the A-P axis into distinct foregut, midgut, and hindgut domains (Zorn & Wells, 2009) (Fig. 2B). In general, these signals are posteriorizing. For example, Wnt and BMP ligands from the posterior mesoderm promote hindgut fate and repress foregut identity, whereas Wnt/BMP-antagonists are required to repress Wnt and BMP signaling in the foregut (McLin et al., 2007). Recent genomic studies have begun to reveal how combinatorial signaling is integrated at the genomic level with β-catenin and Smad1, the effectors of Wnt and BMP respectively, co-binding enhancers to regulate the foregut *versus* hindgut transcriptomes (Mukherjee et al., 2020; Stevens et al., 2017). Disruption of these key signaling pathways or transcription factors required for endoderm axis or organ patterning can lead to multi-organ disorders, particularly when BMP signaling is disrupted early in development—loss of BMP regulators Chordin and Szl disrupts anterior endoderm development and leads to agenesis of foregut organs in *Xenopus* (Bilogan & Horb, 2012; Kenny et al., 2012).

Tissue recombination experiments in *Xenopus* embryos have been instrumental in defining the reciprocal signaling between the mesoderm and endoderm, regulating positional information in the gut tube and progressively defining organ lineages. One observation that has emerged is that the same signals are used over and over again with different context dependent outcomes. For example, while RA, Fgf, Wnt and BMP inhibit foregut fate in the gastrula, just 24 h later they promote liver, lung, and/or pancreatic fate in the foregut (McLin et al., 2007; Rankin et al., 2018; Zhang,

Rankin, & Zorn, 2013) (Fig. 2B). Epistasis experiments have further begun to define the molecular basis of combinatorial signaling in the foregut, with RA signaling from the mesoderm controlling both the competence of the foregut endoderm and the expression of Wnt and BMP ligands that subsequently induce Nkx2-1 positive respiratory progenitors in the endoderm (Rankin et al., 2016). Similarly, early repression of Wnt/β-catenin signaling is necessary for anterior foregut fate but is then later required for differentiation of pancreatic progenitors (Li et al., 2008; McLin et al., 2007). Studies on the context-dependent combinatorial signals of endoderm organogenesis in *Xenopus* continue to inform and optimize protocols to differentiate hPSC into different lineages *in vitro* (Rankin et al., 2016, 2018).

Between neurula (NF15) and tadpole stages (NF40s), the endodermal mass elongates by radial cell intercalation and cavitation to form the classical gut tube (Chalmers & Slack, 2000; Reed et al., 2009). As a result of patterning and mesoderm-endoderm interactions, organ-specific lineages are induced at the appropriate position around NF30-35. Lung, liver, and pancreatic organ buds then emerge from the gut tube around NF40, which then proliferate and differentiate into physiological functioning organs as the tadpole's digestive system becomes active with feeding by stage NF45 (Fig. 1E and F). During metamorphosis, many endoderm-derived organs undergo dramatic remodeling and reorganization to have structural similarities to mammalian postnatal organs, including the pancreas and small intestine (Mukhi, Horb, & Brown, 2009; Pearl, Bilogan, Mukhi, Brown, & Horb, 2009).

4. Mouth and pharynx development

The mouth and pharynx are the most anterior endoderm-derived structures, providing an opening to the external environment, and allowing passage of food into the digestive tract and air into the respiratory system. Many discoveries about vertebrate mouth development were first made in *Xenopus* (reviewed in Chen, Jacox, Saldanha, & Sive, 2017). Mouth development begins in the late neurula stages where the anterior endoderm comes into direct contact with the head ectoderm, without any intervening mesoderm. Signals from this region recruit cranial neural crest cells that will later form the jaws, palate, and facial cartilage (Minoux & Rijli, 2010). A signaling cascade in the primitive mouth involving Hedgehog and the secreted Wnt-antagonists Frzb1 and Crescent regulates the degradation of the basement membrane separating the endoderm and ectoderm through reducing levels of basement membrane components fibronectin and laminin (Dickinson & Sive, 2009; Tabler, Bolger, Wallingford, & Liu, 2014) (Fig. 4A). Endoderm and ectoderm cells then intercalate to form a thin, 1–2 cell layer buccopharyngeal membrane, which ultimately perforates *via* Hedgehog and JNK-dependent mechanisms, and opens the mouth between NF39-40 (Houssin, Bharathan, Turner, & Dickinson, 2017; Tabler et al., 2014).

The pharynx develops from a set of segmented outpouchings on the lateral side of the head named the pharyngeal arches. The pharyngeal arches are composed of tissue from all three germ layers. The inner lining of the arches is endoderm-derived and forms the pharyngeal epithelium and glands. Pharyngeal arch mesoderm becomes musculoskeletal derivatives, while the arch ectoderm forms sensory neurons and epidermis (Grevellec & Tucker, 2010). Each arch contributes to different sets of organs including the thymus, parathyroid and thyroid glands, and ultimobranchial bodies which are derived from the third pharyngeal arch (Lee et al., 2013). Conserved development and function of these pharyngeal endoderm-derived glands between *Xenopus* and mammals have made *Xenopus* a useful model for developmental studies of the immune system and the effects of thyroid-hormone disrupting chemicals (Buchholz, 2015; Colombo, Scalvenzi, Benlamara, & Pollet, 2015). As thyroid hormones (T3/T4) are required for frog metamorphosis, successful *Xenopus laevis* metamorphosis is used as a phenotypic readout of screening endocrine disrupting compounds by the Organization for Economic Co-Operation and Development (OECD) and Environmental Protection Agency (EPA, USA) (Miyata & Ose, 2012).

4.1 Modeling human orofacial defects in *Xenopus*

Orofacial clefts are common congenital defects that occur when structures in the face and oral cavity do not properly fuse during development. While the primitive mouth may have morphological differences between vertebrate species (Chen et al., 2017), the basic phases of development discovered in *Xenopus* are conserved, and human orofacial defects such as oral clefts, choanal atresia, and persistent buccopharyngeal membrane have been modeled in tadpoles (Dickinson, 2016; Dickinson & Sive, 2009; Jacox, Chen, Rothman, Lathrop-Marshall, & Sive, 2016; Jacox et al., 2014). Using loss of function, pharmacological inhibition, and innovative "facial transplants" between control and manipulated embryos, retinoic acid was

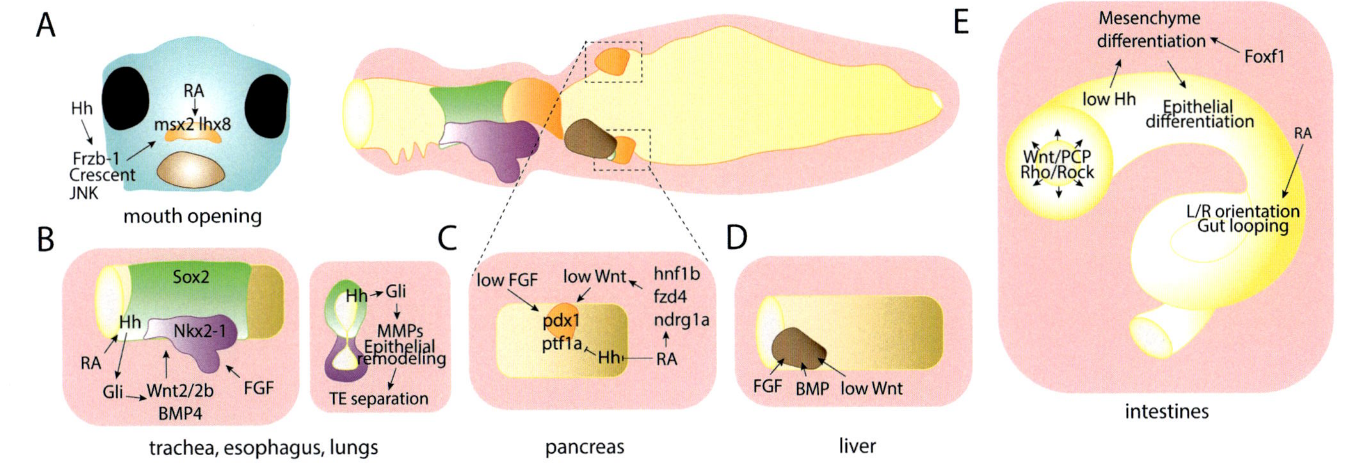

Fig. 4 Patterning and differentiation of endoderm organs. (A) A combination of Hh, JNK, Wnt, and RA signaling pattern and form the embryonic mouth, which eventually perforates to open the gut tube to the external environment. (B) RA, Hh, Wnt, and BMP signaling control trachea-esophageal separation and elongation of the lung buds. (C) The pancreatic transcription factor cascade is regulated by RA and low levels of FGF. (D) The liver is specified by ventral FGF and BMP signaling from the surrounding mesenchyme. (E) Foxf1-dependent mesenchymal differentiation promotes intestinal epithelial differentiation. Wnt/PCP and Rho/Rock signaling within the epithelium promotes differentiation and lumen expansion, while RA regulates left-right orientation and gut looping during intestinal development.

discovered to be a key regulator of *msx2* and *lhx8* in the developing mouth; dysregulation of these critical mouth development genes leads to median facial clefts (Kennedy & Dickinson, 2012). In addition, several candidate genes identified in patients with craniofacial defects have been functionally validated in *Xenopus* including *ism1*, *ccnd1*, *schmd1*, *sf3b4*, *cdh11*, and *adam13*, and experiments have been able to place these genes into a mechanistic frame work, when mutations in humans could result in patient phenotypes (Abbruzzese et al., 2016; Alharatani et al., 2020; Devotta et al., 2016; Gordon et al., 2017; Lansdon et al., 2018).

5. Trachea, lungs, and esophagus development

Evolution of the respiratory system was a key step in adaption to air breathing in amphibians. Studies in the last 5 years have shown that the molecular and cellular basis of early respiratory system development are highly conserved between *Xenopus* and mammals. Respiratory progenitors are first detected at the molecular level by the expression of the transcription factor Nkx2-1 in the ventral foregut endoderm at NF35 in *Xenopus* (equivalent to E9.5 in mice) (Rankin et al., 2015, 2016). Within a day (NF42) the Nkx2-1-expressing epithelium evaginates from the gut tube to form paired lung buds coincident with separation of the trachea and esophagus (Rankin et al., 2015). The tadpole lungs are a simple sac-like structure with a single-layered epithelium surrounded by a thin mesenchymal smooth muscle layer (Rose & James, 2013). During metamorphosis, the *Xenopus* lungs septate to form the rudimentary alveolar structure of the adult amphibian lung, which does not undergo the elaborate branching morphogenesis that mammals do (Rankin et al., 2015).

Despite morphological differences in adult lung structure, the early stages of pulmonary development in *Xenopus* and mice are regulated by conserved FGF, RA, BMP, Hedgehog, and Wnt signaling between the endoderm and surrounding mesoderm (Fig. 4B). In both species Wnt2/2b and Bmp4 signaling from the ventral splanchnic mesoderm are necessary to downregulate Sox2 and induce Nkx2-1 expression in the ventral foregut epithelium. Additional signaling from nearby cardiopulmonary progenitors *via* Tbx5-Wnt2/2b drives pulmonary specification, which then promotes reciprocal signaling back to the developing heart (Steimle et al., 2018). Epistasis experiments possible in *Xenopus* have recently revealed how these different signaling pathways interact in a complex regulatory network initiated by RA (Rankin et al., 2012; Rankin et al., 2016, 2018; Wang et al., 2011).

These studies linking general A-P patterning to lung induction show that RA signaling between NF15-25 patterns both the endoderm and mesoderm promoting the expression of Hedgehog ligands in the endoderm, which in turn stimulate the expression of lung-inducing Wnt and BMP in the mesenchyme. Moreover, RA appears to somehow control the competence of the endoderm to respond to these Wnt and BMP signals. Subsequent studies in mouse embryos and hPSCs indicate that similar mechanisms operate in mammals (Rankin et al., 2016).

Comparative analyses in *Xenopus* and mouse are beginning to reveal the molecular and cellular mechanisms that separate the single foregut tube into the trachea and esophagus (between NF39-42 in *Xenopus* and E10.5–11.5 in mice). After patterning the ventral Nkx2-1 + respiratory domain and a dorsal Sox2 + esophageal domain, the foregut constricts at the midline causing the opposite walls of the foregut endoderm to touch, forming a transient epithelial septum. Upstream of Nkx2-1 expression, Isl1 together with Sox2 and Nkx2-1 mark a unique population in the foregut endoderm where constriction appears to initiate (Kim et al., 2019). As development proceeds, the epithelial cells of the transient septum lose adhesion to one another, regulated by a Hedgehog/Gli/Rab11 dependent process. Basement membrane components between the epithelium and mesenchyme are degraded by matrix metalloproteases, allowing the surrounding mesenchyme to invade and separate the tracheal and esophageal epithelial tubes (Nasr et al., 2019).

5.1 *Xenopus* models of tracheoesophageal defects (TEDs) and airway disease

TEDs, including laryngotracheal clefts, esophageal atresia, and tracheoesophageal fistulas, occur when tracheoesophageal separation is disrupted. With the exception of a handful of mutations in known human genes (GLI3, SOX2) the genetic etiology of TEDs is poorly understood. The above recent studies on trachea-esophageal morphogenesis have begun to reveal how mutations in SOX2 or GLI3 might cause TEDs in patients. CRISPR mediated mutation of *gli3* in *Xenopus* to mimic a dominant mutation in Pallister-Hall syndrome patients disrupted Rab11-mediated epithelial remodeling causing severe laryngotracheal clefts (Nasr et al., 2019). Ongoing genomic studies in *Xenopus* will help discover targets of major transcription factors like SOX2 and GLI3 and reveal new candidate genes that when mutated in human patients may cause TEDs.

The structure and function of the mammalian airway mucociliary epithelium (MCE) can also be modeled using *Xenopus* tadpole skin. By two days of development, the ectoderm-derived epidermis differentiates into MCE containing mucous-secreting cells interspersed by beating multiciliated cells that produce fluid flow, similar to the mammalian airway MCE, which helps keep airways clear of particulates and debris. The *Xenopus* MCE develops from basal cells using conserved signaling pathways like Wnt/β-catenin and BMP, which are also important for development and regeneration of the mammalian airway MCE (Walentek & Quigley, 2017). Because the Xenopus external skin is easy to manipulate and visualize in real time, this has made it an attractive model to study development and function of MCE (Cibois et al., 2015; Haas et al., 2019), including *in vivo* modeling of ciliopathies and diseases affecting the airway that would otherwise be challenging to model in primary cultures of mouse or human tracheal airway tissue (Walentek & Quigley, 2017; Wallmeier et al., 2014). Indeed self-organizing *Xenopus* MCE organoids can now be generated to study the impact of tissue mechanical properties on cell development and differentiation in three dimensions (Kim, Jackson, Stuckenholz, & Davidson, 2020).

6. Pancreas development

With exocrine and endocrine cells, the pancreas is an essential digestive organ required for food digestion and blood glucose homeostasis. Studying how the *Xenopus* pancreas develops has provided insight into the signaling networks and transcription factors necessary for pancreatic cell fate and has informed strategies to generate β-cell specific human stem cell differentiation protocols. The pancreas originates from the posterior foregut endoderm as three separate organ buds growing into the surrounding mesenchyme. One dorsal pancreatic bud derived from archenteron roof endoderm appears at NF35/36 beneath the aorta, while two ventral buds emerge posterior to the hepatic bud at NF37/38 (Kelly & Melton, 2000). Dynamic movements and rearrangements of the gastrointestinal tract cause the pancreatic buds to fuse by NF40 (Kelly & Melton, 2000; Pearl et al., 2009). Lineage tracing transplantation experiments performed using transgenic *Xenopus* embryos have shown that cells from the dorsal and ventral organ buds are transcriptionally distinct and contribute to different regions of the pancreas (Jarikji et al., 2009).

Conserved RA, TGFβ, FGF, and Wnt signaling from the surrounding mesenchyme—pathways that are also used to direct lung and liver development—promote pancreatic fate (Fig. 4C). RA is required in gastrulating embryos to initiate a pancreatic program involving Hnf1β, Fzd4, and Ndrg1a (Ariizumi et al., 2017; Gere-Becker, Pommerenke, Lingner, & Pieler, 2018; Rankin et al., 2018; Zhang, Guo, & Chen, 2013). RA also appears to partly mediate its effects by inhibiting Hedgehog signaling (Chen et al., 2004). Combined with Activin, or VegT and Noggin, RA can reprogram *Xenopus* dorsal lip explants and animal cap cells into pancreatic tissue (Ariizumi et al., 2017; Chen et al., 2004; Moriya, Komazaki, & Asashima, 2000; Moriya, Komazaki, Takahashi, Yokota, & Asashima, 2000). Endoderm-only explants suggest that in *Xenopus*, pancreatic progenitors marked by *pdx1* and *ptf1a* expression are already specified by NF16; prolonged activation of FGF signaling after this stage represses pancreatic fate and promotes liver and lung development (Shifley, Kenny, Rankin, & Zorn, 2012).

6.1 Modeling neonatal diabetes and pancreatic developmental defects in *Xenopus*

Xenopus has been used to model neonatal diabetes, a disease caused by mutations in genes, particularly transcription factors, that lead to loss or dysfunction of pancreatic β-cells in human patients (reviewed in Kofent & Spagnoli, 2016; Salanga & Horb, 2015). Congenital defects including pancreas agenesis, hypoplasia, and annular pancreas have also been modeled in *Xenopus*. CRISPR-Cas9 and TALEN-mediated gene editing of the key pancreatic transcription factors such as *pdx1* and *ptf1a* causes hypoplastic pancreas in *Xenopus* (Guo et al., 2014; Lei et al., 2012). Dorsal-ventral bud fusion defects have also been modeled with *tm4sf3* morphant and overexpression models, with the latter resulting in annular pancreas, a structural defect where a ring of pancreatic tissue grows around the duodenum constricting flow in the intestine (Jarikji et al., 2009). Moreover, by knocking down candidate genes and by injecting *Xenopus* embryos with mutant mRNA constructs analogous to patient-specific mutations, researchers were able to validate that diabetes in Mitchell-Riley syndrome patients is likely to be caused by mutations in the transcription factor RFX6 (Pearl et al., 2011). Likewise, *Xenopus* with mutant Neurog3 (mimicking patients) lacked *neurod1* expression, reenforcing how disruptions to the transcription factor cascade of β-cell specification can lead to diabetes (Wang et al., 2006). The discovery of newly identified factors like Pcbd1, a Hnf1α cofactor, and Ire1α provide insight into the interplay between these transcriptional cascades and endoplasmic reticulum stress in pancreas development (Simaite et al., 2014; Yuan et al., 2014).

Neonatal diabetes can also be caused by defective β-cell electrophysiology and insulin release. Under physiological conditions, β-cells respond to glucose by potassium channel activation and cell depolarization, triggering a calcium influx that stimulates insulin release. Genetic mutations in a number of these ion channels and transporters can impact β-cell function. The *Xenopus* oocyte has been used extensively for structure-function studies to assay the impact of ion channel and glucose transporter mutations. *Xenopus* oocytes have a tremendous translation capacity and can be microinjected with synthetic mRNA to generate high levels of wild type or mutant protein, which can then be assayed electro-physiologically at the single cell level, effectively modeling the predicted defects in pancreatic β-cells. Mutations in ABCC8, KCNJ11, GLUT2 and WFS1 associated with neonatal diabetes have been functionally characterized in *Xenopus* oocytes (Girard et al., 2006; Michau et al., 2013; Osman et al., 2003; Proks et al., 2006; Vedovato et al., 2016). These examples used *in vivo* structure-function studies of patient mutations, which would otherwise be much more challenging and time consuming to perform in mouse genetic models.

7. Liver development

The liver is critical for protein synthesis, bile acid production, metabolizing nutrients, and detoxifying biochemical compounds. The endoderm derived parenchyma includes hepatocytes and cholangiocytes that make up the epithelium of the biliary system, which transports bile to the gall bladder and intestine. The hepato-biliary system originates from the ventral foregut endoderm as a single rudiment, the hepatic diverticulum, budding near where the future stomach and duodenum meet (Chalmers & Slack, 2000; Zorn & Mason, 2001). In *Xenopus*, the hepatic diverticulum grows anteriorly into a round structure with thick walls at NF33/34. At NF37-39, the liver then develops inward folds and the cavity fills with hepatocytes, with the gallbladder and extrahepatic ducts becoming obvious at the anterio-ventral aspect of the liver bud (Womble et al., 2018; Zorn & Mason, 2001). The hepato-biliary system undergoes further maturation, along with many digestive organs, at metamorphosis as the diet of the animal changes (Ueno et al., 2015).

Like other foregut organs, liver development initially requires low Wnt and BMP signaling to maintain Hhex+ progenitors in the anterior endoderm between NF11-20 (McLin et al., 2007; Zhang, Rankin, et al., 2013), but then between NF25-30, Wnt, BMP and FGF signaling from the cardiogenic and septum transversum mesenchyme are required to induce hepatic fate and promote the outgrowth and expansion of the liver bud (Fig. 4D) (McLin et al., 2007; Shifley et al., 2012). *Xenopus* experiments have revealed how the temporally dynamic and reiterative Wnt and BMP signaling impact the expression of key transcription factors Hhex, Gata4, Gata5, and Gata6 that are critical to coordinate development of the liver, biliary system, and ventral pancreas from a common population of ventral foregut endoderm progenitors (Haworth, Kotecha, Mohun, & Latinkic, 2008; McLin et al., 2007; Rankin, Kormish, Kofron, Jegga, & Zorn, 2011, Zhao, Han, Dawid, Pieler, & Chen, 2012).

7.1 Liver developmental defects and disease in *Xenopus*

As the major site of drug metabolism and heavy metal detoxification, liver damage is frequently used as a readout for drug and environmental toxicity screening. *Xenopus* tadpoles treated with increasing doses of paracetamol between NF38-45 exhibit similar drug-induced liver injury phenotypes observed in mammalian models. Interestingly, the *Xenopus* liver expresses many of the same enzymes responsible for paracetamol metabolism in humans (Saide et al., 2019). Because of these similarities, *Xenopus* tadpoles could act as a first pass drug-induced livery injury screen for future preclinical compounds without having to perform similar experiments in mammalian models (Saide et al., 2019). Copper and polycyclic aromatic hydrocarbons have also been tested for liver toxicity after environmental exposure in *Xenopus* tadpoles (Carotenuto et al., 2020; Regnault et al., 2014). Common liver diseases like cirrhosis and hepatitis are less well studied in *Xenopus*, but the physiological similarities of the *Xenopus* liver with mammalian livers provide an opportunity to test environmental factors or genetic predisposition to these diseases, as well as to investigate rare genetic mutations leading to congenital defects like biliary atresia and heterotaxia.

8. Stomach and intestinal development

The stomach originates from the posterior foregut, whereas the midgut and hindgut domains give rise to the small and large intestines, respectively. These progenitor domains can be distinguished by gene expression patterns as early as NF20-25, but it is not until after this stage that the endodermal mass elongates and cavitates to form the GI tract (Chalmers & Slack,

2000). At NF32, the gut tube is filled with large, round, non-polarized epithelial cells. From NF37 onwards, these cells become more radially oriented, organized, polarized, and become more elongated in shape within the gut tube (Reed et al., 2009). Eventually the lumen of the gut tube expands, the gut tube dramatically lengthens, and the epithelium becomes single-layered due to cell rearrangements driven by radial intercalation and convergent extension (Chalmers & Slack, 2000; Reed et al., 2009). Between NF41-47 the gut continues to elongate, the epithelium differentiates, and the gut begins to exhibit a characteristic coiling.

Wnt/Planar Cell Polarity and Rho/ROCK/Myosin II signaling within the gut epithelium and Hedgehog/RA signaling between the epithelium and lateral plate and visceral mesenchyme and are critical for driving this radial cell intercalation-mediated intestinal lengthening and epithelial differentiation (Fig. 4E) (Dush & Nascone-Yoder, 2019; Reed et al., 2009; Zhang, Rosenthal, De Sauvage, & Shivdasani, 2001). Low Hedgehog ligand secretion by the gut epithelium promotes Foxf1 positive mesenchyme differentiation and subsequent epithelial differentiation, while RA ensures correct left-right orientation and looping of the developing gut (Lipscomb, Schmitt, Sablyak, Yoder, & Nascone-Yoder, 2006; Tseng et al., 2004; Zhang et al., 2001). BMP signaling from the posterior visceral mesenchyme promotes the emergence of Satb2positive cells fated to become the distal small intestine and colon in between *Xenopus*, mouse, and humans; hPSCs in fact can be directed toward a colonic fate by promoting high levels of BMP during differentiation (Múnera et al., 2017).

During metamorphosis, *Xenopus* transition from being vegetarian to carnivorous and thyroid hormone (T3), the main driver of metamorphosis, induces a number of changes and maturation in intestinal structure and other GI organs (Schreiber, Cai, & Brown, 2005). T3-mediated changes cause the tadpole gut epithelium to degenerate and Lgr5positive adult epithelial stem cells to develop, remodeling the *Xenopus* intestines to form folding epithelial structures containing absorptive enterocytes and secretory cells with a crypt-villus epithelial axis similar to that found in the postnatal mammalian intestines (Hasebe, Fujimoto, Buchholz, & Ishizuya-Oka, 2020; Ishizuya-Oka et al., 2001; Sterling, Fu, Matsuura, & Shi, 2012). Studies of *Xenopus* metamorphosis have revealed the mechanism of thyroid hormone action, which models similar changes that occur during late prenatal and perinatal mammalian development, and suggests how environmental endocrine disruptions might result in birth defects (reviewed in Buchholz, 2015).

8.1 Modeling intestine defects and disease in *Xenopus*

Xenopus has been used to study congenital intestinal malrotation, as the direction of intestinal coiling can be clearly observed during tadpole development (Dush & Nascone-Yoder, 2013, 2019). Studies have implicated Wnt/PCP, Rho kinase, and JNK signaling as the regulators of cell behaviors controlling intestinal rotation which when disrupted could cause defects. Similarly, experimental manipulation of factors involved in left-right asymmetry including *nodal1* or *pitx2* can result in reversed gut coiling or disrupted left-right organ asymmetry (Davis et al., 2017; Womble et al., 2018) providing mechanistic insight into the etiology of heterotaxy in human patients.

Xenopus has also recently been used to model genetic GI cancers (Naert, Van Nieuwenhuysen, & Vleminckx, 2017). Screening cancer-associated genes with TALEN and CRISPR-Cas9 genome editing is much faster in *Xenopus* compared to rodents as cancer phenotypes can be detected in the F0 generation (Naert et al., 2017). Familial adenomatous polyposis, an inherited genetic condition that causes the accumulation of polyps in the epithelium of the large intestine, was modeled by mutating *apc* in *Xenopus tropicalis*. This genetic model generates tumors faster than mouse cancer models and can be used to screen for small molecules as therapeutic compounds (Van Nieuwenhuysen et al., 2015). *Xenopus* has also been used as a reporter assay to screen colorectal cancer risk variants in human SMAD7 using restriction enzyme-mediated transgenesis (Pittman et al., 2009).

9. Future directions

Research discoveries made in *Xenopus* have provided a fundamental understanding of how the endoderm gives rise to the digestive and respiratory systems. Insight gained from these studies are continuing to inform strategies to direct the differentiation of human organoids from iPSCs for regenerative medicine. With advances in CRISPR genome editing techniques and the ability to create precise mutations analogous to those found in patients, *Xenopus* is increasingly being used to understand the mechanisms of congenital birth defects. Moreover, by combining the classical experimental advantages of *Xenopus* embryos with sophisticated new technologies like single cell genomics and high-resolution imaging, investigators are discovering the gene regulatory networks governing endoderm organogenesis and revealing at a deep mechanistic level how disruptions in these networks can result in disease.

Acknowledgments

The authors thank Scott Rankin and Mardi Nenni for critically reviewing the manuscript. This work is supported in part by grants P41HD054556, P01HD093363, and DK123092 to AMZ. We acknowledge Xenbase the *Xenopus* model organism knowledge base (RRID:SCR_003280) and the National *Xenopus* Resource (RRID:SCR_013731) for supporting the research reviewed here.

References

Abbruzzese, G., Becker, S. F., Kashef, J., & Alfandari, D. (2016). ADAM13 cleavage of cadherin-11 promotes CNC migration independently of the homophilic binding site. *Developmental Biology*, *415*(2), 383–390. https://doi.org/10.1016/j.ydbio.2015.07.018.

Afouda, B. A., Nakamura, Y., Shaw, S., Charney, R. M., Paraiso, K. D., Blitz, I. L., et al. (2020). Foxh1/nodal defines context-specific direct maternal Wnt/β-catenin target gene regulation in early development. *IScience*, *23*(7), 101314. https://doi.org/10.1016/j.isci.2020.101314.

Alharatani, R., Ververi, A., Beleza-Meireles, A., Ji, W., Mis, E., Patterson, Q. T., et al. (2020). Novel truncating mutations in CTNND1 cause a dominant craniofacial and cardiac syndrome. *Human Molecular Genetics*, *29*(11), 1900–1921. https://doi.org/10.1093/hmg/ddaa050.

Ariizumi, T., Michiue, T., & Asashima, M. (2017). In vitro induction of Xenopus embryonic organs using animal cap cells. *Cold Spring Harbor Protocols*, *2017*(12), 982–987. https://doi.org/10.1101/pdb.prot097410.

Asashima, M., Nakano, H., Shimada, K., Kinoshita, K., Ishii, K., Shibai, H., et al. (1990). Mesodermal induction in early amphibian embryos by activin A (erythroid differentiation factor). *Roux's Archives of Developmental Biology*, *198*(6), 330–335. https://doi.org/10.1007/BF00383771.

Beck, C. W., & Slack, J. M. W. (1999). Gut specific expression using mammalian promoters in transgenic Xenopus laevis. *Mechanisms of Development*, *88*(2), 221–227. https://doi.org/10.1016/S0925-4773(99)00217-8.

Bhattacharya, D., Marfo, C. A., Li, D., Lane, M., & Khokha, M. K. (2015). CRISPR/Cas9: An inexpensive, efficient loss of function tool to screen human disease genes in Xenopus. *Developmental Biology*, *408*(2), 196–204. https://doi.org/10.1016/j.ydbio.2015.11.003.

Bilogan, C. K., & Horb, M. E. (2012). Xenopus Staufen2 is required for anterior endodermal organ formation. *Genesis*, *50*(3), 251–259. https://doi.org/10.1002/dvg.22000.

Blitz, I. L., Biesinger, J., Xie, X., & Cho, K. W. Y. (2013). Biallelic genome modification in F0 Xenopus tropicalis embryos using the CRISPR/Cas system. *Genesis*, *51*(12), 827–834. https://doi.org/10.1002/dvg.22719.

Buchholz, D. R. (2015). More similar than you think: Frog metamorphosis as a model of human perinatal endocrinology. *Developmental Biology*, *408*(2), 188–195. https://doi.org/10.1016/j.ydbio.2015.02.018.

Butler, M. T., & Wallingford, J. B. (2017). Planar cell polarity in development and disease. *Nature Reviews Molecular Cell Biology*, *18*(6), 375–388. https://doi.org/10.1038/nrm.2017.11.

Carotenuto, R., Capriello, T., Cofone, R., Galdiero, G., Fogliano, C., & Ferrandino, I. (2020). Impact of copper in Xenopus laevis liver: Histological damages and atp7b downregulation. *Ecotoxicology and Environmental Safety*, *188*(November 2019), 109940. https://doi.org/10.1016/j.ecoenv.2019.109940.

Chalmers, A. D., & Slack, J. M. W. (2000). The Xenopus tadpole gut: Fate maps and morphogenetic movements. *Development*, *127*(2), 381–392.

Charney, R. M., Paraiso, K. D., Blitz, I. L., & Cho, K. W. Y. (2017). A gene regulatory program controlling early Xenopus mesendoderm formation: Network conservation and motifs. *Seminars in Cell and Developmental Biology*, *66*, 12–24. https://doi.org/10.1016/j.semcdb.2017.03.003.
Chen, J., Jacox, L. A., Saldanha, F., & Sive, H. (2017). Mouth development. *Wiley Interdisciplinary Reviews: Developmental Biology*, *6*(5), 1–16. https://doi.org/10.1002/wdev.275.
Chen, Y., Pan, F. C., Brandes, N., Afelik, S., Sölter, M., & Pieler, T. (2004). Retinoic acid signaling is essential for pancreas development and promotes endocrine at the expense of exocrine cell differentiation in Xenopus. *Developmental Biology*, *271*(1), 144–160. https://doi.org/10.1016/j.ydbio.2004.03.030.
Cibois, M., Luxardi, G., Chevalier, B., Thomé, V., Mercey, O., Zaragosi, L. E., et al. (2015). BMP signalling controls the construction of vertebrate mucociliary epithelia. *Development (Cambridge)*, *142*(13), 2352–2363. https://doi.org/10.1242/dev.118679.
Colombo, B. M., Scalvenzi, T., Benlamara, S., & Pollet, N. (2015). Microbiota and mucosal immunity in amphibians. *Frontiers in Immunology*, *6*(Mar), 1–15. https://doi.org/10.3389/fimmu.2015.00111.
Crease, D. J., Dyson, S., & Gurdon, J. B. (1998). Cooperation between the activin and Wnt pathways in the spatial control of organizer gene expression. *Proceedings of the National Academy of Sciences of the United States of America*, *95*(8), 4398–4403. https://doi.org/10.1073/pnas.95.8.4398.
Dale, L., & Slack, J. M. W. (1987). Fate map for the 32-cell stage of Xenopus laevis. *Development*, *99*(4), 527–551.
D'Amour, K. A., Agulnick, A. D., Eliazer, S., Kelly, O. G., Kroon, E., & Baetge, E. E. (2005). Efficient differentiation of human embryonic stem cells to definitive endoderm. *Nature Biotechnology*, *23*(12), 1534–1541. https://doi.org/10.1038/nbt1163.
Davis, A., Amin, N. M., Johnson, C., Bagley, K., Ghashghaei, H. T., & Nascone-Yoder, N. (2017). Stomach curvature is generated by left-right asymmetric gut morphogenesis. *Development (Cambridge)*, *144*(8), 1477–1483. https://doi.org/10.1242/dev.143701.
Devotta, A., Juraver-Geslin, H., Gonzalez, J. A., Hong, C. S., & Saint-Jeannet, J. P. (2016). Sf3b4-depleted Xenopus embryos: A model to study the pathogenesis of craniofacial defects in Nager syndrome. *Developmental Biology*, *415*(2), 371–382. https://doi.org/10.1016/j.ydbio.2016.02.010.
Dickinson, A. J. G. (2016). Using frogs faces to dissect the mechanisms underlying human orofacial defects. *Seminars in Cell and Developmental Biology*, *51*, 54–63. https://doi.org/10.1016/j.semcdb.2016.01.016.
Dickinson, A. J. G., & Sive, H. L. (2009). The Wnt antagonists Frzb-1 and crescent locally regulate basement membrane dissolution in the developing primary mouth. *Development*, *136*(7), 1071–1081. https://doi.org/10.1242/dev.032912.
Dush, M. K., & Nascone-Yoder, N. M. (2013). Jun N-terminal kinase maintains tissue integrity during cell rearrangement in the gut. *Development (Cambridge)*, *140*(7), 1457–1466. https://doi.org/10.1242/dev.086850.
Dush, M. K., & Nascone-Yoder, N. M. (2019). Vangl2 coordinates cell rearrangements during gut elongation. *Developmental Dynamics*, *248*(7), 569–582. https://doi.org/10.1002/dvdy.61.
Foley, A. C., & Mercola, M. (2005). Heart induction by Wnt antagonists depends on the homeodomain transcription factor Hex. *Genes and Development*, *19*(3), 387–396. https://doi.org/10.1101/gad.1279405.
Gentsch, G. E., Spruce, T., Owens, N. D. L., & Smith, J. C. (2019). Maternal pluripotency factors initiate extensive chromatin remodelling to predefine first response to inductive signals. *Nature Communications*, *10*(1). https://doi.org/10.1038/s41467-019-12263-w.
Gere-Becker, M. B., Pommerenke, C., Lingner, T., & Pieler, T. (2018). Retinoic acid-induced expression of hnf1b and fzd4 is required for pancreas development in

Xenopus laevis. *Development (Cambridge)*, *145*(12), dev161372. https://doi.org/10.1242/dev.161372.

Girard, C. A. J., Shimomura, K., Proks, P., Absalom, N., Castano, L., Perez De Nanclares, G., et al. (2006). Functional analysis of six Kir6.2 (KCNJ11) mutations causing neonatal diabetes. *Pflügers Archiv - European Journal of Physiology*, *453*(3), 323–332. https://doi.org/10.1007/s00424-006-0112-3.

Gordon, C. T., Xue, S., Yigit, G., Filali, H., Chen, K., Rosin, N., et al. (2017). De novo mutations in SMCHD1 cause Bosma arhinia microphthalmia syndrome and abrogate nasal development. *Nature Genetics*, *49*(2), 249–255. https://doi.org/10.1038/ng.3765.

Grevellec, A., & Tucker, A. S. (2010). The pharyngeal pouches and clefts: Development, evolution, structure and derivatives. *Seminars in Cell and Developmental Biology*, *21*(3), 325–332. https://doi.org/10.1016/j.semcdb.2010.01.022.

Guo, X., Zhang, T., Hu, Z., Zhang, Y., Shi, Z., Wang, Q., et al. (2014). Efficient RNA/Cas9-mediated genome editing in Xenopus tropicalis. *Development (Cambridge)*, *141*(3), 707–714. https://doi.org/10.1242/dev.099853.

Haas, M., Gómez Vázquez, J. L., Sun, D. I., Tran, H. T., Brislinger, M., Tasca, A., et al. (2019). ΔN-Tp63 mediates Wnt/β-catenin-induced inhibition of differentiation in basal stem cells of mucociliary epithelia. *Cell Reports*, *28*(13), 3338–3352.e6. https://doi.org/10.1016/j.celrep.2019.08.063.

Hasebe, T., Fujimoto, K., Buchholz, D. R., & Ishizuya-Oka, A. (2020). Stem cell development involves divergent thyroid hormone receptor subtype expression and epigenetic modifications in the Xenopus metamorphosing intestine. *General and Comparative Endocrinology*, *292*(November 2019), 113441. https://doi.org/10.1016/j.ygcen.2020.113441.

Haworth, K. E., Kotecha, S., Mohun, T. J., & Latinkic, B. V. (2008). GATA4 and GATA5 are essential for heart and liver development in Xenopus embryos. *BMC Developmental Biology*, *8*, 1–16. https://doi.org/10.1186/1471-213X-8-74.

Hellsten, U., Harland, R. M., Gilchrist, M. J., Hendrix, D., Jurka, J., Kapitonov, V., et al. (2010). The genome of the western clawed frog xenopus tropicalis. *Science*, *328*(5978), 633–636. https://doi.org/10.1126/science.1183670.

Hempel, A., & Kühl, M. (2016). A matter of the heart: The African clawed frog Xenopus as a model for studying vertebrate cardiogenesis and congenital heart defects. *Journal of Cardiovascular Development and Disease*, *3*(2), 21. https://doi.org/10.3390/jcdd3020021.

Horb, M. E., & Slack, J. M. W. (2001). Endoderm specification and differentiation in Xenopus embryos. *Developmental Biology*, *236*(2), 330–343. https://doi.org/10.1006/dbio.2001.0347.

Horb, M., Wlizla, M., Abu-Daya, A., McNamara, S., Gajdasik, D., Igawa, T., et al. (2019). Xenopus resources: Transgenic, inbred and mutant animals, training opportunities, and web-based support. *Frontiers in Physiology*, *10*(Apr), 1–10. https://doi.org/10.3389/fphys.2019.00387.

Houssin, N. S., Bharathan, N. K., Turner, S. D., & Dickinson, A. J. G. (2017). Role of JNK during buccopharyngeal membrane perforation, the last step of embryonic mouth formation. *Developmental Dynamics*, *246*(2), 100–115. https://doi.org/10.1002/dvdy.24470.

Hudson, C., Clements, D., Friday, R. V., Stott, D., & Woodland, H. R. (1997). Xsox17α and -β mediate endoderm formation in Xenopus. *Cell*, *91*, 397–405. Retrieved from http://www.cell.com/cell/pdf/S0092-8674(00)80423-7.pdf.

Hwang, W. Y., Marquez, J., & Khokha, M. K. (2019). Xenopus: Driving the discovery of novel genes in patient disease and their underlying pathological mechanisms relevant for organogenesis. *Frontiers in Physiology*, *10*(Jul), 1–7. https://doi.org/10.3389/fphys.2019.00953.

Ishizuya-Oka, A., Ueda, S., Inokuchi, T., Amano, T., Damjanovski, S., Stolow, M., et al. (2001). Thyroid hormone-induced expression of sonic hedgehog correlates with adult epithelial development during remodeling of the Xenopus stomach and intestine. *Differentiation*, *69*(1), 27–37. https://doi.org/10.1046/j.1432-0436.2001.690103.x.

Jacox, L., Chen, J., Rothman, A., Lathrop-Marshall, H., & Sive, H. (2016). Formation of a "pre-mouth array" from the extreme anterior domain is directed by neural crest and Wnt/PCP signaling. *Cell Reports*, *16*(5), 1445–1455. https://doi.org/10.1016/j.celrep.2016.06.073.

Jacox, L., Sindelka, R., Chen, J., Rothman, A., Dickinson, A., & Sive, H. (2014). The extreme anterior domain is an essential craniofacial organizer acting through Kinin-Kallikrein signaling. *Cell Reports*, *8*(2), 596–609. https://doi.org/10.1016/j.celrep.2014.06.026.

Jarikji, Z., Horb, L. D., Shariff, F., Mandato, C. A., Cho, K. W. Y., & Horb, M. E. (2009). The tetraspanin Tm4sf3 is localized to the ventral pancreas and regulates fusion of the dorsal and ventral pancreatic buds. *Development*, *136*(11), 1791–1800. https://doi.org/10.1242/dev.032235.

Jensen, J. N., Rosenberg, L. C., Hecksher-Sørensen, J., & Serup, P. (2007). Mutant neurogenin-3 in congenital malabsorptive diarrhea. *New England Journal of Medicine*, *356*(17), 1781–1782. https://doi.org/10.1056/NEJMc063247.

Juárez-Morales, J. L., Martinez-De Luna, R. I., Zuber, M. E., Roberts, A., & Lewis, K. E. (2017). Zebrafish transgenic constructs label specific neurons in Xenopus laevis spinal cord and identify frog V0v spinal neurons. *Developmental Neurobiology*, 77(8), 1007–1020. https://doi.org/10.1002/dneu.22490.

Karimi, K., Fortriede, J. D., Lotay, V. S., Burns, K. A., Wang, D. Z., Fisher, M. E., et al. (2018). Xenbase: A genomic, epigenomic and transcriptomic model organism database. *Nucleic Acids Research*, *46*(D1), D861–D868. https://doi.org/10.1093/nar/gkx936.

Keller, R., & Sutherland, A. (2020). Convergent extension in the amphibian, Xenopus laevis. In *Vol. 136. Current topics in developmental biology* (1st ed.). Elsevier Inc. https://doi.org/10.1016/bs.ctdb.2019.11.013.

Kelly, O. G., & Melton, D. A. (2000). Development of the pancreas in Xenopus laevis. *Developmental Dynamics*, *218*(4), 615–627. https://doi.org/10.1002/1097-0177(2000)9999:9999<::AID-DVDY1027>3.0.CO;2-8.

Kennedy, A. E., & Dickinson, A. J. G. (2012). Median facial clefts in Xenopus laevis: Roles of retinoic acid signaling and homeobox genes. *Developmental Biology*, *365*(1), 229–240. https://doi.org/10.1016/j.ydbio.2012.02.033.

Kenny, A. P., Rankin, S. A., Allbee, A. W., Prewitt, A. R., Zhang, Z., Tabangin, M. E., et al. (2012). Sizzled-tolloid interactions maintain foregut progenitors by regulating fibronectin-dependent BMP signaling. *Developmental Cell*, *23*(2), 292–304. https://doi.org/10.1016/j.devcel.2012.07.002.

Kim, H. Y., Jackson, T. R., Stuckenholz, C., & Davidson, L. A. (2020). Tissue mechanics drives regeneration of a mucociliated epidermis on the surface of Xenopus embryonic aggregates. *Nature Communications*, *11*(1), 1–10. https://doi.org/10.1038/s41467-020-14385-y.

Kim, E., Jiang, M., Huang, H., Zhang, Y., Tjota, N., Gao, X., et al. (2019). Isl1 regulation of Nkx2.1 in the early foregut epithelium is required for trachea-esophageal separation and lung lobation. *Developmental Cell*, *51*(6), 675–683.e4. https://doi.org/10.1016/j.devcel.2019.11.002.

Kofent, J., & Spagnoli, F. M. (2016). Xenopus as a model system for studying pancreatic development and diabetes. *Seminars in Cell and Developmental Biology*, *51*, 106–116. https://doi.org/10.1016/j.semcdb.2016.01.005.

Kolm, P. J., & Sive, H. L. (1995). Efficient hormone-inducible protein function in Xenopus laevis. *Developmental Biology*, *171*(1), 267–272. https://doi.org/10.1006/dbio.1995.1279.

Kubo, A., Shinozaki, K., Shannon, J. M., Kouskoff, V., Kennedy, M., Woo, S., et al. (2004). Development of definitive endoderm from embryonic stem cells in culture. *Development*, *131*(7), 1651–1662. https://doi.org/10.1242/dev.01044.

Lansdon, L. A., Darbro, B. W., Petrin, A. L., Hulstrand, A. M., Standley, J. M., Brouillette, R. B., et al. (2018). Identification of isthmin 1 as a novel clefting and craniofacial patterning gene in humans. *Genetics*, *208*(1), 283–296. https://doi.org/10.1534/genetics.117.300535.

Lee, Y. H., Williams, A., Hong, C. S., You, Y., Senoo, M., & Saint-Jeannet, J. P. (2013). Early development of the thymus in Xenopus laevis. *Developmental Dynamics*, *242*(2), 164–178. https://doi.org/10.1002/dvdy.23905.

Lei, Y., Guo, X., Liu, Y., Cao, Y., Deng, Y., Chen, X., et al. (2012). Efficient targeted gene disruption in Xenopus embryos using engineered transcription activator-like effector nucleases (TALENs). *Proceedings of the National Academy of Sciences of the United States of America*, *109*(43), 17484–17489. https://doi.org/10.1073/pnas.1215421109.

Li, Y., Rankin, S. A., Sinner, D., Kenny, A. P., Krieg, P. A., & Zorn, A. M. (2008). Sfrp5 coordinates foregut specification and morphogenesis by antagonizing both canonical and noncanonical Wnt11 signaling. *Genes and Development*, *22*(21), 3050–3063. https://doi.org/10.1101/gad.1687308.

Lipscomb, K., Schmitt, C., Sablyak, A., Yoder, J. A., & Nascone-Yoder, N. (2006). Role for retinoid signaling in left-right asymmetric digestive organ morphogenesis. *Developmental Dynamics*, *235*(8), 2266–2275. https://doi.org/10.1002/dvdy.20879.

Loh, K. M., Ang, L. T., Zhang, J., Kumar, V., Ang, J., Auyeong, J. Q., et al. (2014). Efficient endoderm induction from human pluripotent stem cells by logically directing signals controlling lineage bifurcations. *Cell Stem Cell*, *14*(2), 237–252. https://doi.org/10.1016/j.stem.2013.12.007.

Love, N. R., Thuret, R., Chen, Y., Ishibashi, S., Sabherwal, N., Paredes, R., et al. (2011). pTransgenesis: A cross-species, modular transgenesis resource. *Development*, *138*(24), 5451–5458. https://doi.org/10.1242/dev.066498.

McLin, V. A., Rankin, S. A., & Zorn, A. M. (2007). Repression of Wnt/β-catenin signaling in the anterior endoderm is essential for liver and pancreas development. *Development*, *134*(12), 2207–2217. https://doi.org/10.1242/dev.001230.

Michau, A., Guillemain, G., Grosfeld, A., Vuillaumier-Barrot, S., Grand, T., Keck, M., et al. (2013). Mutations in SLC2A2 gene reveal hGLUT2 function in pancreatic βcell development. *Journal of Biological Chemistry*, *288*(43), 31080–31092. https://doi.org/10.1074/jbc.M113.469189.

Minoux, M., & Rijli, F. M. (2010). Molecular mechanisms of cranial neural crest cell migration and patterning in craniofacial development. *Development*, *137*(16), 2605–2621. https://doi.org/10.1242/dev.040048.

Miyata, K., & Ose, K. (2012). Thyroid hormone-disrupting effects and the amphibian metamorphosis assay. *Journal of Toxicologic Pathology*, *25*(1), 1–9. https://doi.org/10.1293/tox.25.1.

Moody, S. A. (1987a). Fates of the blastomeres of the 16-cell stage Xenopus embryo. *Developmental Biology*, *119*(2), 560–578. https://doi.org/10.1016/0012-1606(87)90059-5.

Moody, S. A. (1987b). Fates of the blastomeres of the 32-cell-stage Xenopus embryo. *Developmental Biology*, *122*(2), 300–319. https://doi.org/10.1016/0012-1606(87)90296-X.

Moriya, N., Komazaki, S., & Asashima, M. (2000). In vitro organogenesis of pancreas in Xenopus laevis dorsal lips treated with retinoic acid. *Development, Growth & Differentiation*, *42*(2), 175–185. https://doi.org/10.1046/j.1440-169X.2000.00498.x.

Moriya, N., Komazaki, S., Takahashi, S., Yokota, C., & Asashima, M. (2000). In vitro pancreas formation from Xenopus ectoderm treated with activin and retinoic acid.

Development, Growth & Differentiation, *42*(6), 593–602. https://doi.org/10.1046/j.1440-169X.2000.00542.x.

Mukherjee, S., Chaturvedi, P., Rankin, S., Fish, M., Wlizla, M., Paraiso, K., et al. (2020). Sox17 and β-catenin co-occupy Wnt-responsive enhancers to govern the endodermal gene regulatory network. *eLife*, *9*(e58029). https://doi.org/10.7554/eLife.58029.

Mukhi, S., Horb, M. E., & Brown, D. D. (2009). Remodeling of insulin producing β-cells during Xenopus laevis metamorphosis. *Developmental Biology*, *328*(2), 384–391. https://doi.org/10.1016/j.ydbio.2009.01.038.

Muller, J. K., Prather, D. R., & Nascone-Yoder, N. M. (2003). Left-right asymmetric morphogenesis in the Xenopus digestive system. *Developmental Dynamics*, *228*(4), 672–682. https://doi.org/10.1002/dvdy.10415.

Múnera, J. O., Sundaram, N., Rankin, S. A., Hill, D., Watson, C., Mahe, M., et al. (2017). Differentiation of human pluripotent stem cells into colonic organoids via transient activation of BMP signaling. *Cell Stem Cell*, *21*(1), 51–64.e6. https://doi.org/10.1016/j.stem.2017.05.020.

Naert, T., Van Nieuwenhuysen, T., & Vleminckx, K. (2017). TALENs and CRISPR/Cas9 fuel genetically engineered clinically relevant Xenopus tropicalis tumor models. *Genesis*, *55*(1–2), 1–10. https://doi.org/10.1002/dvg.23005.

Nakayama, T., Fisher, M., Nakajima, K., Odeleye, A. O., Zimmerman, K. B., Fish, M. B., et al. (2015). Xenopus pax6 mutants affect eye development and other organ systems, and have phenotypic similarities to human aniridia patients. *Developmental Biology*, *408*(2), 328–344. https://doi.org/10.1016/j.ydbio.2015.02.012.

Nascone, N., & Mercola, M. (1995). An inductive role for the endoderm in Xenopus cardiogenesis. *Development*, *121*(2), 515–523.

Nasr, T., Mancini, P., Rankin, S. A., Edwards, N. A., Agricola, Z. N., Kenny, A. P., et al. (2019). Endosome-mediated epithelial remodeling downstream of hedgehog-gli is required for tracheoesophageal separation. *Developmental Cell*, *51*(6), 665–674.e6. https://doi.org/10.1016/j.devcel.2019.11.003.

Nenni, M. J., Fisher, M. E., James-Zorn, C., Pells, T. J., Ponferrada, V., Chu, S., et al. (2019). XenBase: Facilitating the use of Xenopus to model human disease. *Frontiers in Physiology*, *10*(Feb), 1–13. https://doi.org/10.3389/fphys.2019.00154.

Okabayashi, K., & Asashima, M. (2003). Tissue generation from amphibian animal caps. *Current Opinion in Genetics and Development*, *13*(5), 502–507. https://doi.org/10.1016/S0959-437X(03)00111-4.

Osman, A. A., Saito, M., Makepeace, C., Permutt, M. A., Schlesinger, P., & Mueckler, M. (2003). Wolframin expression induces novel ion channel activity in endoplasmic reticulum membranes and increases intracellular calcium. *Journal of Biological Chemistry*, *278*(52), 52755–52762. https://doi.org/10.1074/jbc.M310331200.

Paraiso, K. D., Blitz, I. L., Coley, M., Cheung, J., Sudou, N., Taira, M., et al. (2019). Endodermal maternal transcription factors establish super-enhancers during zygotic genome activation. *Cell Reports*, *27*(10), 2962–2977.e5. https://doi.org/10.1016/j.celrep.2019.05.013.

Pearl, E. J., Bilogan, C. K., Mukhi, S., Brown, D. D., & Horb, M. E. (2009). Xenopus pancreas development. *Developmental Dynamics*, *238*(6), 1271–1286. https://doi.org/10.1002/dvdy.21935.

Pearl, E. J., Jarikji, Z., & Horb, M. E. (2011). Functional analysis of Rfx6 and mutant variants associated with neonatal diabetes. *Developmental Biology*, *351*(1), 135–145. https://doi.org/10.1016/j.ydbio.2010.12.043.

Pittman, A. M., Naranjo, S., Webb, E., Broderick, P., Lips, E. H., Van Wezel, T., et al. (2009). The colorectal cancer risk at 18q21 is caused by a novel variant altering SMAD7 expression. *Genome Research*, *19*(6), 987–993. https://doi.org/10.1101/gr.092668.109.

Popov, I. K., Kwon, T., Crossman, D. K., Crowley, M. R., Wallingford, J. B., & Chang, C. (2017). Identification of new regulators of embryonic patterning and morphogenesis in Xenopus gastrulae by RNA sequencing. *Developmental Biology*, *426*(2), 429–441. https://doi.org/10.1016/j.ydbio.2016.05.014.

Proks, P., Arnold, A. L., Bruining, J., Girard, C., Flanagan, S. E., Larkin, B., et al. (2006). A heterozygous activating mutation in the sulphonylurea receptor SUR1 (ABCC8) causes neonatal diabetes. *Human Molecular Genetics*, *15*(11), 1793–1800. https://doi.org/10.1093/hmg/ddl101.

Rankin, S. A., Gallas, A. L., Neto, A., Gómez-Skarmeta, J. L., & Zorn, A. M. (2012). Suppression of Bmp4 signaling by the zinc-finger repressors Osr1 and Osr2 is required for Wnt/β-catenin-mediated lung specification in xenopus. *Development (Cambridge)*, *139*(16), 3010–3020. https://doi.org/10.1242/dev.078220.

Rankin, S. A., Han, L., McCracken, K. W., Kenny, A. P., Anglin, C. T., Grigg, E. A., et al. (2016). A retinoic acid-hedgehog cascade coordinates mesoderm-inducing signals and endoderm competence during lung specification. *Cell Reports*, *16*(1), 66–78. https://doi.org/10.1016/j.celrep.2016.05.060.

Rankin, S. A., Kormish, J., Kofron, M., Jegga, A., & Zorn, A. M. (2011). A gene regulatory network controlling hhex transcription in the anterior endoderm of the organizer. *Developmental Biology*, *351*(2), 297–310. https://doi.org/10.1016/j.ydbio.2010.11.037.

Rankin, S. A., McCracken, K. W., Luedeke, D. M., Han, L., Wells, J. M., Shannon, J. M., et al. (2018). Timing is everything: Reiterative Wnt, BMP and RA signaling regulate developmental competence during endoderm organogenesis. *Developmental Biology*, *434*(1), 121–132. https://doi.org/10.1016/j.ydbio.2017.11.018.

Rankin, S. A., Thi Tran, H., Wlizla, M., Mancini, P., Shifley, E. T., Bloor, S. D., et al. (2015). A molecular atlas of Xenopus respiratory system development. *Developmental Dynamics*, *244*(1), 69–85. https://doi.org/10.1002/dvdy.24180.

Reed, R. A., Womble, M. A., Dush, M. K., Tull, R. R., Bloom, S. K., Morckel, A. R., et al. (2009). Morphogenesis of the primitive gut tube is generated by Rho/ROCK/myosin II-mediated endoderm rearrangements. *Developmental Dynamics*, *238*(12), 3111–3125. https://doi.org/10.1002/dvdy.22157.

Regnault, C., Worms, I. A. M., Oger-Desfeux, C., MelodeLima, C., Veyrenc, S., Bayle, M. L., et al. (2014). Impaired liver function in Xenopus tropicalis exposed to benzo[a]pyrene: Transcriptomic and metabolic evidence. *BMC Genomics*, *15*(1), 1–16. https://doi.org/10.1186/1471-2164-15-666.

Rose, C. S., & James, B. (2013). Plasticity of lung development in the amphibian, Xenopus laevis. *Biology Open*, *2*(12), 1324–1335. https://doi.org/10.1242/bio.20133772.

Saide, K., Sherwood, V., & Wheeler, G. N. (2019). Paracetamol-induced liver injury modelled in Xenopus laevis embryos. *Toxicology Letters*, *302*(July 2018), 83–91. https://doi.org/10.1016/j.toxlet.2018.09.016.

Salanga, M. C., & Horb, M. E. (2015). Xenopus as a model for GI/pancreas disease. *Current Pathobiology Reports*, *3*(2), 137–145. https://doi.org/10.1007/s40139-015-0076-0.

Schreiber, A. M., Cai, L., & Brown, D. D. (2005). Remodeling of the intestine during metamorphosis of Xenopus laevis. *Proceedings of the National Academy of Sciences of the United States of America*, *102*(10), 3720–3725. https://doi.org/10.1073/pnas.0409868102.

Session, A. M., Uno, Y., Kwon, T., Chapman, J. A., Toyoda, A., Takahashi, S., et al. (2016). Genome evolution in the allotetraploid frog Xenopus laevis. *Nature*, *538*(7625), 336–343. https://doi.org/10.1038/nature19840.

Shifley, E. T., Kenny, A. P., Rankin, S. A., & Zorn, A. M. (2012). Prolonged FGF signaling is necessary for lung and liver induction in Xenopus. *BMC Developmental Biology*, *12*(27), 1–14. https://doi.org/10.1186/1471-213X-12-27.

Simaite, D., Kofent, J., Gong, M., Rüschendorf, F., Jia, S., Arn, P., et al. (2014). Recessive mutations in PCBD1 cause a new type of early-onset diabetes. *Diabetes*, *63*(10), 3557–3564. https://doi.org/10.2337/db13-1784.

Sinner, D., Kirilenko, P., Rankin, S., Wei, E., Howard, L., Kofron, M., et al. (2006). Global analysis of the transcriptional network controlling Xenopus endoderm formation. *Development*, *133*(10), 1955–1966. https://doi.org/10.1242/dev.02358.

Sinner, D., Rankin, S., Lee, M., & Zorn, A. M. (2004). Sox17 and β-catenin cooperate to regulate the transcription of endodermal genes. *Development*, *131*(13), 3069–3080. https://doi.org/10.1242/dev.01176.

Smith, J. C., & Green, J. B. (1990). Graded changes in dose of a Xenopus activin A homologue elicit stepwise transitions in embryonic cell fate. *Nature*, *34*(September), 391–394.

Sparrow, D. B., Cai, C., Kotecha, S., Latinkic, B., Cooper, B., Towers, N., et al. (2000). Regulation of the tinman homologues in Xenopus embryos. *Developmental Biology*, *227*(1), 65–79. https://doi.org/10.1006/dbio.2000.9891.

Steimle, J. D., Rankin, S. A., Slagle, C. E., Bekeny, J., Rydeen, A. B., Chan, S. S. K., et al. (2018). Evolutionarily conserved Tbx5-Wnt2/2b pathway orchestrates cardiopulmonary development. *Proceedings of the National Academy of Sciences of the United States of America*, *115*(45), E10615–E10624. https://doi.org/10.1073/pnas.1811624115.

Sterling, J., Fu, L., Matsuura, K., & Shi, Y. B. (2012). Cytological and morphological analyses reveal distinct features of intestinal development during Xenopus tropicalis metamorphosis. *PLoS One*, 7(10), 1–10. https://doi.org/10.1371/journal.pone.0047407.

Sterner, Z. R., Rankin, S. A., Wlizla, M., Choi, J. A., Luedeke, D. M., Zorn, A. M., et al. (2019). Novel vectors for functional interrogation of Xenopus ORFeome coding sequences. *Genesis*, *57*(10), 1–11. https://doi.org/10.1002/dvg.23329.

Stevens, M. L., Chaturvedi, P., Rankin, S. A., Macdonald, M., Jagannathan, S., Yukawa, M., et al. (2017). Genomic integration of Wnt/β-catenin and BMP/Smad1 signaling coordinates foregut and hindgut transcriptional programs. *Development (Cambridge)*, *144*(7), 1283–1295. https://doi.org/10.1242/dev.145789.

Sun, L., Bertke, M. M., Champion, M. M., Zhu, G., Huber, P. W., & Dovichi, N. J. (2015). Quantitative proteomics of Xenopus laevis embryos: Expression kinetics of nearly 4000 proteins during early development. *Scientific Reports*, *4*, 1–9. https://doi.org/10.1038/srep04365.

Tabler, J. M., Bolger, T. G., Wallingford, J., & Liu, K. J. (2014). Hedgehog activity controls opening of the primary mouth. *Developmental Biology*, *396*(1), 1–7. https://doi.org/10.1016/j.ydbio.2014.09.029.

Tseng, H. T., Shah, R., & Jamrich, M. (2004). Function and regulation of FoxF1 during Xenopus gut development. *Development*, *131*(15), 3637–3647. https://doi.org/10.1242/dev.01234.

Ueno, T., Ishihara, A., Yagi, S., Koike, T., Yamauchi, K., & Shiojiri, N. (2015). Histochemical analyses of biliary development during metamorphosis of *Xenopus laevis* tadpoles. *Zoological Science*, *32*(1), 88–96. https://doi.org/10.2108/zs140104.

Van Nieuwenhuysen, T., Naert, T., Tran, H. T., Van Imschoot, G., Geurs, S., Sanders, E., et al. (2015). TALEN-mediated apc mutation in Xenopus tropicalis phenocopies familial adenomatous polyposis. *Oncoscience*, *2*(5), 555–566. https://doi.org/10.18632/oncoscience.166.

Vedovato, N., Cliff, E., Proks, P., Poovazhagi, V., Flanagan, S. E., Ellard, S., et al. (2016). Neonatal diabetes caused by a homozygous KCNJ11 mutation demonstrates that tiny changes in ATP sensitivity markedly affect diabetes risk. *Diabetologia*, *59*(7), 1430–1436. https://doi.org/10.1007/s00125-016-3964-x.

Walentek, P., & Quigley, I. K. (2017). What we can learn from a tadpole about ciliopathies and airway diseases: Using systems biology in Xenopus to study cilia and mucociliary epithelia. *Genesis*, *55*(1–2), 1–12. https://doi.org/10.1002/dvg.23001.

Wallmeier, J., Al-Mutairi, D. A., Chen, C. T., Loges, N. T., Pennekamp, P., Menchen, T., et al. (2014). Mutations in CCNO result in congenital mucociliary clearance disorder with reduced generation of multiple motile cilia. *Nature Genetics*, *46*(6), 646–651. https://doi.org/10.1038/ng.2961.

Wang, J., Cortina, G., Wu, S. V., Tran, R., Cho, J.-H., Tsai, M.-J., et al. (2006). Mutant neurogenin-3 in congenital malabsorptive diarrhea. *New England Journal of Medicine*, *355*(3), 270–280. https://doi.org/10.1056/NEJMoa054288.

Wang, J. H., Deimling, S. J., D'Alessandro, N. E., Zhao, L., Possmayer, F., & Drysdale, T. A. (2011). Retinoic acid is a key regulatory switch determining the difference between lung and thyroid fates in *Xenopus laevis*. *BMC Developmental Biology*, *11*(75). https://doi.org/10.1186/1471-213X-11-75.

Wen, J. W. H., & Winklbauer, R. (2017). Ingression-type cell migration drives vegetal endoderm internalisation in the Xenopus gastrula. *eLife*, *6*, 1–35. https://doi.org/10.7554/eLife.27190.

Winklbauer, R. (2020). Mesoderm and endoderm internalization in the Xenopus gastrula. In *Vol. 136. Current topics in developmental biology* (1st ed.). Elsevier Inc. https://doi.org/10.1016/bs.ctdb.2019.09.002.

Womble, M., Amin, N. M., & Nascone-Yoder, N. (2018). The left-right asymmetry of liver lobation is generated by Pitx2c-mediated asymmetries in the hepatic diverticulum. *Developmental Biology*, *439*(2), 80–91. https://doi.org/10.1016/j.ydbio.2018.04.021.

Wühr, M., Freeman, R. M., Presler, M., Horb, M. E., Peshkin, L., Gygi, S. P., et al. (2014). Deep proteomics of the *Xenopus laevis* egg using an mRNA-derived reference database. *Current Biology*, *24*(13), 1467–1475. https://doi.org/10.1016/j.cub.2014.05.044.

Xanthos, J. B., Kofron, M., Wylie, C., & Heasman, J. (2001). Maternal VegT is the initiator of a molecular network specifying endoderm in *Xenopus laevis*. *Development*, *128*(2), 167–180.

Yuan, L., Li, X., Feng, J., Yin, C., Yuan, F., & Wang, X. (2014). IRE 1 α is essential for Xenopus pancreas development. *Journal of Biomedical Research*, *28*(2), 123–131. https://doi.org/10.7555/JBR.28.20130076.

Zhang, T., Guo, X., & Chen, Y. (2013). Retinoic acid-activated Ndrg1a represses Wnt/β-catenin signaling to allow xenopus pancreas, oesophagus, stomach, and duodenum specification. *PLoS One*, *8*(5), e65058. https://doi.org/10.1371/journal.pone.0065058.

Zhang, Z., Rankin, S. A., & Zorn, A. M. (2013). Different thresholds of Wnt-Frizzled 7 signaling coordinate proliferation, morphogenesis and fate of endoderm progenitor cells. *Developmental Biology*, *378*(1), 1–12. https://doi.org/10.1016/j.ydbio.2013.02.024.

Zhang, J., Rosenthal, A., De Sauvage, F. J., & Shivdasani, R. A. (2001). Downregulation of hedgehog signaling is required for organogenesis of the small intestine in Xenopus. *Developmental Biology*, *229*(1), 188–202. https://doi.org/10.1006/dbio.2000.9953.

Zhao, H., Han, D., Dawid, I., Pieler, T., & Chen, Y. (2012). Homeoprotein hhex-induced conversion of intestinal to ventral pancreatic precursors results in the formation of giant pancreata in Xenopus embryos. *Proceedings of the National Academy of Sciences of the United States of America*, *109*(22), 8594–8599. https://doi.org/10.1073/pnas.1206547109.

Zorn, A. M., Barish, G. D., Williams, B. O., Lavender, P., Klymkowsky, M. W., & Varmus, H. E. (1999). Regulation of Wnt signaling by Sox proteins: XSox17α/β and XSox3 physically interact with β-catenin. *Molecular Cell*, *4*(4), 487–498. https://doi.org/10.1016/S1097-2765(00)80200-2.

Zorn, A. M., & Mason, J. (2001). Gene expression in the embryonic Xenopus liver. *Mechanisms of Development*, *103*(1–2), 153–157. https://doi.org/10.1016/S0925-4773(01)00341-0.

Zorn, A. M., & Wells, J. M. (2009). Vertebrate endoderm development. *Annual Review of Cell Biology*, *25*, 221–251. https://doi.org/10.1146/annurev.cellbio.042308.113344. Vertebrate.

CHAPTER FOUR

From egg to embryo in marsupial frogs

Eugenia M. del Pino*
Escuela de Ciencias Biológicas, Pontificia Universidad Católica del Ecuador, Quito, Ecuador
*Corresponding author: e-mail address: edelpino@puce.edu.ec

Contents

Abstract

Marsupial frogs (Hemiphractidae) evolved exceptional mechanisms for the conquest of terrestrial life. These adaptations include very large eggs. In some species eggs reach 10 mm in diameter, and are considered to be the largest in frogs. Females have reproductive modifications for the incubation of embryos in their bodies. Modifications of embryos include adaptations for development inside the body of the mother, and changes in the developmental pattern. Moreover, in some species, oocytes are multinucleated instead of having a single germinal vesicle as in most vertebrates. This chapter provides an overview of the adaptations of marsupial frogs associated with terrestrial life, with a discussion of gastrulation and multinucleated oogenesis.

1. Introduction

The conquest of terrestrial life by the amniotes, i.e. reptiles, birds and mammals, was facilitated by the enormous reserves of yolk of their large eggs (in reptiles and birds), and by the evolution of the amnion, chorion, allantois, and yolk sac (Arendt & Nübler-Jung, 1999; Elinson & Beckham, 2002;

Current Topics in Developmental Biology, Volume 145
ISSN 0070-2153
https://doi.org/10.1016/bs.ctdb.2020.10.008

Elinson, Stewart, Bonneau, & Blackburn, 2014; Gilbert, 2014). The amniotic sac encloses the embryo within a fluid environment. Embryos of reptiles and birds are additionally enclosed within an eggshell. The eggshell is a key adaptation for development in dry areas as it encloses the aquatic fluid needed for embryonic development and it provides physical protection to the embryo. In addition the eggshell functions in gas exchange, and is a calcium reserve for embryonic development (Hallmann & Griebeler, 2015).

The life cycles of the majority of anamniotes, i.e. fishes and amphibians, include aquatic feeding–larvae. Eggs of anamniotes lack eggshells and do not develop the amnion, chorion, and allantois. Fishes and most amphibians develop in water or in water associated environments (Duellman & Trueb, 1986). Some frogs have terrestrial eggs, such as the frog without tadpoles, *Eleutherodactylus coqui* (Eleutherodactylidae: Terrarana) or the poison–arrow dendrobatid frogs (Dendrobatidae). To prevent desiccation of their eggs, development of these frogs is associated with a humid environment (Duellman & Trueb, 1986). In contrast, embryos of marsupial frogs (Hemiphractidae) do not require an external humid environment, as the body of the mother provides the required humidity. As in other frogs, embryos of marsupial frogs are enclosed inside jelly capsules, and lack eggshells (Schmid et al., 2012). The topics of this chapter are gastrulation and oogenesis as mechanisms associated with terrestrial life in frogs of the family Hemiphractidae, with an overview of reproduction and development in these frogs.

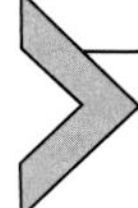

2. Overview of reproduction and development in marsupial frogs

The family Hemiphractidae (commonly known as hemiphractid frogs) includes about 100 species of Latin American frogs (Duellman, 2015; Schmid et al., 2012). Hemiphractid frogs develop from large eggs, including the largest eggs of anuran amphibians (Fig. 1C). Development occurs in the body of the mother until the birth of froglets or of advanced tadpoles instead of development in an external aquatic or humid–environment (del Pino, 2018; del Pino & Escobar, 1981). In the genera *Flectonotus* and *Gastrotheca*, embryos are enclosed in a pouch of integument located in the back of the frog–mother (Fig. 1A). These hemiphractid frogs are commonly known as marsupial frogs. In the remaining genera, *Cryptobatrachus, Hemiphractus, Fritziana* and *Stefania*, the eggs are attached to the dorsal skin of the frog–mother, and are exposed to the external environment (Schmid et al., 2012).

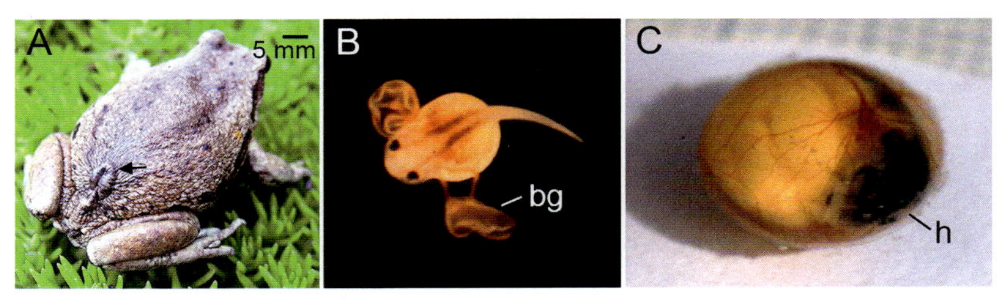

Fig. 1 Reproduction of marsupial frogs. (A) Adult female of *G. riobambae* during the brooding period. The pouch is a derivative of the dorsal skin and opens through an aperture, anterior to the cloaca (arrow). (B) *G. riobambae* embryo removed from the maternal pouch. The egg capsule was removed, and the bell gills collapsed on each side of the head. (C) Embryo of *Gastrotheca orophylax* taken from the maternal pouch. The embryo is within the egg capsule, and it is surrounded by a vascularized sac formed by the bell gills. Embryos of *G. riobambae* and *G. orophylax* develop from eggs of 3 mm and 6 mm diameter, respectively. In other species eggs are larger, reaching 10 mm in diameter. bg, bell gill; h, head.

The adaptations of the mother and embryos for terrestrial reproduction have been analyzed in detail in the marsupial frog *Gastrotheca riobambae* (Hemiphractidae), a frog from Ecuador (Fig. 1A) (del Pino, 1989, 2018, 2019; del Pino et al., 2007; Elinson & del Pino, 2012; Schmid et al., 2012). *G. riobambae* females incubate about 100 eggs in a pouch derived from the dorsal skin (Fig. 1A). Each egg measures 3 mm in diameter, representing almost 16 times the volume of a *Xenopus laevis* egg (del Pino, 2019). Large eggs provide the needed nourishment to reach advanced tadpole stages at the time of birth in *G. riobambae* or to frog stages, in other species of hemiphractid frogs (del Pino & Escobar, 1981; Schmid et al., 2012).

The maternal pouch of *G. riobambae* is a derivative of the dorsal skin (Fig. 1A). The integument of the pouch resembles frog skin during the non-reproductive period, and changes dramatically for the incubation of embryos. The pouch is under hormonal control. Estrogen triggers the development of the pouch in juvenile females (del Pino, 1980; Jones, Gerrard, & Roth, 1973). Moreover, under the influence of progesterone, the pouch becomes highly vascularized and develops partitions that envelop each embryo in a vascularized chamber during the incubatory period (del Pino, 1980). The hormonal control of reproduction by sex steroid hormones in *G. riobambae* resembles mammalian reproduction.

Adaptations of embryos for development inside the maternal pouch include a change in the morphology and function of the primary gills. The primary gills are transformed into disk-shaped vascularized structures, called bell gills (Fig. 1B). Bell gills envelop each embryo in a chamber filled with fluid (Fig. 1C) (del Pino, Galarza, de Albuja, & Humphries, 1975). The enveloping bell gills resemble the amniotic sac of amniotes, although the embryonic origin of bell gills differs from development of the amnion (del Pino et al., 1975; Elinson & del Pino, 2012).

The maternal pouch and the bell gills of embryos are separated by a thin egg–capsule, formed by the fertilization envelope and several layers of egg–jelly (del Pino et al., 1975; Schmid et al., 2012). Thus, the maternal circulation of the pouch and the embryonic blood vessels of the bell gills are closely associated for exchanges between the mother and embryos (del Pino et al., 1975). The association of the bell gills of embryos with the pouch of the mother parallels the association of embryos in the uterus of placental mammals (del Pino, 2018).

Another unusual adaptation to terrestrial life is the excretion of urea in *G. riobambae* (Alcocer, Santacruz, Steinbeisser, Thierauch, & del Pino, 1992). Tadpoles of frogs with aquatic reproduction excrete ammonia, and nitrogen–waste excretion changes to urea at metamorphosis. In contrast, embryos, tadpoles and adults of *G. riobambae* excrete urea. In *G. riobambae*, urea accumulates in the capsular fluid of embryos, and functions as an osmolyte for the retention of water (Alcocer et al., 1992; del Pino, Alcocer, & Grunz, 1994). The accumulation of urea in the capsular fluid allows the progression of embryonic development under the conditions of water–stress found in the maternal pouch (del Pino et al., 1994). As an outcome of these studies, a saline solution, containing urea, was designed for the in vitro culture of *G. riobambae* embryos, allowing the study of gastrulation in this frog (del Pino et al., 1994).

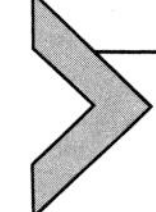

3. Patterns of gastrulation in frogs

3.1 Early development and gastrulation in *G. riobambae*

The large and pale eggs of *G. riobambae*, as in *X. laevis* and other frogs, develop small blastomeres in the animal region, and very large blastomeres in the vegetal region (Fig. 2A). However, in contrast to the regular cleavage pattern of *X. laevis, G. riobambae* cleavage is irregular and becomes asynchronous after a few cell divisions (del Pino & Loor-Vela, 1990; Nieuwkoop & Faber, 1994). In the blastula, the blastocoel roof consists of a monolayer of transparent cells, derived from animal blastomeres (Fig. 2B) (Elinson & del Pino, 1985, 2012). During gastrulation, the blastocoel roof gradually becomes opaque, as it becomes covered with yolky cells of the likely

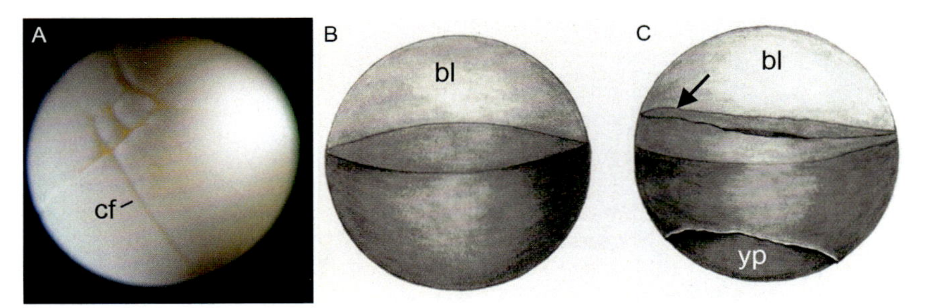

Fig. 2 From cleavage to the gastrula of *G. riobambae* embryos. (A) Cleavage stage embryo. Animal blastomeres are small. (B) Diagram of the blastula. The blastocoel roof consists of a transparent cell–monolayer, and allows the observation of internal cell movements. (C) Diagram of the gastrula. The blastocoel roof is invaded by yolky cells, beginning on the dorsal side (arrow). The blastopore surrounds a large yolk plug. The animal region faces to the front in (A) and to the top in (B), (C). bl, blastocoel; cf., cleavage furrow; yp, yolk plug.

mesendoderm starting on the dorsal side. This change is a visible landmark of the onset of gastrulation in living embryos (Fig. 2C) (Elinson & del Pino, 1985, 2012). At about the same time, an inconspicuous dorsal blastopore lip develops in the vegetal region of *G. riobambae* embryos (Fig. 2C).

In *G. riobambae*, as in other frogs, the morphogenetic processes of gastrulation include epiboly, involution, convergence, the formation of the dorsal blastopore lip and development of the blastopore around a prominent yolk plug (del Pino & Elinson, 1983; Moya, Alarcón, & del Pino, 2007; Solnica-Krezel, 2005). The blastopore of *G. riobambae* embryos develop in the subequatorial region, as in embryos of *X. laevis* and other frogs (Fig. 3A). In contrast with *X. laevis,* involuted cells accumulate in the blastopore lip. When the blastopore closes, the blastopore lip generates a thick circumblastoporal collar of small cells surrounding the closed blastopore, on top of a tiny archenteron (Fig. 3A). Small cells also occur in the surface of the circumblastoporal collar, delineating an embryonic disk. In comparison, cells of the vegetal region are large. In dendrobatid frogs and in *E. coqui*, a thick circumblastoporal collar develops around the closing blastopore, and elongation of the notochord occurs after blastopore closure, as in embryos of *G. riobambae*. However, these frogs do not develop an embryonic disk (del Pino et al., 2007; Moya et al., 2007). Development from an embryonic disk has not been found in frogs with exception of the family Hemiphractidae (del Pino, 2019; del Pino & Elinson, 1983).

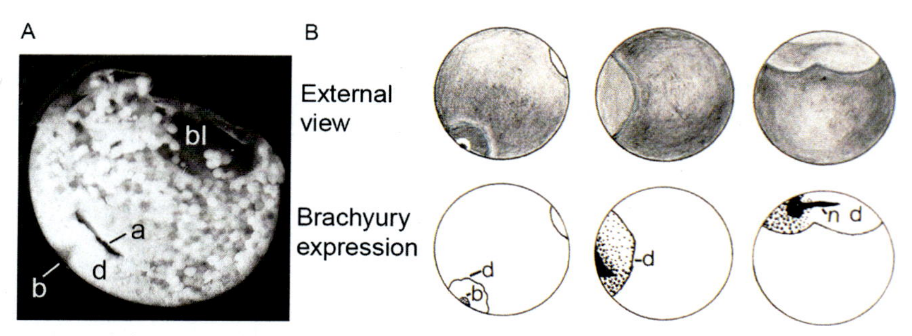

Fig. 3 Development of the embryonic disk in *G. riobambae*. (A) Bisection of a late gastrula. The thick blastopore lip surrounds the closed blastopore and generates a thick circumblastoporal collar and the embryonic disk. The archenteron is very small. The blastocoel is visible. The animal region faces to the top. (B) Diagrams of embryos after blastopore closure. *Top row:* External views. *Lower row*: Expression of Brachury in the notochord, depicted in black. *Left images*: An embryonic disk of small cells develops around the blastopore. Internally, there is the thick circumblastoporal collar and a small archenteron, as shown in (A). The borders of the embryonic disk are indicated by a solid line. The Brachyury signal is faint, not shown. *Middle images*: The archenteron and embryonic disk enlarge and the embryo undergoes an upward rotation. The blastocoel is obliterated by yolky cells. The Brachury signal is strong in the area of the closed blastopore and in the dorsal side. The elongated Brachury–positive area indicates the elongation of the notochord. *Right images*: Rotation has been completed, and the archenteron and embryonic disk face upward. The Brachury signal is strong and restricted to the notochord and tail bud. a, archenteron; b, blastopore; bl, blastocoel; d, disk; n, notochord. *Diagrams in (B) are reprinted from del Pino, E. M. (1996). The expression of brachyury during gastrulation in the marsupial frog* Gastrotheca riobambae. Developmental Biology, 177, *64–72, Fig. 1C–E′.*

A weak distiction of the dorsal side can be detected in the embryonic disk of *G. riobambae* embryos before elongation (Moya et al., 2007). Elongation of the embryonic disk, notochord and archenteron occur after blastopore closure in an antero–posterior direction marking the antero–posterior and dorsal–ventral axis of the embryo. Simultaneously, the embryo rotates, the embryonic disk enlarges, and faces upwards. The early embryos have an extraordinary planar orientation on the embryonic disk. The streams of cranial neural crest cells are conspicuous and develop in the same plane as the developing body, facilitating the identification of their relationships with the central nervous system (del Pino & Medina, 1998). The embryo develops from the embryonic disk, bearing resemblance to development of the chick embryo (Fig. 3B) (del Pino & Elinson, 1983). The embryonic disks of birds and reptiles have a different embryonic origin (del Pino & Elinson, 1983).

A comparison with *X. laevis* aids in the understanding of the *G. riobambae* pattern of gastrulation. At the beginning of gastrulation in *X. laevis*, convergence without extension, called convergent thickening, occurs in the marginal zone, with concomitant thickening of the blastopore lip (Keller & Shook, 2004; Shook, Kasprowicz, Davidson, & Keller, 2018). This movement is followed in the mid–gastrula by convergent extension. Convergent extension of the dorsal marginal zone contributes to blastopore closure and to elongation of the notochord, archenteron, and body of *X. laevis* embryos (Keller & Danilchik, 1988; Keller et al., 2000).

Variation in the timing of events allows development of an embryonic disk in the *G. riobambae* gastrula (del Pino & Elinson, 1983). In *G. riobambae,* the blastopore closes apparently by convergent thickening (Keller & Shook, 2004). Extension occurs after blastopore closure, evidenced by expression of the protein Brachyury in the notochord (Fig. 3B) (del Pino, 1996). The tiny archenteron of *G. riobambae* elongates simultaneously with the notochord after blastopore closure (Fig. 3A and B) (del Pino, 1996). Therefore, extension and notochord elongation have been naturally separated from gastrulation in *G. riobambae* (del Pino, 1996). Experimentally, extension can be separated from other gastrulation processes in embryos of *X. laevis* and the zebrafish (Ewald, Peyrot, Tyszka, Fraser, & Wallingford, 2004; Roszko, Sawada, & Solnica-Krezel, 2009). It is interesting that while gene expression patterns are highly conserved during development of different frog species (Sudou, Garcés-Vásconez, López-Latorre, Taira, & del Pino, 2016), it appears to be the modular nature of gastrulation that accounts for the diversity of frog gastrulation strategies (Ewald et al., 2004).

3.2 Reproductive strategies and frog gastrulation modes

The analysis of gastrulation in 15 species of frogs revealed two modes of gastrulation according to the timing of notochord elongation (Table 1; Fig. 4A and B) (del Pino, 2019; del Pino et al., 2007; Vargas & del Pino, 2017). In gastrulation mode 1, exemplified by *X. laevis*, elongation of the notochord begins in the mid–gastrula (Fig. 4A). Gastrulation mode 1 occurs in frogs that deposit eggs in or proximal to water or in water associated moist environments. Eggs are small and develop rapidly in comparison with species following gastrulation mode 2. Eggs measure 2 mm diameter or less, with few exceptions and gastrulation takes 1–4 times the *X. laevis* gastrulation time (Table 1; Fig. 4A) (del Pino, 2018). Gastrulation times give indications of the speed of early development in the analyzed frogs (Table 1).

Table 1 Characteristics associated with gastrulation modes in several frog species[a].

Characteristic	Gastrulation mode 1	Gastrulation mode 2
Egg diameter (mm)	1.2–2.9	1.6–3.5
Egg volume (as times *X. laevis* egg volume)[b]	1–14	2–25
Gastrulation time (as times *X. laevis* time)[b]	1–4	4–31
Species analyzed	7	8

[a]Species analyzed are listed in Vargas and del Pino (2017) and del Pino (2019).
[b]*X. laevis* egg diameter: 1.2 mm, gastrulation time 5.5 h (Nieuwkoop & Faber, 1994).

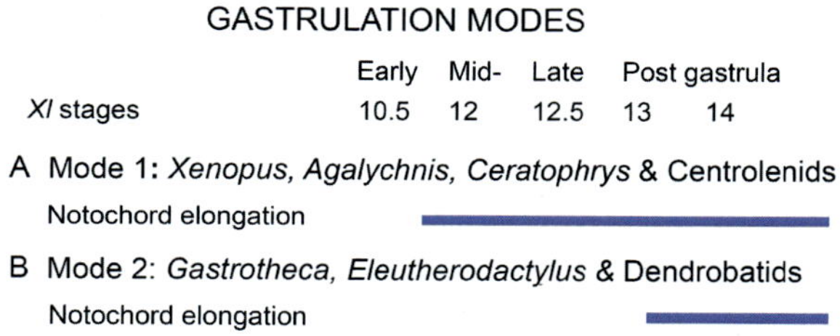

GASTRULATION MODES

Xl stages — Early 10.5, Mid- 12, Late 12.5, Post gastrula 13, 14

A Mode 1: *Xenopus, Agalychnis, Ceratophrys* & Centrolenids
Notochord elongation

B Mode 2: *Gastrotheca, Eleutherodactylus* & Dendrobatids
Notochord elongation

Fig. 4 Frog gastrulation modes. (A) Gastrulation mode 1. (B) Gastrulation mode 2. Gastrulation mode 1 was detected in seven species, representative of four genera. Eight species, belonging to three genera, displayed gastrulation mode 2 (del Pino, 2019). Table 1 summarizes the egg size and gastrulation times associated with gastrulation modes. *X laevis* stages according to Nieuwkoop and Faber (1994). *Modified from Moya, I. M., Alarcón, I., & del Pino, E. M. (2007). Gastrulation of* Gastrotheca riobambae *in comparison with other frogs.* Developmental Biology, 304, *467–478, Fig. 7.*

The overlap between gastrulation and body elongation can help the rapid development of free–living tadpoles, allowing the survival of frog embryos derived from eggs laid in water, especially from eggs with smaller amounts of yolk (Table 1; Fig. 4A).

In contrast, in frogs with gastrulation mode 2, the notochord elongates after blastopore closure, as shown for *G. riobambae* (Table 1; Figs. 3A, B and 4B) (del Pino, 2019; del Pino et al., 2007). *G. riobambae*, dendrobatid frogs, and the frog without tadpoles, *E. coqui*, follow gastrulation mode 2 (Table 1; Fig. 4B) (del Pino, 2019). Eggs are housed in the body of the mother in marsupial frogs or are deposited in the humid soil in dendrobatid frogs and in *E. coqui*. Free–living tadpole stages are reduced in *G. riobambae*, and in dendrobatid frogs (del Pino, 2019; del Pino et al., 2007), whereas in *E. coqui*, the tadpole stages are missing completely, and embryos develop directly into tiny frogs.

The shift of extension to the post-gastrula in frogs with gastrulation mode 2 is associated with the evolution of large eggs (del Pino, 2018; Elinson & del Pino, 2012; Venegas-Ferrín, Sudou, Taira, & del Pino, 2010). The volume of an *E. coqui* egg, with 3.5 mm diameter, represents 25 times the volume of the *X. laevis* egg (Table 1). Moreover, eggs of some hemiphractids are enormous by frog standards, with diameters of 10 mm, i.e., 580 times the volume of a *X. laevis* egg (Fig. 1C) (del Pino & Humphries, 1978; Elinson, del Pino, Townsend, Cuesta, & Eichhorn, 1990). In contrast, eggs of 1–2.9 mm diameter, typical of gastrulation mode 1, are 1–14 times the volume of the *X. laevis* egg (Table 1). Overlap in egg diameters occurs between species with the two modes of gastrulation as the analyzed species with gastrulation mode 2 have eggs that range in diameter from 1.6–3.5 mm (Table 1) (Vargas & del Pino, 2017).

The two features associated with terrestrial life in frogs with gastrulation mode 2 are the increase in the yolk supply of the egg, and slow development. The yolk–rich eggs allow development in absence of external nourishment, resembling the terrestrial reproduction strategy of birds and reptiles. In contrast with birds and reptiles, the yolk of frog embryos is partitioned into cells. Eggs develop slowly in comparison with species following gastrulation mode 1 (Table 1). Gastrulation in *G. riobambae* takes 31 times, and dendrobatid frogs take 7–13 times the *X. laevis* gastrulation time. *E. coqui* is the exception, as its embryos take only 4 times the *X. laevis* gastrulation time, a time period comparable to that of frogs with gastrulation mode 1 (Table 1) (del Pino, 2018). Development in terrestrial environments in frogs with gastrulation mode 2 allowed a delay in the elongation of the body, as the rapid elongation of a free–living tadpole was no longer necessary. Therefore, separation of extension from other gastrulation mechanisms is a developmental mechanism of frogs associated with the conquest of terrestrial life.

4. Multinucleated oogenesis and nurse cells

Two types of oogenesis have been detected in frogs, the mononucleated and the multinucleated modes of oogenesis (del Pino & Humphries, 1978). Oocytes of the majority of frogs and other vertebrates are mononucleated, with a single nucleus, called the germinal vesicle, in each oocyte (Fig. 5A) (Schmid et al., 2018). More than one nucleus per oocyte is uncommon, and has been detected in the tailed frog of North America, *Ascaphus truei,* and in 12, of a total of 35 species of hemiphractid frogs analyzed (Fig. 5B and C) (del Pino & Humphries, 1978; Elinson et al., 1990;

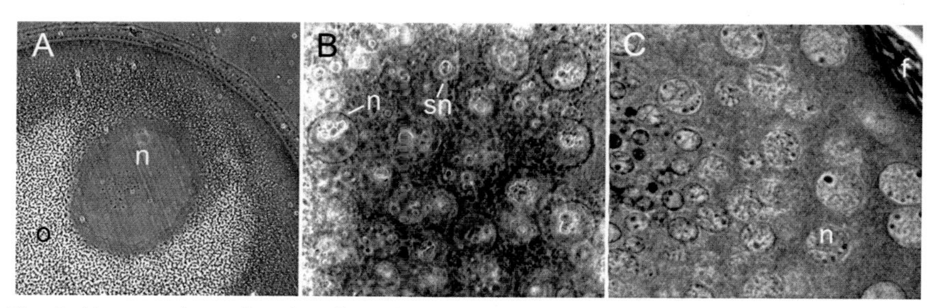

Fig. 5 Mononucleated and multinucleated oocytes. (A) Histological section of a mononucleated oocyte of *G. riobambae*. (B–C) Multinucleated oocytes of *F. pygmaeus*. (B) A massive amount of nuclei was released after opening a living oocyte. This oocyte had nuclei of various sizes. (C) Histological section of an oocyte with nuclei of various sizes. The central region of the oocyte is visible toward the left, and the oocyte periphery is located toward the right. The centrally located nuclei are small, and nuclei located toward the periphery are large. f, follicle wall; n, nucleus; o, oocyte; sn, small nucleus. *Micrograph in (C) is reprinted from Elinson, R. P., & del Pino, E. M. (2012). Developmental diversity of amphibians.* WIREs Developmental Biology, 1, *345–369, Fig. 4C.*

Macgregor & Kezer, 1970). This mode of oogenesis was named multinucleated oogenesis (del Pino & Humphries, 1978). Oocytes have 8–nuclei in *A. truei* (Macgregor & Kezer, 1970). In contrast, nuclear numbers range from one to thousands of nuclei in the oocytes of hemiphractid frogs, according to species (Fig. 5A–C) (del Pino & Humphries, 1978; Elinson et al., 1990).

Mononucleated and multinucleated modes of oogenesis occur in hemiphractid frogs (del Pino & Humphries, 1978; Elinson et al., 1990). Oocytes are mononucleated in species of *Fritziana* (Hemiphractidae), and *Cryptobatrachus* (Hemiphractidae). Both, mononucleated and multinucleated oogenesis typify *Hemiphractus* and *Gastrotheca* (del Pino & Humphries, 1978; Elinson et al., 1990), and multinucleated oogenesis characterizes the analyzed species of *Flectonotus* and *Stefania* (del Pino & Humphries, 1978; Elinson et al., 1990). Multinucleated oogenesis is a derived character that apparently evolved several times during the evolution of hemiphractid frogs (Schmid et al., 2012).

Eggs with final diameters of 2.5–6.0mm characterize species of hemiphractid frogs with mononucleated oocytes, and eggs of 7–10mm diameter occur in species with multinucleated oocytes (del Pino & Humphries, 1978; Elinson et al., 1990). However, there are exceptions. The oocytes of *Flectonotus* and *Fritziana* are multinucleated, although the final diameter of oocytes is 3–4mm (Elinson et al., 1990). Most species of the genus *Hemiphractus* have mononucleated oocytes, despite reaching a final oocyte–diameter of 7–10 mm (del Pino & Humphries, 1978; Elinson et al., 1990). In spite of these exceptions, multinucleated oogenesis is the tendency of hemiphractid frogs with very large eggs (del Pino & Humphries, 1978; Elinson et al., 1990).

Hemiphactid frogs with multinucleated oocytes are rarely collected in nature, a fact that limits the analysis of this mode of oogenesis. The study of multinucleated oogenesis, reviewed in this chapter, was based on a few living specimens of *F. pygmaeus,* a marsupial frog from Venezuela, and museum specimens of other hemiphractid frogs (del Pino & Humphries, 1978; Macgregor & del Pino, 1982). The characteristics of the multinucleated oocytes *F. pygmaeus* are summarized in Table 2 (del Pino & Humphries, 1978; Macgregor & del Pino, 1982). For a comparison of the multinucleated oocytes of *F. pygmaeus* with the mononucleated oocytes of *G. riobambae* see del Pino (1989) and Schmid et al. (2012).

Table 2 Multinucleated oogenesis of the marsupial frog *F. pygmaeus* (Hemiphractidae)[a].

Oocyte diameter (μm)	Nuclear characteristics
<500	Previtellogenic oocytes with nuclei of uniform size (Fig. 6A). Oocytes contain 1000–3000 meiotic nuclei. There is variation in the level of ribosomal RNA amplification between nuclei. Nuclei are active in the synthesis of RNA. Chromosomes are visible as fine threads. Nuclei contain nucleolar–like bodies
500–800	Previtellogenic oocytes with nuclei of various sizes (Figs. 5B, C and 6B). Nuclei located toward the periphery are larger than the centrally located nuclei. Large and medium size nuclei contain lampbrush chromosomes. Nucleoli are clustered in a few large masses of irregular shape. Nuclei are active in the synthesis of RNA
>800	Vitellogenic oocytes with large nuclei located in the periphery (Fig. 6C). Small nuclei are clustered in the central region of the oocyte. Inside nuclei, nucleolar aggregates are large. Centrally located nuclei disintegrate
1200–3000	Vitellogenic oocytes with a single nucleus (Fig. 6D). The nucleus has the typical appearance of the single germinal vesicle of *X. laevis* oocytes. Numerous round nucleoli are located in the central part of the nucleus

[a]According to del Pino and Humphries (1978) and Macgregor and del Pino (1982).

4.1 Multinucleated oocytes of *F. pygmaeus*

The number of nuclei of a multinucleated oocyte of *F. pygmaeus* ranges from 2000 to 3000 nuclei (del Pino & Humphries, 1978). In the ovarian cysts, multinucleated oocytes derive from cellular aggregates of about 100 μm diameter, and result from nuclear divisions without cytokinesis (del Pino & Humphries, 1978). Nuclei of a multinucleated oocyte have a nuclear envelope, chromosomes, and extrachromosomal nucleoli. All nuclei incorporate ^{3}H-uridine into RNA and there is amplification of rDNA in each nucleus (del Pino & Humphries, 1978; Macgregor & del Pino, 1982). The changes of multinucleated oocytes of *F. pygmaeus* during oogenesis are summarized in Table 2.

In oocytes with diameters of 500 μm or less, the nuclei have similar sizes, and the chromosomes have the appearance of pachytene bivalents (Fig. 6A). In oocytes of 500–800 μm diameter, the oocyte contains an outer shell of large nuclei and an inner core of smaller nuclei with variable chromatin density. All nuclei contain 5–20 nucleolus-like bodies (Figs. 5B, C and 6B). The mass of nuclei that emerge after puncturing an oocyte of *F. pygmaeus* gives a visual evidence of the high density of nuclei found in each oocyte (Fig. 5B). In oocytes larger than 800 μm diameter, the internal mass of nuclei disintegrates, and the outer shell of large nuclei have numerous nucleoli and lampbrush chromosomes (Fig. 6C). The differentiation of these two classes of nuclei may depend on the positions the nuclei occupy within the oocyte, with larger nuclei located in the periphery and small nuclei in the central part of the oocyte (Fig. 6B and C) (Macgregor & del Pino, 1982). It may be that

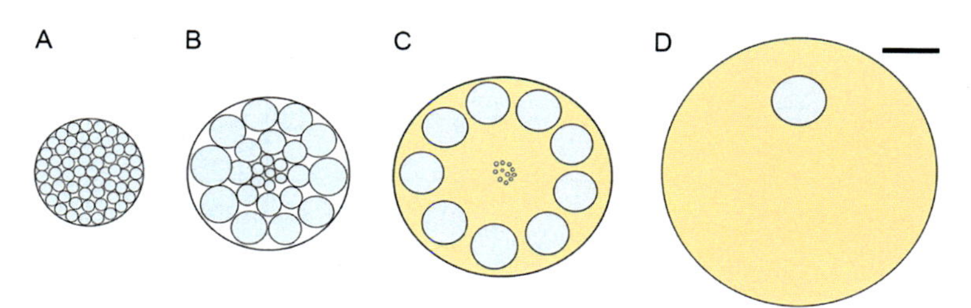

Fig. 6 Diagrams of the progression of multinucleated oogenesis in *F. pygmaeus*. (A) Previtellogenic oocyte with nuclei of uniform size. (B) Previtellogenic oocyte with large nuclei in the periphery and smaller nuclei in internal regions. (C) Vitellogenic oocyte with reduction in the number of nuclei. Peripheral nuclei are large. Central nuclei are small. (D) Vitellogenic oocyte with a single germinal vesicle. Nuclei are shown in blue, and vitellogenic oocytes are highlighted in yellow. Bar, 250 μm. The characteristics of *F. pygmaeus* oocytes are given in Table 2.

the transcriptional activity of lampbrush chromosomes in peripheral nuclei contributes to enrich the cortex of the oocyte with transcripts of developmental value. The number of nuclei is reduced in oocytes of 1200 μm diameter, and the large nuclei that remain have lampbrush chromosomes and many nucleoli. In full–grown oocytes of 3000 μm, only one germinal vesicle remains with all the properties of the single germinal vesicle of the amphibian oocyte (Fig. 6D; Table 2) (del Pino & Humphries, 1978; Macgregor & del Pino, 1982).

Each nucleus of a multinucleated oocyte is a meiotic nucleus with four copies of the haploid genome, and variable amounts of rDNA. The level of rDNA amplification varies between nuclei (Macgregor & del Pino, 1982). The rDNA amplification level of each nucleus is modest in comparison with the level of rDNA amplification of the single germinal vesicle of *X. laevis* oocytes (Macgregor & del Pino, 1982). However, due to the enormous number of nuclei, the total level of rDNA amplification of a multinucleated oocyte exceeds greatly the level of rDNA amplification detected in the single germinal vesicle of *X. laevis* oocytes (Macgregor & del Pino, 1982).

During oogenesis, the *X. laevis* oocyte, with its single germinal vesicle, accumulates vast amounts of ribosomes, various kinds of RNA molecules, and proteins required for the initial advancement of embryonic development. The *X. laevis* germinal vesicle has four copies of the haploid genome. In contrast, there are 12,000 copies of the haploid genome in an oocyte *F. pygmaeus* with 3000 nuclei. The high degree of endopolyploidy, the amplification of ribosomal genes, and the transcriptional activity of all nuclei of a *F. pygmaeus* oocyte must contribute large amounts of gene products and proteins needed to support early development.

Incubation in the pouch of *F. pygmaeus* is considerably abbreviated in comparison with *G. riobambae*, a marsupial frog with mononucleated oocytes. Development in the pouch of *F. pygmaeus* lasts 29 days, and at birth, non–swimming and non–feeding tadpoles with hind limbs are released from the pouch. In nature, tadpoles complete metamorphosis in 13 days (del Pino & Escobar, 1981; del Pino & Humphries, 1978). Whereas, *G. riobambae* development in the pouch lasts nearly 4 months, and birth occurs at a feeding–tadpole stage. Free–living tadpoles of *G. riobambae* require several months to reach metamorphosis (del Pino et al., 1975). However, early development is slow in both frogs in comparison with *X. laevis*. Seven hours after fertilization, embryos of *X. laevis* reach the midblastula transition, and have around 4000 blastomeres. In the same period, the embryos of *F. pygmaeus* and *G. riobambae* have not yet completed the first cleavage cycle (del Pino & Escobar, 1981; Schmid et al., 2012). Therefore, the relevance of multinucleated oogenesis to provide gene products to support early development needs additional observations and analysis.

The numerous nuclei of *F. pygmaeus* oocytes may accelerate the processes of oogenesis. However, the time required to complete oogenesis is unknown. (del Pino & Humphries, 1978; Macgregor & del Pino, 1982; Schmid et al., 2012). Other features of multinucleated oogenesis are similarly unknown and require analysis. For example, the events that result in the placement of many meiotic nuclei inside a multinucleated oocyte are unknown. The features that allow enlargement of peripheral nuclei in oocytes is puzzling and requires analysis. The mechanisms of nuclear degradation, and the protection from degradation of the final germinal vesicle that remains in the oocyte are also unknown.

4.2 Nurse cells and large egg size

Nurse cells can be of germline or of somatic origin. Nurse cells have a germline origin in *Caenorhabditis elegans, Drosophila,* the mouse, and other animals. In these organisms, the founder cell of the each germline cyst, the cystoblast, undergoes incomplete cell–divisions. As a result, the oocyte is connected with its sister cells, the nurse cells (Cox & Spradling, 2003; Lei & Spradling, 2013, 2016; Pepling, de Cuevas, & Spradling, 1999). In *Drosophila*, four incomplete divisions of each cystoblast result in the formation of the oocyte, and of 15 nurse cells (Pepling et al., 1999). Nurse cells in *Drosophila* polyploidize their genome and transfer gene products, cytoplasm, and organelles, enriching the cytoplasm of the oocyte. Later in oogenesis, the nurse cells undergo apoptosis (Cooley, 1995). The multiple nuclei of an oocyte of *F. pygmaeus* and of other hemiphractid frogs are probably derived from germline nurse cells that were incorporated into the cytoplasm of the oocyte (del Pino, 2018).

Reptiles, like the Italian lizard *Podarcis sicula* (Lacertidae), enrich the oocyte with transcripts, proteins and organelles by way of nurse cells of somatic origin, called pyriform cells (Andreuccetti, Taddei, & Filosa, 1978). The oocytes of *P. sicula* are mononucleated, and are connected with pear–shaped pyriform cells of the follicular epithelium (Andreuccetti et al., 1978; Maurizii et al., 2009; Maurizii & Taddei, 2012). The follicular epithelium of previtellogenic oocytes is multi–layered. Small follicle–cells connect with the oocyte through intercellular bridges, and differentiate into pyriform cells (Andreuccetti et al., 1978). Pyriform cells transfer transcripts and organelles to the oocyte during the previtellogenic period. (Maurizii et al., 2009). At the end of the previtellogenic period, pyriform cells undergo apoptosis. In larger oocytes, the follicular epithelium consists of a single layer of small cells (Motta, Filosa, & Andreuccetti, 1996). The contribution of pyriform cells to the stores of gene products, proteins and organelles must be significant given the large number of pyriform cells observed in the follicular wall of *P. sicula* oocytes.

Besides the lizards, pyriform cells occur in the ovarian follicle of Elasmobranch *Raya asterias*, and *Torpedo marmorata*, and in the cobra *Naja naja* (Andreuccetti, Iodice, Prisco, & Gualtieri, 1999; Lance & Lofts, 1977; Prisco, Liguoro, Ricchiari, Del Giudice, & Andreuccetti, 2007). Pyriform–like cells occur in ovarian follicles of birds, and of the frog *Ceratophrys cranwelli* (Callebaut, 1991; Callebaut, Van Nassauw, & Harrisson, 1997; Villecco, Monaco, & Sánchez, 2007). Pyriform–like cells are cells of the follicular epithelium that do not acquire the pyriform morphology but function as nurse cells. Pyriform cells and pyriform–like cells of fish, amphibians, reptiles, and bird ovaries may enrich the large oocytes of these organisms with transcripts, ribosomes and organelles required to support early development.

Increase in egg size in vertebrates is due to yolk, made in the liver, transported via blood to the ovarian follicles, and incorporated into the growing oocytes. The synthesis and accumulation of yolk in oocytes is not discussed in this chapter. Nonetheless, an important question that should be answered is: How much active cytoplasm (non–yolk and non–oil storage inclusions) is needed in the large eggs of different organisms such as frogs or different reptiles or birds? The answer to this question is an important consideration that might involve the need to have numerous nuclei providing gene products, in addition to the germinal vesicle of the oocyte.

5. Conclusion

In amniotes (reptiles and birds), the enrichment of the oocyte with transcripts, provided by nurse cells, is likely required to support development from very large eggs. Large egg size in reptiles and birds is associated with meroblastic cleavage, development from a disk of cells and the formation of extra–embryonic structures, such as the amnion. These developmental adaptations were important for the conquest of terrestrial life. Hemiphractid frogs apparently use multinucleated oogenesis to enrich the cytoplasm of oocytes with nuclear products, allowing for an extraordinary increase in egg size. Large egg size in hemiphractid frogs allowed modifications of early development, including the separation of extension from other gastrulation mechanisms, and the conquest of terrestrial life by parallel mechanisms to those of the amniotes, as discussed in this chapter. The function of nurse cells in oogenesis apparently provided the basis for the evolution of terrestrial life in amniotes, paralleled by the multinucleated oogenesis of hemiphractid frogs. It remains important to determine the concentration and localization of transcripts and molecules of developmental importance in the large eggs of hemiphractid frogs and in the amniotes as well as the possible contribution of nurse cells to this process.

Acknowledgments

I would like to acknowledge the collaboration of former students and colleagues. I thank Igor B. Dawid, Richard P. Elinson and Allan C. Spradling for critical revision of the text, and José García–Arrarás for important comments and language revision. I gratefully acknowledge the Pontificia Universidad Católica del Ecuador for supporting the research of my laboratory.

References

Alcocer, I., Santacruz, X., Steinbeisser, H., Thierauch, K.-H., & del Pino, E. M. (1992). Ureotelism as the prevailing mode of nitrogen excretion in larvae of the marsupial frog *Gastrotheca riobambae* (Fowler) (Anura, Hylidae). *Comparative Biochemistry and Physiology*, *101A*(2), 229–231.

Andreuccetti, P., Iodice, M., Prisco, M., & Gualtieri, R. (1999). Intercellular bridges between granulosa cells and the oocyte in the elasmobranch *Raya asterias*. *The Anatomical Record*, *255*(2), 180–187.

Andreuccetti, P., Taddei, C., & Filosa, S. (1978). Intercellular bridges between follicle cells and oocyte during the differentiation of follicular epithelium in *Lacerta sicula* Raf. *Journal of Cell Science*, *33*, 341–350.

Arendt, D., & Nübler-Jung, K. (1999). Rearranging gastrulation in the name of yolk: Evolution of gastrulation in yolk-rich amniote eggs. *Mechanisms of Development*, *81*(1–2), 3–22.

Callebaut, M. (1991). Pyriform-like and holding granulosa cells in the avian ovarian follicle wall. *European Archives of Biology*, *102*, 135–145.

Callebaut, M., Van Nassauw, L., & Harrisson, F. (1997). Comparison between oogenesis and related ovarian structures in a reptile, *Pseudemys scripta elegans* (turtle) and in a bird *Coturnix coturnix japonica* (quail). *Reproduction, Nutrition, Development*, *37*(3), 233–252.

Cooley, L. (1995). Oogenesis: Variations on a theme. *Developmental Genetics*, *16*(1), 1–5.

Cox, R. T., & Spradling, A. C. (2003). A Balbiani body and the fusome mediate mitochondrial inheritance during *Drosophila* oogenesis. *Development*, *130*(8), 1579–1590.

del Pino, E. M. (1980). Morphology of the pouch and incubatory integument in marsupial frogs (Hylidae). *Copeia*, *1980*(1), 10–17.

del Pino, E. M. (1989). Modifications of oogenesis and development in marsupial frogs. *Development*, *107*(2), 169–187.

del Pino, E. M. (1996). The expression of brachyury (T) during gastrulation in the marsupial frog *Gastrotheca riobambae*. *Developmental Biology*, *177*(1), 64–72.

del Pino, E. M. (2018). The extraordinary biology and development of marsupial frogs (Hemiphractidae) in comparison with fish, mammals, birds, amphibians and other animals. *Mechanisms of Development*, *154*, 2–11.

del Pino, E. M. (2019). Embryogenesis of marsupial frogs (Hemiphractidae), and the changes that accompany terrestrial development in frogs. *Results and Problems in Cell Differentiation*, *68*, 379–418.
del Pino, E. M., Alcocer, I., & Grunz, H. (1994). Urea is necessary for the culture of embryos of the marsupial frog *Gastrotheca riobambae*, and is tolerated by embryos of the aquatic frog *Xenopus laevis*. *Development, Growth & Differentiation*, *36*(1), 73–80.
del Pino, E. M., & Elinson, R. P. (1983). A novel development pattern for frogs: Gastrulation produces an embryonic disk. *Nature*, *306*, 589–591.
del Pino, E. M., & Escobar, B. (1981). Embryonic stages of *Gastrotheca riobambae* (Fowler) during maternal incubation and comparison of development with that of other egg-brooding hylid frogs. *Journal of Morphology*, *167*(3), 277–295.
del Pino, E. M., Galarza, M. L., de Albuja, C. M., & Humphries, A. A. (1975). The maternal pouch and development in the marsupial frog *Gastrotheca riobambae* (Fowler). *The Biological Bulletin*, *149*, 480–491.
del Pino, E. M., & Humphries, A. A. (1978). Multiple nuclei during early oogenesis in *Flectonotus pygmaeus* and other marsupial frogs. *The Biological Bulletin*, *154*, 198–212.
del Pino, E. M., & Loor-Vela, S. (1990). The pattern of early cleavage of the marsupial frog *Gastrotheca riobambae*. *Development*, *110*(3), 781–789.
del Pino, E. M., & Medina, A. (1998). Neural development in the marsupial frog *Gastrotheca riobambae*. *The International Journal of Developmental Biology*, *4*(1), 723–731.
del Pino, E. M., Venegas-Ferrín, M., Romero-Carvajal, A., Montenegro-Larrea, P., Sáenz-Ponce, N., Moya, I. M., et al. (2007). A comparative analysis of frog early development. *Proceedings of the National Academy of Sciences of the United States of America*, *104*(29), 11882–11888.
Duellman, W. E. (2015). *Marsupial frogs: Gastrotheca and allied genera*. Baltimore: Johns Hopkins University Press.
Duellman, W. E., & Trueb, L. (1986). *Biology of amphibians*. New York, USA: McGraw Hill, Inc.
Elinson, R. P., & Beckham, Y. (2002). Development in frogs with large eggs and the origin of amniotes. *Zoology*, *105*(2), 105–117.
Elinson, R. P., & del Pino, E. M. (1985). Cleavage and gastrulation in the egg-brooding, marsupial frog, *Gastrotheca riobambae*. *Journal of Embryology and Experimental Morphology*, *90*, 223–232.
Elinson, R. P., & del Pino, E. M. (2012). Developmental diversity of amphibians. *WIREs Developmental Biology*, *1*(3), 345–369.
Elinson, R. P., del Pino, E. M., Townsend, D. S., Cuesta, F. C., & Eichhorn, P. (1990). A practical guide to the developmental biology of terrestrial-breeding frogs. *The Biological Bulletin*, *179*, 163–177.
Elinson, R. P., Stewart, J. R., Bonneau, L. J., & Blackburn, D. G. (2014). Amniote yolk sacs: Diversity in reptiles and a hypothesis on their origin. *The International Journal of Developmental Biology*, *58*(10 – 12), 889–894.
Ewald, A. J., Peyrot, S. M., Tyszka, J. M., Fraser, S. E., & Wallingford, J. B. (2004). Regional requirements for dishevelled signaling during *Xenopus* gastrulation: Separable effects on blastopore closure, mesendoderm internalization and archenteron formation. *Development*, *131*(24), 6195–6209.
Gilbert, S. F. (2014). *Developmental biology* (10th ed.). Sunderland, MA: Sinauer Associates, Inc.
Hallmann, K., & Griebeler, E. M. (2015). Eggshell types and their evolutionary correlation with life-history strategies in squamates. *PLoS One*, *10*(9), 1–20.
Jones, R. E., Gerrard, A. M., & Roth, J. J. (1973). Estrogen and brood pouch formation in the marsupial frog *Gastrotheca riobambae*. *The Journal of Experimental Zoology*, *184*(2), 177–184.

Keller, R., & Danilchik, M. (1988). Regional expression, pattern and timing of convergence and extension during gastrulation of *Xenopus laevis*. *Development*, *103*(1), 193–209.
Keller, R., Davidson, L., Edlund, A., Elul, T., Ezin, M., Shook, D., et al. (2000). Mechanisms of convergence and extension by cell intercalation. *Philosophical Transactions of the Royal Society of London. Series B, Biological Sciences*, *355*(1399), 897–922.
Keller, R., & Shook, D. (2004). Gastrulation in amphibians. In C. D. Stern (Ed.), *Gastrulation from cells to embryo* (pp. 171–203). New York, USA: Cold Spring Harbor Laboratory Press.
Lance, V., & Lofts, B. (1977). Studies on the reproductive cycle of the female cobra, *Naja naja*. II in vitro ovarian steroid synthesis. *General and Comparative Endocrinology*, *32*(3), 279–288.
Lei, L., & Spradling, A. C. (2013). Mouse primordial germ cells produce cysts that partially fragment prior to meiosis. *Development*, *140*(10), 2075–2081.
Lei, L., & Spradling, A. C. (2016). Mouse oocytes differentiate through organelle enrichment from sister cyst germ cells. *Science*, *352*(6281), 95–99.
Macgregor, H. C., & del Pino, E. M. (1982). Ribosomal gene amplification in multinucleate oocytes of the egg brooding hylid frog *Flectonotus pygmaeus*. *Chromosoma*, *85*(4), 475–488.
Macgregor, H. C., & Kezer, J. (1970). Gene amplification in oocytes with 8 germinal vesicles from the tailed frog *Ascaphus truei* Stejneger. *Chromosoma*, *29*(2), 189–206.
Maurizii, M. G., Cavaliere, V., Gamberi, C., Lasko, P., Gargiulo, G., & Taddei, C. (2009). Vasa protein is localized in the germ cells and in the oocyte-associated pyriform follicle cells during early oogenesis in the lizard *Podarcis sicula*. *Development Genes and Evolution*, *219*(7), 361–367.
Maurizii, M. G., & Taddei, C. (2012). Microtubule organization and nucleation in the differentiating ovarian follicle of the lizard *Podarcis sicula*. *Journal of Morphology*, *273*(10), 1089–1095.
Motta, C. M., Filosa, S., & Andreuccetti, P. (1996). Regression of the epithelium in late previtellogenic follicles of *Podarcis sicula*: A case of apoptosis. *The Journal of Experimental Zoology*, *276*(3), 233–241.
Moya, I. M., Alarcón, I., & del Pino, E. M. (2007). Gastrulation of *Gastrotheca riobambae* in comparison with other frogs. *Developmental Biology*, *304*(2), 467–478.
Nieuwkoop, P. D., & Faber, J. (1994). *Normal table of Xenopus laevis (Daudin)*. New York, USA: Garland Publishing.
Pepling, M. E., de Cuevas, M., & Spradling, A. C. (1999). Germline cysts: A conserved phase of germ cell development? *Trends in Cell Biology*, *9*(7), 257–262.
Prisco, M., Liguoro, A., Ricchiari, L., Del Giudice, G., & Andreuccetti, P. (2007). Oogenesis in the spotted ray *Torpedo marmorata*. *Reviews in Fish Biology and Fisheries*, *17*, 1–10.
Roszko, I., Sawada, A., & Solnica-Krezel, L. (2009). Regulation of convergence and extension movements during vertebrate gastrulation by the Wnt/PCP pathway. *Seminars in Cell & Developmental Biology*, *20*(8), 986–997.
Schmid, M., Steinlein, C., Bogart, J. P., Feichtinger, W., Haaf, T., Nanda, I., et al. (2012). The Hemiphractid frogs phylogeny, embryology, life history, and cytogenetics. *Cytogenetic and Genome Research*, *138*(2–4), 1–384.
Schmid, M., Steinlein, C., Haaf, T., Feichtinger, W., Guttenbach, M., Bogart, J. P., et al. (2018). The Arboranan frogs. Evolution, biology, cytogenetics. *Cytogenetic and Genome Research*, *155*(1), 1–325.
Shook, D. R., Kasprowicz, E. M., Davidson, L. A., & Keller, R. (2018). Large, long range tensile forces drive convergence during *Xenopus* blastopore closure and body axis elongation. *eLife*, 7, e26944.
Solnica-Krezel, L. (2005). Conserved patterns of cell movements during vertebrate gastrulation. *Current Biology*, *15*(6), 213–228.

Sudou, N., Garcés-Vásconez, A., López-Latorre, M. A., Taira, M., & del Pino, E. M. (2016). Transcription factors Mix1 and VegT, relocalization of vegt mRNA, and conserved endoderm and dorsal specification in frogs. *Proceedings of the National Academy of Sciences of the United States of America, 113*(20), 5628–5633.

Vargas, A., & del Pino, E. M. (2017). Analysis of cell size in the gastrula of ten frog species reveals a correlation of egg with cell sizes, and a conserved pattern of small cells in the marginal zone. *Journal of Experimental Zoology. Part B, Molecular and Developmental Evolution, 328*(1–2), 88–96.

Venegas-Ferrín, M., Sudou, N., Taira, M., & del Pino, E. M. (2010). Comparison of Lim1 expression in embryos of frogs with different modes of reproduction. *The International Journal of Developmental Biology, 54*(1), 195–202.

Villecco, E. I., Monaco, M. E., & Sánchez, S. S. (2007). Ultrastructural changes in the follicular epithelium of *Ceratophrys cranwelli* previtellogenic oocytes. *Zygote, 15*(3), 273–283.

SECTION II

Systems biology approaches in amphibians

CHAPTER FIVE

LIM homeodomain proteins and associated partners: Then and now

Yuuri Yasuoka[a,*] and Masanori Taira[b,*]

[a]Laboratory for Comprehensive Genomic Analysis, RIKEN Center for Integrative Medical Sciences, Yokohama, Japan
[b]Department of Biological Sciences, Faculty of Science and Engineering, Chuo University, Bunkyo-ku, Tokyo, Japan
*Corresponding authors: e-mail addresses: yuuriyasuoka@gmail.com (Yuuri Yasuoka); m-taira.183@g.chuo-u.ac.jp (Masanori Taira)

Contents

Abstract

The field of molecular embryology started around 1990 by identifying new genes and analyzing their functions in early vertebrate embryogenesis. Those genes encode transcription factors, signaling molecules, their regulators, etc. Most of those genes are relatively highly expressed in specific regions or exhibit dramatic phenotypes when

Current Topics in Developmental Biology, Volume 145
ISSN 0070-2153
https://doi.org/10.1016/bs.ctdb.2021.04.003

ectopically expressed or mutated. This review focuses on one of those genes, *Lim1/Lhx1*, which encodes a transcription factor. Lim1/Lhx1 is a member of the LIM homeodomain (LIM-HD) protein family, and its intimate partner, Ldb1/NLI, binds to two tandem LIM domains of LIM-HDs. The most ancient LIM-HD protein and its partnership with Ldb1 were innovated in the metazoan ancestor by gene fusion combining LIM domains and a homeodomain and by creating the LIM domain-interacting domain (LID) in ancestral Ldb, respectively. The LIM domain has multiple interacting interphases, and Ldb1 has a dimerization domain (DD), the LID, and other interacting domains that bind to Ssbp2/3/4 and the boundary factor, CTCF. By means of these domains, LIM-HD-Ldb1 functions as a hub protein complex, enabling more intricate and elaborate gene regulation. The common, ancestral role of LIM-HD proteins is neuron cell-type specification. Additionally, Lim1/Lhx1 serves crucial roles in the gastrula organizer and in kidney development. Recent studies using *Xenopus* embryos have revealed Lim1/Lhx1 functions and regulatory mechanisms during development and regeneration, providing insight into evolutionary developmental biology, functional genomics, gene regulatory networks, and regenerative medicine. In this review, we also discuss recent progress at unraveling participation of Ldb1, Ssbp, and CTCF in enhanceosomes, long-distance enhancer–promoter interactions, and *trans*-interactions between chromosomes.

1. Introduction

Since the 1980s, genetics and molecular biology of *Drosophila melanogaster*, and later *Caenorhabditis elegans*, have identified many important genes involved in embryogenesis. Around 1990, molecular embryology commenced by identifying new genes and analyzing their functions in early vertebrate embryogenesis, using *Xenopus*, mice, and zebrafish, and became an established field during the 1990s. Using *Xenopus* embryos, subtraction and functional screening worked very well to identify newly discovered genes. Most such genes encode transcription factors (TFs), signaling molecules, and signal transducers or modifiers, including secreted inhibitors and inhibitory cytoplasmic components. When new types of genes were identified, homologous genes, such as paralogs, ohnologs, and orthologs, are relatively easily identified by homology screening. Ohnologs are paralogs that were created by two rounds of whole genome duplication in the common ancestor of vertebrates. Since about 2000, whole genome sequences of various organisms, including humans, have dominated gene cloning strategies with in silico screening to identify homologous genes of interests and subtraction screening using regional RNA-seq data. Once identified, loss-of-function analysis is performed by targeting gene knockout in mice, antisense morpholino oligos (MOs) in other species, or more recently the

CRISPR/Cas9 system, and gain-of-function analysis is done using DNA transfection or mRNA injection.

Homeobox genes were first identified in *Drosophila* as the genes responsible for homeotic mutations, such as Antennapedia and Ultrabithorax. The homeobox is a conserved nucleotide sequence of about 180 base pairs, that encodes a ~60-amino acid sequence of a DNA-binding domain, called the homeodomain. Homeodomain (HD) TFs or homeobox genes constitute a large gene family and are classified into several subfamilies. One of them is the LIM-HD protein, or the LIM homeobox gene. Like other homeobox gens, LIM homeobox genes are involved in various developmental pathways.

In this review, we first focus on the LIM homeobox gene *Lim1/Lhx1*, which was initially identified as a Spemann-Mangold organizer-specific gene (Taira, Jamrich, Good, & Dawid, 1992). Then we summarize the long history of molecular studies of all LIM-HD proteins in neuron cell-type specification, regionalization, organogenesis, regeneration, and tumorigenesis. We also focus on the LIM-domain binding protein, Ldb1/NLI, which functions not only as a cofactor of LIM-HD, but also interacts with single strand DNA-binding proteins (Ssbp2/3/4) and the CCCTC-binding factor (CTCF) to function as a key factor for enhanceosome and long-distance enhancer–promoter communications in *cis* and even in *trans*. We will also discuss recent analyses of genomic and evolutionary aspects of LIM and Ldb proteins from plants to animals to argue that innovation of LIM-HD contributed significantly to the diversity and complexity of metazoans.

2. History and an overview of LIM-HD genes

2.1 Identification of LIM-HD and the LIM protein family

2.1.1 LIM-HD proteins

Entirely new discoveries are often completely unexpected, because complete novelty has no relevance to available information or hypotheses. Identification of the LIM domain or motif illustrates this principle. The first LIM-HD discovered was Mec-3 in *Caenorhabditis elegans*, the gene responsible for mechanosensory mutant 3 (*mec-3*) (Way & Chalfie, 1988). At that time, however, the LIM domain had not yet been recognized, and its classification as a LIM-HD occurred later. In 1990, Islet-1 and Lin-11 were identified in rats as regulators of the insulin gene (Karlsson, Thor, Norberg, Ohlsson, & Edlund, 1990) and from *C. elegans* as the gene responsible for lineage mutant 11 (*lin-11*) (Freyd, Kim, & Horvitz, 1990).

Alignment of the three amino acid sequences revealed a new conserved motif in addition to the homeodomains, called the LIM motif or domain, after the first letters of the three gene names (Freyd et al., 1990; Karlsson et al., 1990).

The LIM motif consists of eight dispersed cysteine and histidine residues, and based on this primary structure, the LIM domain was expected to be a zinc-finger-like structure (Fig. 2A). Discovery of a new class of homeobox gene enabled researchers to identify new members of this class. However, because of less conserved amino acid sequences of LIM domains, except for the LIM motifs, it was difficult to use them for PCR cloning with degenerate oligos. Instead, using homeodomain amino acid sequences specific to lin-11 and Ceh-14 to design degenerate oligos, *Xlim-1* (also called *lim1* or *lhx1*; Table 1) was identified as an ortholog of *lin-11* from an embryonic cDNA library of the African clawed frog, *Xenopus laevis* (Taira et al., 1992). With this screening, *Xlim-3* (*lim3* or *lhx3*) and *lim5* (*lhx5*) were also identified (Taira, Hayes, Otani, & Dawid, 1993; Toyama et al., 1995). The other LIM-HD genes identified in vertebrates by homology screening or other methods from *Xenopus*, mice, and rats are *lmx1* (*lmx1a*) (German, Wang, Chadwick, & Rutter, 1992), *LH-2* (*lhx2*) (Xu et al., 1993), *Gsh-4* (*lhx4*) (Li et al., 1994), *isl2* (Tsuchida et al., 1994), *lmx1b* (Chen et al., 1998), *lhx6* (Grigoriou, Tucker, Sharpe, & Pachnis, 1998), *lhx8* (*L3*, the same as *lhx7*) (Matsumoto et al., 1996), *lhx9* (Bertuzzi et al., 1999) (Table 1). These genes are classified into six subfamilies of the LIM class homeobox gene family (Fig. 1A; Table 1 for gene nomenclature). In vertebrates, each subfamily has two ohnologs (Fig. 1A and Table 1).

LIM only protein 1 (LMO1, also called Rhombotin-1 or Ttg-1) has two tandem LIM domains that are closely related to those of LIM-HD proteins, but they share almost no other attributes (Boehm, Foroni, Kaneko, Perutz, & Rabbitts, 1991; McGuire, Davis, & Korsmeyer, 1991). This LMO family has four vertebrate members (LMO1–4), and was initially thought to comprise natural dominant negative proteins against LIM-HD proteins, but LMO proteins have their own functions as adaptor proteins, as described below, in addition to serving as negative regulators.

2.1.2 Other LIM proteins

LIM domains are associated not only with the homeodomain, but also with known domains, uncharacterized regions, or almost nothing else, as exemplified by LIM kinase, focal adhesion-associated adaptor proteins (Paxillin, Prickle, etc.), and four and half LIM protein 2 (FHL2) (Tran, Kurakula,

Table 1 Protein and gene names and synonyms for LIM-HD, LMO, Ldb, and Ssbp.

Protein family	Protein name	Gene name[a]	Gene symbol	Synonyms or orthlogs				
				Vertebrates	*C. elegans*	*Drosophila*	*Arabidopsis*	Yeast
LIM-HD	Lhx1	LIM homeobox 1	*lhx1*	*Xlim-1, Xlim1, lim-1, lim1*	*lin-11, mec-3*[b]	*lim1*		
	Lhx2	LIM homeobox 2	*lhx2*	*LH-2*	*ttx-3*	*Apterous*		
	Lhx3	LIM homeobox 3	*lhx3*	*Xlim-3, lim3, P-lim*	*ceh-14*	*lim3*		
	Lhx4	LIM homeobox 4	*lhx4*	*gsh4, cphd4*				
	Lhx5	LIM homeobox 5	*lhx5*	*Xlim-2, Xlim-5, Xlim5, lim5*	*lin-11, mec-3*[b]	*lim1*		
	Lhx6	LIM homeobox 6	*lhx6*		*lim-4*[c]	*Arrowhead*		
	Lhx8	LIM homeobox 8	*lhx8*	*lhx7, L3*				
	Lhx9	LIM homeobox 9	*lhx9*		*ttx-3*	*Apterous*		
	Islet1	ISL LIM homeobox 1	*isl1*		*lim-7*[c]	*Tailup*		
	Islet2	ISL LIM homeobox 2	*isl2*					
	Lmx1a	LIM homeobox transcription factor 1 alpha	*lmx1a*	*lmx-1 , lmx1.1*	*lim-6*[c]	*lmx1a, CG4328*[d]		
	Lmx1b	LIM homeobox transcription factor 1 beta	*lmx1b*	*Xlmx1b, lmx1.2, nps1, GENA 191, lcst*				

Continued

Table 1 Protein and gene names and synonyms for LIM-HD, LMO, Ldb, and Ssbp.—cont'd

				Synonyms or orthlogs				
Protein family	**Protein name**	**Gene name[a]**	**Gene symbol**	**Vertebrates**	***C. elegans***	***Drosophila***	***Arabidopsis***	**Yeast**
LMO	LMO1	LIM domain only 1	*lmo1*	*Rbtn1, Rbtn-1, Ttg1*	e	*Beadex*		
	LMO2	LIM domain only 2	*lmo2*	*Rbtn2, Rbtn-2, Ttg2*				
	LMO3	LIM domain only 3	*lmo3*	*Rbtn3, Rbtn-3, RBTNL2, RHOM3*				
	LMO4	LIM domain only 4	*lmo4*					
Ldb	Ldb1	LIM domain binding 1	*ldb1*	*NLI, CLIM2, ldb1a and ldb1b (teleost paralogs; ldb1a = ldb4, ldb1b = ldb1)*	*ldb-1*	*Chip*	*SEUSS, SEUSS-LIKE*	*adn1*[f]
	Ldb2	LIM domain binding 2	*ldb2*	*clim1, ldb2a and ldb2b (teleost paralogs; ldb2a = ldb2, ldb2b = ldb3)*				
Ssbp	Ssbp2	single stranded DNA binding protein 2	*ssbp2*	*ssdp2*	*ssdp*	*ssdp, dSsdp*	*LEUNIG, LEUNIG_HOMOLOG*	*Flo8, Mss11*[g]
	Ssbp3	single stranded DNA binding protein 3	*ssbp3*	*csdp, ssdp, ssdp1*				
	Ssbp4	single stranded DNA binding protein 4	*ssbp4*	*Ssdp3*				

[a]According to the Human Gene Nomenclature Committee (HGNC).
[b]*lin-11* and *mec-3* are paralogs, not synonym.
[c]A confusing name (be careful to use it).
[d]*lmx1a* and *CG4328* are paralogs.
[e]Not found by BLAST search using the GenBank nr database.
[f]*Schizosaccharomyces pombe.*
[g]*Saccharomyces cerevisiae, Flo8* and *Mss11* are paralogs.

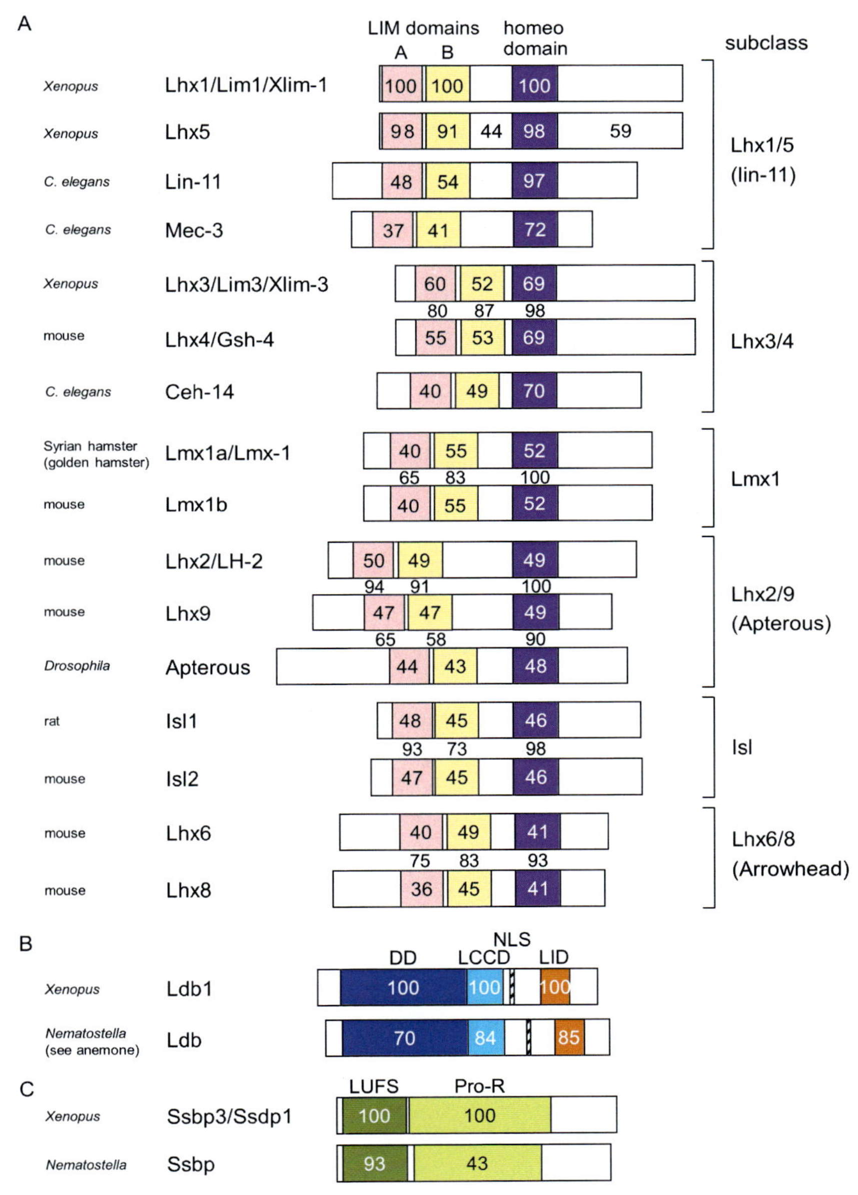

Fig. 1 Classification and sequence homology among LIM-HD, Ldb, and Ssbp. (A) Representatives of six LIM-HD subclasses are schematically represented with protein domains. Numbers in each domain indicate sequence identities (%) with Xlim-1/Lhx1. Numbers between proteins indicate sequence identities (%) with the protein on the top in each subclass. Each subclass has two ohnologs and some of *C. elegans* or *Drosophila* proteins. (B) Ldb proteins from *Xenopus* and *Nematostella* are schematically presented with sequence identities (%) in each domain. DD, dimerization domain; LCCD, Ldb1-Chip conserved domain; NLS, nuclear localization signal; LID, LIM domain-interacting domain. (C) sbp proteins from *Xenopus* and *Nematostella* are schematically presented with sequence identities (%) in each domain. LUFS, LUG/LUH, Flo8, and SSBP conserved domain; Pro-R, proline-rich transactivation domain.

Koenis, & de Vries, 2016). Those proteins, called LIM domain proteins or LIM proteins, possess one or more LIM domains, and are classified into several groups, according to sequence homology of LIM domains and domain compositions (Dawid, Breen, & Toyama, 1998; Sala & Ampe, 2018; Taira, Evrard, Steinmetz, & Dawid, 1995). While LIM-HD and LMO are localized in the nucleus, other LIM proteins are localized in either the cytoplasm or the nucleus, or both, or are shuttled between them (Bach, 2000; Kadrmas & Beckerle, 2004; Sala & Ampe, 2018; references therein).

2.2 The function of tandem LIM domains of LIM-HD

2.2.1 Inhibitory role of LIM domains in DNA binding and activity of LIM-HD

The first indications of LIM domain function in LIM-HD proteins are the inhibitory roles of LIM-HD proteins, such as DNA binding (Sánchez-García, Osada, Forster, & Rabbitts, 1993; Xue, Tu, & Chalfie, 1993) and biological activity (Taira, Otani, Saint-Jeannet, & Dawid, 1994). DNA-binding assays using bacterially produced Mec-3 (Xue et al., 1993) and Islet-1 (Sánchez-García et al., 1993) proteins have shown that DNA-binding ability of LIM HD proteins is enhanced when LIM domains are deleted or denatured by chelation of the zinc ion.

The role of LIM domains in biological activity of LIM-HD was revealed by an analysis of *Xlim-1/lhx1*. Once *Xlim-1* was identified by homology screening, whole mount in situ hybridization (WISH) using *Xenopus* embryos, which was newly invented at that time (Hemmati-Brivanlou et al., 1990), revealed that this gene is specifically expressed in the dorsal marginal zone of gastrulae, called the Spemann-Mangold organizer region (Taira et al., 1992). This discovery prompted analyses of the "organizer activity" of Xlim-1. Organizer activity means that ventral misexpression of a given gene(s) mimics transplantation of the dorsal blastopore lip (the organizer) into the ventral marginal zone, leading to secondary axis formation, as shown by Spemann and Mangold (1924). The first gene that showed organizer activity is *goosecoid* (*gsc*), a homeobox gene expressed in the organizer (Cho, Blumberg, Steinbeisser, & De Robertis, 1991). In contrast to *gsc*, ventral injection of *Xlim-1* mRNA did not show any apparent organizer activity. However, mutations or deletion of LIM domains activated Xlim-1 (3m and ΔNA in Fig. 3A and B) to exert organizer activity. That is, mutant Xlim1 causes formation of a secondary axis and induces the *gsc* gene in the animal explant (Taira, Otani, Saint-Jeannet, & Dawid, 1994). Thus, Xlim-1 functions as an "organizer gene," and LIM domains suppress organizer activity of Xlim-1, suggesting that protein–protein interactions of the LIM domain with a cofactor may activate Xlim-1. This idea led to identification of Ldb1 and discovery of its function, as described below. The organizer function of Xlim-1 was also supported by a headless phenotype of a mouse null mutant of Lim1/Lhx1, suggesting that Lim1 is required for head organizer functions (Shawlot & Behringer, 1995).

What is the biological significance of auto-suppression of LIM-HD by LIM domains? One possibility is that LIM domains suppress unnecessary/ unproductive DNA binding of LIM-HD proteins in the nucleus, allowing them to function only in association with interacting proteins, as discussed below. Interaction-dependent DNA binding of LIM-HD proteins may be useful to build more elaborate gene regulatory mechanisms in various contexts of cell-type specification. This idea was supported by functional analyses of Xlim-1 axis-inducing activity in *Xenopus* embryos (Hiratani et al., 2001; Taira, Otani, Saint-Jeannet, & Dawid, 1994), and by gel-shift DNA-binding assays (Mochizuki et al., 2000; Sánchez-García et al., 1993; Xue et al., 1993; Yamamoto et al., 2003). However, intramolecular interaction through LIM domains in LIM-HD was revealed later (Gadd et al., 2013; Robertson et al., 2018), because this interaction was very weak and could not be detected by the yeast two-hybrid (Y2H) system (see below).

2.2.2 3D structure of the LIM domain

The LIM motif is similar to the Zn finger motif (Fig. 2A), but a clear difference from Zn fingers in the secondary structure is that the sequence of linker amino acids between two fingers is exactly two amino acids, with no exceptions. Therefore, the central motif, HxxCxxCxxC (x = any amino acid) is specific to LIM domains. Later the 3D structure of the LIM domain of Cysteine-Rich Protein (CRP) and Cysteine-Rich Intestinal Protein (CRIP) were revealed by X-ray crystallography (Pérez-Alvarado et al., 1994, 1996), both proteins showing GATA1- or steroid hormone receptor-type Zn finger structures. The 3D structure of the N-terminal LIM domain of a LIM only protein, LMO2, which is the closest to that of LIM-HD, was also determined (Fig. 2B) (Deane et al., 2003), showing a result similar to CRP and CRIP.

Primary structures or motifs of RING fingers and PHD domains are very similar to those of LIM domains, but their secondary structures are very different (Capili, Schultz, Rauscher, & Borden, 2001). LIM, RING, and PHD domains are all thought to belong to Zn-finger domains, which are defined

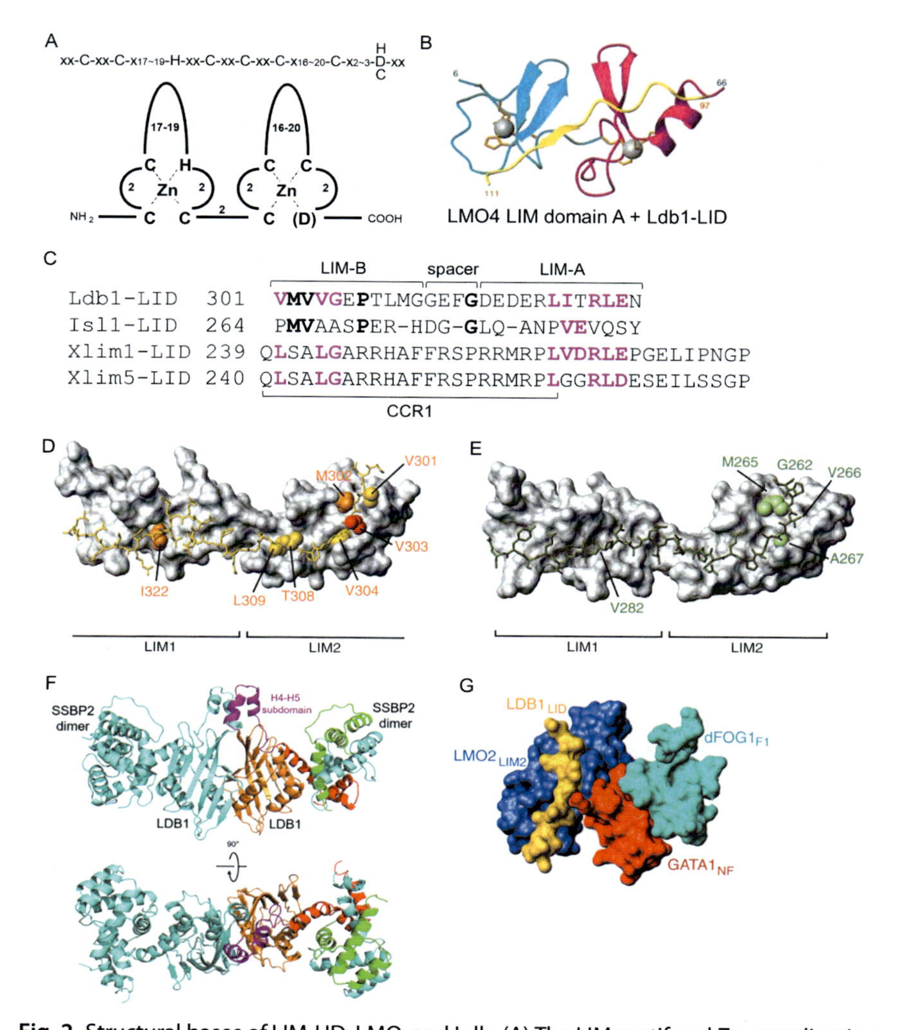

Fig. 2 Structural bases of LIM-HD, LMO, and Ldb. (A) The LIM motif and Zn-coordination spheres of LIM domains. (B) A ribbon diagram of a fusion protein of the N-terminal LIM domain of LMO4 (blue and red) and Ldb1-LID (yellow). Spheres, zinc ion. The diagram is reproduced from Deane et al. (2003). (C) Amino acid sequence comparisons with Ldb1-LID and Isl1-LID revealed LID motifs around CCR1 (C-terminal conserved region 1) of Xlim-1/Lhx1 and Xlim-5/Lhx5. Bold black letters indicate conserved residues between Ldb1-LID and Isl1-LID. Bold magenta letters indicate conservative residues of Xlim-1 and Xlim-5 with either Ldb1-LID or Isl1-LID. (D) Surface models of tandem LIM domains of Lhx3 are reproduced from Bhati et al. (2008). Left, interaction with Ldb1-LID (yellow); right, that with Isl1-LID (green). For Ldb1-LID, the side chains of key residues from alanine scanning mutagenesis screens are classified as having a strong (red), moderate (orange), or weak (yellow) effect, whereas for Isl1-LID the indicated residues (green) all have a strong effect. (E) Ribbon diagrams of the human LDB1/SSBP2 complex are reproduced from Wang et al. (2020). Each molecule of a LDB1 dimer1 interacts with a SSBP2 dimer. (F) A surface model of the LMO2/LDB1/GATA1/FOG1 complex is reproduced from Wilkinson-White et al. (2011). (G) The C-terminal LIM domain of LMO2 (blue) interacts with LDB1-LID (yellow) as domain-motif interaction and with GATA1 (red) of a GATA1-FOG1(cyan) complex as domain-domain interaction through different interaction surfaces of the LIM domain.

as Zn-dependent folding domains or Zn-coordination spheres (Laitaoja, Valjakka, & Jänis, 2013). Functions of Zn-finger domains include binding of DNA/RNA, proteins, or lipids, and in many cases, single Zn-finger domains, such as C4 or C2H2 Zn-finger domains, bind to DNA or protein (Krishna, Majumdar, & Grishin, 2003). By contrast, LIM, RING, and PHD domains always comprise two tandemly repeated Zn-coordination spheres and bind only to proteins.

2.3 Discovery of cofactors of LIM-HD

2.3.1 Ldb1, LIM domain-binding 1

LIM domain-binding protein 1 or LIM-domain binding 1 (Ldb1) (Agulnick et al., 1996), also known as nuclear interactor (NLI) (Jurata, Kenny, & Gill, 1996) or CLIM2 (Bach, Carrière, Ostendorff, Andersen, & Rosenfeld, 1997), was identified by protein–protein interaction screening using the LIM domains of Xlim-1/Lhx1 (Agulnick et al., 1996), Rbtn2/LMO2 (Jurata et al., 1996), or P-Lim/Lhx3 (Bach et al., 1997) as a probe, or with the Y2H system, using the LIM domains of Xlim-1/Lhx1 as a bait (Breen, Agulnick, Westphal, & Dawid, 1998). Ldb1, 375 amino acids (aa) in length, is extraordinarily conservative evolutionarily, with a sequence identity between mice and *Xenopus* of 98% (Agulnick et al., 1996). Ldb1/NLI/CLIM2 has a vertebrate paralog (an ohnolog), named Ldb2/CLIM1 (Bach et al., 1997), with an amino acid identity to Ldb1 of 75%. Subfunctionalization may have occurred between them. In fact, *Ldb1*, but not *Ldb2*, is expressed in mature olfactory sensory neurons of mice (Monahan et al., 2019). However, functional analysis of Ldb2 is very limited, and awaits further investigation. Note that LIM domain-binding protein 3 (Ldb3), also called Zasp, Cypher, or Oracle, is not a paralog of Ldb1, but a protein containing a PDZ domain and three LIM domains (Faulkner et al., 1999; Passier, Richardson, & Olson, 2000; Zhou, Ruiz-Lozano, Martone, & Chen, 1999). Confusingly, *ldb3* and *ldb4* were first identified as paralogs of *ldb2* and *ldb1*, respectively, in zebrafish (Toyama, Kobayashi, Tomita, & Dawid, 1998). Because these genes were generated by the teleost-specific whole genome duplication, *ldb1*, *ldb2*, *ldb3*, and *ldb4* in teleosts are currently named *ldb1b*, *ldb2a*, *ldb2b*, and *ldb1a*, respectively.

Once Ldb1 was identified as a LIM domain-binding protein, an experiment was performed to examine whether Ldb1 could activate Xlim-1/Lhx1 to exhibit organizer activity. As expected, co-injection of mRNAs for Xlim-1 and Ldb1 into the ventral region of *Xenopus* embryos induced a secondary axis, suggesting that Ldb1 activates Xlim-1 by binding to LIM domains to relieve autosuppression by LIM domains (Agulnick et al., 1996). This function is widely conserved among cnidarians and bilaterians, as demonstrated by secondary axis formation in embryos injected with mRNAs for Xlim-1 and *Nematostella vectensis* (sea anemone) Ldb (Fig. 1B) (Yasuoka et al., 2009).

Protein–protein interaction assays involving an amino acid domain of about 200 aa (called the dimerization domain: DD in Fig. 1B) (Jurata & Gill, 1997) and Y2H screening (Breen et al., 1998) revealed homodimerization of Ldb1. This led to the idea that dimerized Ldb1 can form a tetrameric complex with the same or a different combination of LIM-HD proteins, increasing the variety of TF combinations (Fig. 3C).

2.3.2 Ssbp as a Ldb1-interacting cofactor

Ldb1-interacting partners were subsequently sought using protein–protein interaction assays. Co-immunoprecipitation with FLAG-tagged Ldb1, followed by spectrometric analysis, identified Ssdp1/Ssbp3 and Ssdp3/Ssbp4 proteins (Chen et al., 2002). Note that the gene name, *Ssdp* (sequence-specific single stranded DNA-binding protein), was unified to the official name *Ssbp* (single stranded DNA-binding protein) with different numbers. Y2H screening with Ldb1 as a probe also identified Ssdp2/Ssbp2 (van Meyel, 2003). In a functional assay, Ssdp1/Ssbp3 further enhances the axis-inducing activity of Xlim1 and Ldb1 (Chen et al., 2002). In addition, genetic interactions in *Drosophila* have been shown between Ssdp/Ssbp, Chip/Ldb, and Beadex/LMO (Chen et al., 2002), or between Ssbp, Chip, and Apterous/Lhx2 (van Meyel, 2003) (see Table 1 for gene nomenclature).

Ssbp2, Ssbp3, and Ssbp4 have an N-terminal conserved region, named the LUFS domain (van Meyel, 2003) or a FORWARD domain (Enkhmandakh, Makeyev, & Bayarsaihan, 2006), that is conserved among the three proteins LUG/LUH, Flo8, and SSBP/Ssdp (Fig. 1C). The LUFS domain interacts with an LDB/Chip Conserved Domain, named LCCD, of Ldb1 (Fig. 1B) (van Meyel, 2003). In mice, Ssdp1/Ssbp3 was identified as the gene responsible for a mouse mutant, *headshrinker*, in which the embryonic head is dramatically reduced (Nishioka et al., 2005), resembling the *Lim1/Lhx1* knockout headless phenotype (Shawlot & Behringer, 1995). Furthermore, genetic interactions of Ssdp1/Ssbp3 with Lim1 and Ldb1 have been shown in mice (Nishioka et al., 2005). A reporter assay using culture cells showed that a *gsc* reporter gene, which is driven by

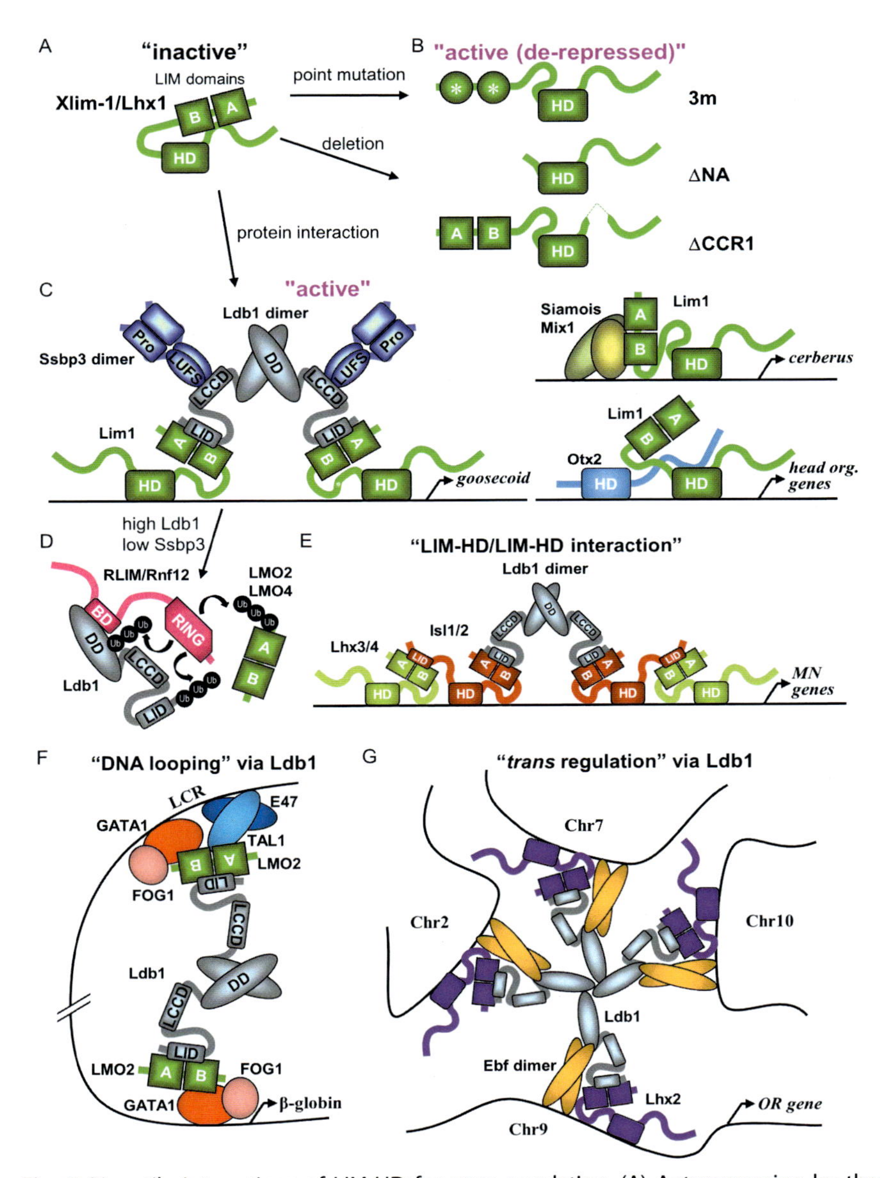

Fig. 3 Versatile interactions of LIM-HD for gene regulation. (A) Autorepression by the intramolecular interaction in Xlim-1/Lhx1. (B) De-repression forms of Xlim-1/Lhx1 mutants: 3 m, a point mutant, in which both LIM domains are mutated by changing a zinc ion-binding cysteine to glycine; ΔNA, a deletion mutant, in which both LIM domains are deleted (Taira, Otani, Jamrich, & Dawid, 1994; Taira, Otani, Saint-Jeannet, & Dawid, 1994); ΔCCR1, a deletion mutant, in which CCR1 is deleted (Hiratani, Mochizuki, Tochimoto, & Taira, 2001). (C) Ldb1/Ssbp3 (left), Siamois/Mix1 (right, top), and Otx2 (right, bottom)

(Continued)

Xlim1/Lhx1-responsible enhancer elements of the *gsc* gene (Mochizuki et al., 2000), is synergistically activated by Xlim-1, Ldb1, and Ssdp1/Ssbp3 (Nishioka et al., 2005). These findings suggest ternary complex formation of LIM-HD, Ldb1, and Ssbp3 to activate target genes (Fig. 3C).

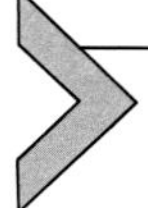

3. Molecular interactions between LIM-HD and cofactors to form a regulatory complex

3.1 Domain–motif interactions of LIM domains

3.1.1 The LIM domain-interacting domain (LID) as an intrinsically disordered region

Since two tandem LIM domains of LIM-HD and LMO proteins were found to bind to Ldb1, the mode of protein–protein interactions has been analyzed in detail. First, the LIM domains of Isl-1/Isl1 or Xlim-1/Lhx1 were shown to bind to a short stretch of about 30 amino acids (called an LIM interaction domain; LID or Ldb1-LID) in the C-terminal region of Ldb1 (Fig. 1B) (Breen et al., 1998; Jurata & Gill, 1997). In addition, LIM domains of Lhx3 were found to bind to a short sequence of Isl1, called the Lhx3-binding domain (LBD) (Jurata, Pfaff, & Gill, 1998). Note that the LBD is also called Isl1-LID (Robertson et al., 2018) to indicate that LBD is an "LID." Although the LID was initially defined as a specific sequence of Ldb1, here we use "LIM domain-interacting domains (LIDs)" as a generic term.

Fig. 3—Cont'd interact with Lim1/Lhx1 to activate target genes, such as *goosecoid*, *cerberus*, and head organizer genes, respectively (Agulnick et al., 1996; Mochizuki et al., 2000; Nakano, Murata, Matsuo, & Aizawa, 2000; Yamamoto, Hikasa, Ono, & Taira, 2003; Yasuoka et al., 2014). (D) RLIM/Rnf12 ubiquitinates free Ldb1 under the condition of high Ldb1 and/or low Ssbp3 (Bach et al., 1999; Hiratani, Yamamoto, Mochizuki, Ohmori, & Taira, 2003; Xu et al., 2007). RLIM/Rnf12 also ubiquitinates LMO2 (Ostendorff et al., 2002). (E) LIM domains of Lhx3/4 bind Isl1/2-LID and form a ternary complex with Ldb1 to activate motor neuron (MN) genes (Bhati et al., 2008; Thaler, Lee, Jurata, Gill, & Pfaff, 2002). (F) The erythroid Ldb1 complex (GATA1, FOG1, Tal1, E47, LMO2, and Ldb1) binds to the locus control region (LCR) and forms an enhanceosome to activate *β-globin* gene. LCR and the proximal enhancer of the *β-globin* gene is bridged by Ldb1 homodimerization through DNA looping (Deng et al., 2012; Krivega, Dale, & Dean, 2014; Love, Warzecha, & Li, 2014; Wilkinson-White et al., 2011). (G) Lhx2, Ldb1, and a Ebf1 dimer bind to Greek islands to activate an olfactory receptor (OR) gene. Greek islands on the same (*cis*) or different (*trans*) chromosomes are assembled through Ldb1 interactions (details are unclear) to form an enhanceosome (Monahan et al., 2017, 2018; Monahan, Horta, & Lomvardas, 2019).

Curiously, Ldb1-LID and Isl1-LID have rather different amino acid sequences (Fig. 2C). Nevertheless, 3D structures of a complex of the LIM domains of Lhx3 or LMO4 with Ldb1-LID or Isl1-LID (LBD) showed that the N-terminal (1st) LIM domain (called LIM-1 or LIM-A) and the C-terminal (2nd) LIM domain (called LIM-1 or LIM-B) bind to the C- and N-terminal halves of LID/LBD, respectively, in an antiparallel fashion (Fig. 2D and E) (Bhati et al., 2008; Deane et al., 2004)).

In general, there are two types of protein–protein interactions: domain–motif interactions (DMIs) and domain–domain interactions (DDIs) (Garamszegi, Franzosa, & Xia, 2013), in which "domain" denotes a structural domain in the protein. In DMIs, protein interaction domains usually recognize short linear motifs of 4–8 amino acids, such as NPxY^{P} (Y^{P}, phosphotyrosine) by the SH2 domain, [RKY]xPxxP by the SH3 domain, and Qxx[*ILM*]xx [DHFM][FMY] by PCNA (proliferating cell nuclear antigen) (Neduva & Russell, 2006). Compared to those short linear motifs, two tandem LIM domains recognize a linear sequence of 26–28 residues (Bhati et al., 2008) (Fig. 2C), which is much longer that those seen in other DMIs, but the mode of protein–protein interactions of LIM domains clearly belong to DMIs.

Ldb1-LID and Isl1-LID (LBD) actually are not "structural domains," but unstructured regions, generally called "intrinsically disordered regions (IDRs)" (Robertson et al., 2018) (see also references therein). IDRs exhibit a remarkable feature, called "coupled folding and binding," which means that IDRs "fold" upon binding with a partner protein (Berlow, Dyson, & Wright, 2015; Robustelli, Piana, & Shaw, 2020). This coupled folding and binding mechanism provides flexibility to recognize partner proteins so that a single IDR can interact with various domains of partner proteins, thereby conferring the hub function on IDR-containing proteins to create protein–protein interaction networks (Berlow et al., 2015; Robertson et al., 2018; Robustelli et al., 2020). The hub function resides in both IDRs and IDR-interacting domains such that a single IDR-interacting domain can interact with multiple IDRs, and a single IDR can interact with multiple IDR-interacting domains, thereby forming an intricate protein–protein interaction network. LIDs are likely to be such IDRs undergoing folding coupled with binding to tandem LIM domains. The following lines of evidence strongly suggest LIDs as IDRs.

3.1.2 Variable interacting interface of LIM domains and LIDs

Tandem LIM domains and LID interactions have been analyzed by comparing 3D structures of the complexes in several combinations, that is, LIM domains from Lhx3, Isl1, or LMO4 and LIDs from Ldb1, Isl1, or Isl2 (Bhati et al., 2008; Deane et al., 2004; Gadd et al., 2011). Results were as follows. (i) Overall 3D structures of the complexes are similar in all combinations, despite low sequence homology between LIM domains (Lhx3, Isl1, and LMO4) and between LIDs (Ldb1 and Isl1) (Fig. 2C–E). (ii) Amino acid residues of LIM domains that contact the LID, vary among Lhx3, Isl1, and LMO4, indicating that distinct tandem LIM domains recognize the same sequence in different ways. (iii) LIM domains of Lhx3 interact similarly with LIDs, such that only 4 of 28 residues are identical (Fig. 2C), indicating that LIM domains can recognize different sequences in a similar ways. (iv) Alanine-scanning mutagenesis analysis of LIDs using the Y2H system identified several critical residues for interaction with LIM domains, but those critical residues differ depending on the LIM domain in question (Bhati et al., 2008). For example, disruption effects of mutated residues in Ldb1-LID on complex formation are V303 > M302 = I322 > V301 = V304 = T308 = L309 for Lhx3-LIM domains, and I322 > M302 = V303 = R320 = V301 for Isl1. Thus, the mechanism of LIM domain–LID interactions is very flexible in terms of interacting residues of both LIM domains and LIDs, suggesting that LIM domain–LID interactions fit with the coupled folding and binding mechanism of IDRs (Robertson et al., 2018).

Even though LIM domain–LID interactions are flexible, a notable feature of Ldb1-LID is extremely conserved evolutionarily. It is perfectly matched in vertebrates and highly conserved even in cnidarians and bilaterians, comprising just one conservative amino acid change in the 28 residues between vertebrates and *Nematostella* or *Drosophila* (see Fig. 4B). This constraint upon LID sequences in cnidarians and bilaterians probably comes from the fact that LID interacts with all six LIM-HDs and one LMO protein in *Nematostella*, and with all 12 LIM-HDs and four LMO proteins in vertebrates. In other words, a change in any of 28 amino acids in the sequence would in most cases ruin an interaction with one or more LIM-HDs or LMO proteins, leading to exclusion of such variants from the species. This speculation seems to contradict alanine-scanning mutagenesis data using the Y2H system, in which mutations of the spacer sequence of Lidb1-LID do not affect interaction with LIM domains (Bhati et al., 2008). However, Y2H assays essentially do not discriminate subtle effects of binding affinity between bait and prey proteins. This suggests that weakly interacting amino acids of LIDs are also very important for physiological functions.

Compared to Ldb1-LID, evolutionary conservation of two tandem LIM domains of LIM-HD proteins is not so high, for example, 62% identity

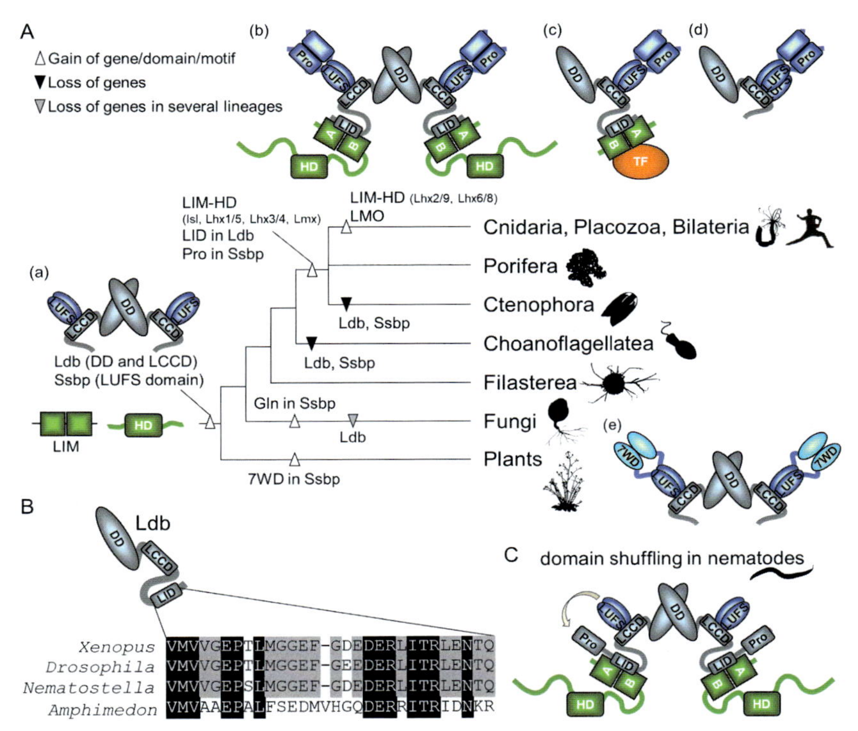

Fig. 4 Evolutionary origins of LIM-HD, LMO, Ldb, and Ssbp. (A) Comparative genomics suggests possible domain creating or shuffling during evolution of eukaryotes. (a) A putative ancestral state of Ldb, Ssbp, and LIM-HD in eukaryotes; (b) A ternary complex of LIM-HD/Ldb/Ssbp in metazoans; (c) A complex of LMO, Ldb, and other TFs in cnidarians, placozoans, and bilaterians; (d) A complex of Ssbp and a splicing isoform of Ldb1 lacking the LID in cnidarians and bilaterians; (e) A Ldb/Ssbp complex in plants. Fungi acquired glutamine-rich transactivation domain (Gln) independently. Note that current missing genes or motifs may be found if genomes of those species are revised or if genomes of more species are decoded for those taxa. (B) Ldb-LID sequences are highly conserved among metazoans. (C) Proline-rich transactivation domain was translocated from Ssbp to Ldb in nematodes.

between orthologs of Lhx1 in *Xenopus* (frog) and a cnidarian, *Nematostella* (sea anemone), or 46% between metazoan paralogs, Lhx3 and Isl1 in mice. This is probably because Ldb1-LID-interacting amino acids of LIM domains need not be well conserved between Lhx3 and Isl1 (Bhati et al., 2008; Deane et al., 2003). Another reason is that Lhx3 LIM domains can similarly interact with both Ldb1-LID and Isl1-LID (LBD), but through different residues of the LIM domains, as there are only 4 identical amino acids out of 29 between Ldb1-LID and Isl1-LID (Fig. 2D and E) (Bhati et al., 2008). Notably,

Lhx3-LIM domains efficiently bind to Isl1-LID (LBD) such that Lhx3 forms a ternary Ldb1/Lhx3/Isl1 complex instead of a binary Ldb1/Lhx3 complex by competing with the Isl1-LID to activate target genes in motor neurons (Fig. 3E) (Bhati et al., 2008; Lee & Pfaff, 2003). Thus, tandem LIM domains interact equally or similarly with very distinct LIDs (Fig. 2C), suggesting that tandem LIM domains function as hub domains through versatile interactions with proteins containing various LIDs.

3.1.3 Quantitative analysis of LIM domain–LID interactions

Initial studies of LIM domain-LID interactions were performed using conventional protein–protein interaction assays, coimmunoprecipitation assays (Agulnick et al., 1996), the Y2H system (Breen et al., 1998; Jurata et al., 1998), and gel shift assays (Yamamoto et al., 2003), as mentioned above. However, to compare binding affinities of LIM domains to LIDs, quantitative analysis using binding kinetics of protein–protein interactions was required. Using a FRET-based method, it was shown that tandem LIM domains of Isl1, Isl2, Lhx3, Lhx4, or Lmo4 have relatively high affinities for Ldb1-LID (dissociation constant (Kd) values from 10^{-10} to 10^{-8} M) (Robertson et al., 2018), showing similar binding characteristics in these LIM-LID combinations, as expected from conventional protein–protein assays. By contrast, those LIM domains interact with Isl1-LID (LBD) with very different affinities (Kd from 10^{-10} to 10^{-4} M), showing distinct binding characteristics among the five LIM proteins for Isl1-LID (Robertson et al., 2018). Notably, Kd values of the LIM domains of Lhx4 for Ldb1-LID and Isl1-LID are 1.0×10^{-9} and 5.4×10^{-8} M, respectively, which are quite similar, suggesting that Ldb1-LID and Isl1-LID compete for Lhx4 as well as Lhx3, as shown before (Thaler et al., 2002). These data verify that the same surface of LIM domains can bind to very different amino acid sequences (only 4 out of 29 residues are the same) with high affinity. Thus, tandem LIM domains target diverse linear sequence motifs with high affinity and also with moderate to low affinities, depending on requirements of their functions, highlighting the versatility of LIM domains through domain–motif interactions. However, this remarkable feature of LIM domains makes it difficult for researchers to identify LIDs from their amino acid sequences in entire or even partial proteins that interact with LIM domains.

3.1.4 The autosuppression mechanism of LIM-HD

As mentioned in the section "Inhibitory role of LIM domains," two tandem LIM domains in LIM-HD proteins autosuppress their own activity or DNA binding. Therefore, it was initially hypothesized that LIM domains bind to an LID in the C-terminal region to mask the HD. However, using conventional protein–protein interaction assays, such intramolecular LIDs had not been identified. For example, Lhx3-LIM domains interact with Isl1-LID (LBD), but Isl1-LIM domains did not interact with their own LIDs, when assayed using Y2H (Bhati et al., 2008). However, it was still possible that Isl1-LIM domains could bind weakly to their own LIDs, which would not be detected using the Y2H system. This possibility reminds us of the result of Xlim-1/Lhx1 (Hiratani et al., 2001), in which a small deletion of C-terminal conserved region 1 (CCR1) activates Xlim-1 to induce a secondary axis in *Xenopus* embryos. Notably, we now found in CCR1 a sequence similar to Ldb1- and Isl1-LIDs (Fig. 2C) and this may explain the autorepression mechanism of Xlim-1, as reported (Taira, Otani, Saint-Jeannet, & Dawid, 1994).

Later, nuclear magnetic resonance revealed that Ist1-LIM domains weakly, but specifically interact with their own LIDs (Gadd et al., 2013), which was previously not detected using the Y2H system, as mentioned above. This provided the first biochemical evidence for the intramolecular interaction of LIM domains in the LIM-HD protein. Quantitative interaction analysis verified that the Kd is 2.9×10^{-8} M for the Isl1-LIM domains and Ldb1-LID, whereas the Kd is 1.7×10^{-4} M for the Isl1-LIM domains and Isl1-LID (Robertson et al., 2018). This 10,000-fold difference between them makes sense because the net concentration of LIM domains and an LID in the same molecule is significantly higher than in intermolecular interactions. Thus, a net higher concentration for intramolecular interactions compensates for lower affinity (larger Kd) to bind efficiently, and this lower affinity (Kd $= 1.7 \times 10^{-4}$ M) may prevent aggregation through intermolecular interactions of Isl1. Furthermore, in ternary complex formation with Lhx3 (or Lhx4), Isl1, and Ldb1 (Bhati et al., 2008; Thaler et al., 2002) (Fig. 3E), the Lhx3/4-LIM domains need to efficiently bind Isl1-LID with higher affinity (Kd $\cong 2.7 \times 10^{-7}$ or 5.4×10^{-8} M) by competing with the intramolecular interaction; hence, intramolecular interactions should be of lower affinity (Kd $\cong 1.7 \times 10^{-4}$ M) (Gadd et al., 2013; Robertson et al., 2018).

Thus, initial speculation about the biological significance of autosuppression by the intramolecular interaction of tandem LIM domains and a LID in LIM-HD is now more likely based on biochemical data, as mentioned above. It is therefore probable that the intramolecular interaction of LIM-HD suppresses unnecessary DNA binding and thereby inactivates

TF functions to prevent leaky gene activation/repression if expressed alone, as has been demonstrated by studies of Mec-3 (Xue et al., 1993) and Xlim-1/Lhx1 (Hiratani et al., 2001; Mochizuki et al., 2000; Taira, Otani, Saint-Jeannet, & Dawid, 1994; Yamamoto et al., 2003).

3.2 Domain–domain interactions of Ldb1 or LIM domains

3.2.1 Interactions between Ldb and Ssbp

Ldb1 has the DD (dimerization domain), the LCCD (LDB/Chip conserved domain), a nuclear localization signal (NLS), another interacting domain (OID), and the LID. We have already discussed the Ldb1–Ldb1 interaction (homodimerization) through the DD (Breen et al., 1998; Jurata & Gill, 1997), and the Ldb1–Ssbp interaction through the LCCD with the LUFS domain of Ssbp (van Meyel, 2003). Size exclusion chromatography coupled with multi-angle light scattering revealed that the LUFS domain alone (roughly 90 aa) forms a stable tetramer, and the tetramer binds to an Ldb1/Chip dimer (Fiedler et al., 2015). More detailed 3D structure analyses further demonstrated that the LUFS domain dimer binds to the LCCD through its surface for tetramerization (Renko et al., 2019; Wang et al., 2020) (Fig. 2F). These interactions are thought to be domain–domain interactions, as shown by 3D structural analysis of dimerized Chip/Ldb1 with two dimerized LUFS domains of Ssdp/Ssbp (Fig. 2F) (Renko et al., 2019; Wang et al., 2020). The LUFS domain includes the LisH (Lis-homology) domain, known from more than 100 proteins (Emes & Ponting, 2001). LisH domains form thermodynamically very stable homodimers and their Kds are reported to be $\sim 10^{-15}$ M (Kim et al., 2004). As expected, LUFS domains efficiently form homodimers so that Ssdp monomers are not detectable (Fiedler et al., 2015). As a result, an Ldb dimer binds two molecules of an Ssdp dimer to form a hexameric binary complex, and this complex functions as a core module for Wnt-responsive enhanceosomes, as described below (Renko et al., 2019).

3.2.2 Interactions between LIM domains and other TFs

LIM domains also interact with various proteins other than Ldb1, some of which occur via domain–domain interactions. An example is LMO2, which interacts with GATA1, FOG1, Tal1 (the same as Scl), and E47, as well as Ldb1 to form a so-called erythroid Ldb1 complex (see Fig. 3F). 3D analysis revealed that the C-terminal LIM domain alone interacts with a GATA1–FOG1 complex through a GATA1 surface via domain–domain interaction (Fig. 2G) (Wilkinson-White et al., 2011), and the N-terminal

LIM domain interacts with the bHLH protein Tal1/Scl, thereby forming a TF complex. Thus, while the two tandem LIM domains of LMO2 bind to a single LID sequence, each LIM domain has two interacting interfaces. One is to interact with part of the Ldb1-LID through DMI (domain–motif interaction), and the other interacts with a TF through DDI (domain–domain interaction).

3.2.3 Regulation of stoichiometry for complex formation

The critical principle of protein complex formation is that the stoichiometry of components needs to be properly maintained. This is true for the LIM-HD/Ldb1 complex and the Ldb1/Ssbp complex. For example, genetically in *Drosophila*, the phenotype of an excess amount of Chip/Ldb is the same as that of null mutation of *Chip* or *apterous/lhx2* (Fernández-Fúnez, Lu, Rincón-Limas, García-Bellido, & Botas, 1998). In *Xenopus*, an excess amount of Ldb1 inhibits Xlim1/Lhx1 functions in the gastrula embryo (Hiratani et al., 2003). To prevent harmful effects of excess Ldb1, the RING finger protein, Rnf12, the same as RLIM (Bach et al., 1999), functions as a ubiquitin ligase only for free Ldb1, thereby leading to proteasome degradation of free Ldb1, but not of Ldb1 associated with Xlim1 (Fig. 3D) (Hiratani et al., 2003). RLIM/Rnf12 also functions as a ubiquitin ligase for LMO2 (Fig. 3D) (Ostendorff et al., 2002). For the Ssbp/Ldb1/LMO complex, ubiquitination-dependent degradation of Ldb1 and LMO2 by RLIM/Rnf12 is suppressed by Ssbp to maintain the stoichiometry between Ldb1, LMO, and Ssbp (Xu et al., 2007). For formation of the erythroid Ldb1 complex, consisting of Ldb1, Ssbp2/3/4, LMO2, GATA1, FOG1, Tal1, and E47 (Fig. 3F), Rnf12 targets Ldb1, Ssbp2/3/4, LMO2, and Tal1 to degrade them when they are not incorporated into the complex (Layer et al., 2020). In other words, the complex itself is stable against a degradation machinery. Because both LMO2 and Tal1 (T-cell acute lymphocytic leukemia protein 1 as the name suggests) act as oncoproteins, persistent stabilization of this complex could lead to the onset of T-cell acute lymphoblastic leukemia (T-ALL) (Layer et al., 2020).

Compared to LMOs and cofactors of LIM-HD, LIM-HD proteins do not appear to be quantitatively regulated by the ubiquitination-proteasome degradation pathway. This may be because the auto-suppression mechanism of LIM-HD proteins by intramolecular interaction between the LIM domains and the LID prevents promiscuous actions of free LIM-HDs from binding to their partner proteins through their LIM domains and binding to their target DNA elements through the HD. In fact, the Lim1/Lhx1 protein is fairly stable in the notochord in *Xenopus* tailbud stage embryos (see Fig. 8A) (Sudou et al., 2012), even after *Xlim-1/lhx1* mRNA disappears in the neurula embryo, as assayed by WISH (Taira, Otani, Jamrich, & Dawid, 1994).

3.3 Other interactions of LIM-HD to be examined

Complex formation with Ldb is the best known mechanism for activating LIM-HD, but other partner proteins also do. A good example is the activation mechanism of *cerberus* by Xlim-1/Lhx1 in the Spemann-Mangold organizer (Sudou et al., 2012; Yamamoto et al., 2003). In *Xenopus* animal explants, the LIM domain mutant of Xlim-1 (3m mutant in Fig. 3B) upregulates *cerberus* expression, but the combination of wild type Xlim-1 and Ldb1 does not (Yamamoto et al., 2003), indicating that Xlim-1 activates *cerberus* in an Ldb1-independent manner. Instead, Xlim-1, Siamois (Sia or Sia1), and Mix.1 (Mix1) synergistically activate the *cerberus* gene in animal explants, and this activation is LIM-domain-dependent. Furthermore, gel shift DNA-binding assays demonstrated that Xlim-1 interacts with Sia1 and Mix1 through LIM domains on the *cerberus* enhancer (Yamamoto et al., 2003) (Fig. 3C). Further exploration may determine how these three TFs interact through the tandem LIM-domains and what kinds of protein–protein interactions are involved in this complex formation. Another example is that Isl1 and a bHLH TF, BETA2, synergistically activate the insulin gene and the LIM domains of Isl1 directly interact with the bHLH domain of BETA2 (Zhang et al., 2009).

Other physical or functional interactions of LIM-HDs with TFs are Lmx1/Lmx1a and Pan-1/E47 (bHLH) (German et al., 1992), Mec-3 and Unc86 (POU-HD) (Xue et al., 1993), and Isl1 and Pannier (GATA) (Biryukova & Heitzler, 2005). Based on analogies from the interaction of LMO2 and GATA1 or Tal1 (bHLH TF), these interactions may occur with the other side of LIM domains of LIM-HD proteins (see Fig. 3F). However, it may be that those interactions occur through cryptic LIDs of those TFs, or occur indirectly via mediating adaptor proteins. Relative to the latter, it was reported that *Drosophila* Chip/Ldb interacts with Pannier (GATA) (Heitzler, Vanolst, Biryukova, & Ramain, 2003) and Da/E47 (bHLH) (Biryukova & Heitzler, 2005), but it is possible that these interactions may mediate the interaction of Beadex/LMO with GATA and bHLH TFs as described above for LMO2 (Wadman et al., 1997; Wilkinson-White et al., 2011).

Direct interaction of Lim1/Lhx1 and Otx2 (a paired-like HD) was shown using GST pull-down assays (Nakano et al., 2000). This interaction was mediated by the homeodomain of Lim1/Lhx1 and the C-terminal region of Otx2 (Fig. 3C). Luciferase reporter analysis and animal explant assays demonstrated that Lim1/Lhx1, Ldb1, and Otx2 synergistically activate *goosecoid* (Mochizuki et al., 2000). How Lim1/Lhx1, Ldb1, and Otx2 form a complex also awaits investigation, because this synergy between Lim1/Lhx1 and Otx2 is crucial for vertebrate head development, as discussed below. In addition, five tyrosine residues in the C-terminal conserved region 2 (CCR2) of Xlim-1/Lhx1 are indispensable for the organizer activity, suggesting that an unidentified coactivator(s) binds to CCR2 (Hiratani et al., 2001). Thus, protein–protein interactions, dependent on or independent of LIM domains, are both involved in LIM-HD functions to activate certain target genes, indicating versatile functions of LIM-HDs.

3.4 Enhanceosomes and long-range enhancer–promoter interactions via an Ldb1-TF complex

The enhancer is defined as a nucleotide sequence that activates transcription from the promoter of a gene, and that functions upstream or downstream of, and near or far from the promoter. Sometimes the enhancer can activate a gene at a distance of several tens of kb or even more than 1 Mb from the promoter, called a remote enhancer. Conversely, the silencer is defined as a nucleotide sequence that represses transcription. In both cases, enhancers or silencers exert functions via binding by activator or repressor TFs, respectively. Usually, enhancers and silencers form a module, called a *cis*-regulatory module (CRM) or *cis*-regulatory element (CRE), spanning several hundred nucleotides. Individual genes have multiple CRMs between the upstream and downstream chromosomal boundaries, which are defined by binding of a boundary protein, CTCF (CCCTC-binding factor). "Enhancers" sometimes have the same meaning as "CRMs" if the CRMs function as enhancers. To activate a gene, multiple TFs bind to a CRM/enhancer to form an enhanceosome, together with various cofactors, such as coactivators (p300, CBP etc.) and chromatin remodeling factors (Brahma, Pygopus, Osa/Arid etc.) (Carey, 1998). The enhanceosome interacts through a Mediator complex with a transcription pre-initiation complex (PIC), which contains RNA polymerase II and general TFs, to activate transcription of the gene (Poss, Ebmeier, & Taatjes, 2013). In the case of a remote enhancer, an enhancer and the promoter approach through DNA looping. This basic molecular biology of gene regulation is often described in textbooks, but only recently have details of enhanceosomes and remote enhancers been gradually documented.

3.4.1 Wnt enhanceosome

Wnt/β-catenin signaling is transduced by binding of nuclear β-catenin to a Tcf/Lef TF to form an enhanceosome on Tcf/Lef DNA-binding sites together with Legless/BCL9 and Pygopus (Pygo), thereby activating a Wnt target gene. In this complex, BCL9 bridges β-catenin to Pygo through an interaction between the HD1 domain of BCL9 and the PHD finger of Pygo (Clevers, 2006). To identify other components of the Wnt enhanceosome, a proteomics approach using HA-tagged Pygo-expressing *Drosophila* Schneider 2 culture cells identified Chip/Ldb1, Ssdp/Ssbp, Beadex/LMO, and Apterous/Lhx2 as components of the Wnt-response enhanceosome (Fiedler et al., 2015). Co-immunoprecipitation assays with the Chip/Ldb–SSDP complex, named "ChiLS," showed that the NPF or NPFxD (Asn-Pro-Phe-X-Asp) motif of Pygo, which is essential for Pygo function, is necessary to bind to ChiLS. Other NPF motif-containing proteins, such as *Drosophila* Osa/Arid1 and Runx and human Pygo2, can also bind to ChiLS (Fiedler et al., 2015). 3D analysis of ChiLS showed a symmetrical $SSDP_2$-LDB_2-$SSDP_2$ architecture that has two pockets to bind the NPFxD motif, suggesting that as a core complex of enhanceosomes, ChiLS can bind two molecules of the same or different NPFxD motif-containing proteins, in addition to two molecules of the same or different LIM-HDs and/or LMO proteins, through Ldb1-LID (Fiedler et al., 2015; Renko et al., 2019). Thus, the Wnt enhanceosome can integrate four types of TFs, some of which may be downstream of various signaling pathways, thereby integrating several signaling outcomes as well.

3.4.2 Long-distance gene activation in the β-globin gene cluster by DNA looping

It has long been known that one of the β-globin cluster genes is activated by the locus control region (LCR) several 10s of kb upstream of the β-globin genes, leading to sequential activation from embryonic to adult genes during development, called "globin gene switching." The β-globin cluster consists of the ε, $G\gamma$, $A\gamma$, δ, and β genes in humans, or the εy, $\beta H1$, βmaj, and βmin in mice. During mouse development, globin gene switching occurs from the embryonic εy to the adult βmaj gene. To activate the βmaj gene, the LCR interacts with the βmaj gene promoter by mediating an erythroid TF complex containing Ldb1, LMO2, and TFs, the GATA1/FOG heterodimer and the Tal1/E47 heterodimer, sometimes called "the erythroid Ldb1 complex" (Love et al., 2014), which is thought to form an enhanceosome. Several lines of evidence from reporter assays and chromosome conformation capture

(3C) analysis suggest that the erythroid Ldb1 complex on the LCR and that on the proximal enhancer of the *βmaj* gene are bridged by Ldb1 homodimerization with DNA looping (Deng et al., 2012; Krivega et al., 2014) (Fig. 3F).

The CCCTC-binding factor named CTCF (11 C2H2-type zinc fingers) was first identified as an insulator or boundary factor of chromatin (Bell, West, & Felsenfeld, 1999). CTCF sites are present upstream and downstream of the globin locus, and CTCF mediates DNA looping between them (Splinter et al., 2006). HiC analysis, which comprehensively identifies proximal to long-distant DNA interactions using next-generation sequencing, reveals "topologically associating domains (TADs)" (Acemel, Maeso, & Gómez-Skarmeta, 2017). TAD organization is created by CTCF and cohesin. CTCF and cohesin also mediate enhancer–promoter interactions to activate genes, leading to intra-TAD formation (Ren et al., 2017). In the case of erythroid-lineage cells, the aforementioned erythroid Ldb1 complex activates numerous erythroid-specific genes beside the β-globin gene, but unlike the β-globin gene, in most genes, the erythroid Ldb1 complex binds only to the remote enhancer, not near the promoter. Analysis of one such gene, *carbonic anhydrase 2* (*Car2*), revealed that the LCCD-OID region of Ldb1 interacts with CTCF and this interaction is required for DNA looping and gene activation, indicating another function of Ldb1 for long-distance enhancer–promoter communication with CTCF (Lee, Krivega, Dale, & Dean, 2017).

3.4.3 Trans *interactions of olfactory receptor gene clusters on different chromosomes*

There are thousands of olfactory receptor (OR) genes in vertebrates. Such numerous OR genes are separated by several OR gene clusters and these clusters are scattered on the same or different chromosomes, but only a single OR gene is selected to function in a single olfactory sensory neuron (Lomvardas et al., 2006). How this works has been a mystery. Each cluster has an enhancer, called "the Greek island," which is bound by Lhx2 and Ebf1/COE1/Olf1 (Monahan et al., 2017). During differentiation of olfactory sensory neurons, Greek islands on the same or different chromosomes are assembled to form a super-enhancer (Monahan et al., 2017, 2019). Analysis using an Lhx2-Ebf1 fusion protein as a dominant negative construct has shown that displacement of Lhx2 and Ebf1 by this construct inhibits *trans* interactions of Greek islands (Monahan et al., 2017). Because Lhx2 interacts with Ldb1 and because Ldb1 apparently mediates a long-range enhancer–promoter interactions, the authors tested the involvement of Ldb1 in *trans* interactions of Greek islands and confirmed it (Fig. 3G) (Monahan et al., 2019). Thus, Ldb1 is also involved in interchromosomal interactions of enhancers, possibly by forming multimeric complexes, of which further investigation is required.

3.5 Evolutionary origins of LIM-HD, Ldb, and Ssbp

3.5.1 LIM-HD is an innovation of metazoans

LIM-HD genes are found in basal metazoan genomes (Cnidaria, Placozoa, Porifera, and Ctenophora), but not in non-metazoans (Fig. 4A) (Koch, Ryan, & Baxevanis, 2012; Larroux et al., 2008; Putnam et al., 2007; Simmons, Pang, & Martindale, 2012; Srivastava et al., 2010). Plants have LIM genes encoding one or two LIM domains (Srivastava & Verma, 2017) and homeobox genes separately, indicating that LIM and homeobox genes are ancient, and that a LIM-HD gene was created by gene fusion in the common ancestor of metazoans (Larroux et al., 2008; Putnam et al., 2007). Comparative genomics of various metazoans have suggested that subsequent gene duplication occurred to create four LIM-HD subfamilies, Lhx1/5 (lin-11), Lhx3/4, Islet, and Lmx before diversification of the Ctenophora and Porifera. Then, additional gene duplications created two LIM-HD subfamilies, Lhx2/9 (Apterous) and Lhx6/8 (Arrowhead) in the last common ancestor of the Cnidaria, Placozoa, and Bilateria (Srivastava et al., 2010) (Fig. 4A). In addition, an *LMO* gene is thought to have been created from a *LIM-HD* gene by gene duplication and deletion in the common ancestor of the Cnidaria, Placozoa, and Bilateria (Koch et al., 2012; Simmons et al., 2012) (Fig. 4A).

3.5.2 Interactions between Ldb and Ssbp homologs are present in plants and fungi

Homologous genes for LIM-HD cofactors, *Ldb* and *Ssbp*, are also present in plants and fungi (Fig. 4A). In plants, *SEUSS* (*SEU*) and *SEUSS-LIKE* (*SLK*) genes share sequence similarity with metazoan Ldb (Bao, Azhakanandam, & Franks, 2010; Franks, Wang, Levin, & Liu, 2002) in which sequences corresponding to the DD (dimerization domain) and LCCD, but not a LID, were identified. Plants also possess Ssbp-like genes, *LEUNIG* (*LUG*) and *LEUNIG_HOMOLOG* (*LUH*), which encode proteins containing the LUFS domain. Curiously, LUG and LUH are transcriptional corepressors that interact with histone deacetylases (HDACs) via seven WD domains in the C-terminal region (Fig. 4A-e) (Gonzalez, Bowen,

Carroll, & Conlan, 2007). Therefore, the SEU–LUG complex functions as a transcriptional corepressor, in contrast to the Ldb–Ssbp complex, which serves as a coactivator (Sridhar, Surendrarao, Gonzalez, Steven Conlan, & Liu, 2004).

No Ldb homolog has been found in the budding yeast, *Saccharomyces cerevisiae*, whereas an Ldb-like gene is present in the fission yeast, *Schizosaccharomyces pombe* (Franks et al., 2002), suggesting that the budding yeast secondarily lost an Ldb-like gene (Fig. 4A). On the other hand, Ssbp homologs, Flo8 and Mss11, have been identified in budding yeast and the fungal pathogen, *Candida* (Misas, Escandón, McEwen, & Clay, 2019; Su, Li, Lu, & Chen, 2009). They have the LUFS domain and function as transcriptional coactivators like the metazoan Ssbp (Fig. 4A), but their glutamine-rich transactivation domains do not resemble the proline-rich transactivation domain of metazoan Ssbp, suggesting that Ssbp homologs acquired the transactivation domain independently in each lineage.

3.5.3 Evolutionary scenario of Lhx/Ldb/Ssbp complex functions

When did Ldb start to interact with LIM domains? Here we identified an Ldb gene in a demosponge, *Amphimedon queenslandica* (Accession No. XP_019849410), although it was not found in a previously reported sponge genome (Simmons et al., 2012). On the other hand, we still could not find an Ldb homolog in the Ctenophora (Simmons et al., 2012). Furthermore, no Ldb genes are found in the choanoflagellate genomes currently available, whereas an Ldb gene was found in *Capsaspora* (Accession No. XP_ 004348054), a sister group to choanoflagerates and metazoans. Because *Amphimedon* Ldb, but not *Capsaspora* Ldb, possesses an LID-like sequence (Fig. 4B), we supposed that the LID was acquired somewhere in the common ancestor of the Porifera, Cnidaria, Placozoa, and Bilateria (Fig. 4A).

Notably, there are splicing isoforms of Ldb1 and Ldb2 lacking the LID, which inhibit the transactivation complex formation on the Protein 4.2 promoter (Tran et al., 2006). This type of isoform also exists in *Nematostella* (Yasuoka et al., 2009), suggesting that Ldb-mediated protein interactions for gene regulation have been modulated by splicing activity since the common ancestor of the Cnidaria and Bilateria arose. Given the metazoan origin of the LID of Ldb, this isoform seems to be an ancient form of Ldb, present among all animals and plants, (Adl et al., 2012). To determine what kinds of TFs were incorporated into the ancestral Ldb–Ssbp complex, it is important to identify partnering TFs of Ldb-like SEU and SLK in plants.

On the basis of current understanding, we speculate that the ancestral Ldb–Ssbp complex was present in the last common ancestor of eukaryotes, and that it served a hub function for gene regulation. Ancient Ssbp may have functioned as a LUFS domain-only protein, and then may have acquired a transactivation or transrepression domain in each lineage. Domain shuffling between Ldb and Ssbp in nematodes supports this idea, in which the transactivation domain of Ssbp is translocated to the C-terminus of Ldb (Enkhmandakh et al., 2006) and Ssbp is truncated to the LUFS domain alone (Fig. 4C). Thus, nematode-type Ssbp may function as the ancestral state of Ssbp proteins.

What is the ancestral role of the Ldb–Ssbp complex? As mentioned above, the *Drosophila* ChiLS complex or the human LDB1–SSBP2 complex forms two binding pockets for NPFxD motifs (Renko et al., 2019; Wang et al., 2020). Without the LID, an ancient Ldb–Ssbp complex may also have had two pockets for interaction with proteins, including TFs, to form enhanceosomes. If so, the ancestral role may have been as a core complex for an enhanceosome, like the Wnt enhanceosome (Fiedler et al., 2015; Renko et al., 2019). Although there is no clear evidence that Ssbp is involved in remote enhancers, long-range and *trans* regulation through Ldb (Fig. 3F) implies that the Ldb–Ssbp core complex is widely used for chromatin looping for enhancer–promoter communications in eukaryotes. In terms of chromosomal folding for long-range gene regulation, TADs and their boundary factors, CTCFs, have never been found beyond bilaterians (Acemel et al., 2017; Heger, Marin, Bartkuhn, Schierenberg, & Wiehe, 2012; Schwaiger et al., 2014; Zimmermann et al., 2020). On the other hand, chromatin looping for enhancer–promoter interactions through CTCF-independent cohesin binding has been reported in mammalian cells (Faure et al., 2012; Kagey et al., 2010; Schmidt et al., 2010). Therefore, the Ldb–Ssbp complex probably serves as a CTCF-independent mechanism for long-distance enhancer–promoter communications among eukaryotes, if remote enhancers are general in eukaryotes. In bilaterians, CTCF was newly invented and co-opted to the Ldb–Ssbp complex as an alternative enhancer looping mechanism (Lee et al., 2017).

In summary, in the metazoan lineage, LIM-HD was innovated and Ldb acquired the LID to interact with LIM-HD. The LIM-HD/Ldb/Ssbp complex may have contributed to formation of more sophisticated gene regulatory networks to generate diverse cell types, through combinatorial regulation by multiple TFs, enhanceosome formation with super enhancer, and long-range/*trans* regulation for target genes.

4. Overview of the role of LIM-HD proteins in development

Mouse gene knockout analyses of LIM-HD genes have revealed interesting phenotypes, including Lim1/Lhx1 (the head organizer) (Shawlot & Behringer, 1995), Lim3/Lhx3 (the pituitary) (Sheng et al., 1996), Isl1 (motoneurons and pancreas) (Ahlgren, Pfaff, Jessell, Edlund, & Edlund, 1997; Pfaff, Mendelsohn, Stewart, Edlund, & Jessell, 1996), Lmx1b (the dorsal limb) (Chen et al., 1998), Lim5/Lhx5 (the hippocampus) (Zhao et al., 1999), and so on, as summarized in Bach (2000), which includes papers published from 1995 to 1999. To date, knockout experiments of LIM-HD genes in mice exhibit remarkable embryonic phenotypes, without exceptions. Notably, not all LIM-HD genes are responsible for human genetic diseases, implying their essential roles in development in humans (Hunter & Rhodes, 2005). Hereafter, we review LIM-HD functions during embryonic development in vertebrates and their evolutionary origins.

4.1 "LIM code" for neuron specification

Founder members, *mec-3*, *lin-11*, and *isl1*, are thought to be involved in neuronal differentiation, cell identity, and tissue-specific gene regulation, respectively (Freyd et al., 1990; Karlsson, Thor, Norberg, Ohlsson, & Edlund, 1990; Way & Chalfie, 1988). The fourth gene identified is *Xlim-1*, which is expressed in the organizer and exhibits organizer activity (Agulnick et al., 1996; Taira et al., 1992; Taira, Otani, Saint-Jeannet, & Dawid, 1994). Roles of these genes initially did not appear related, but later it was shown that most LIM-HD genes, including the aforementioned genes, are commonly expressed in various neurons, in addition to other regional expression domains. More generally speaking, LIM-HD genes are commonly involved in cell-type specification or determination.

A typical example is "LIM code," in which combinations of LIM-HD genes specify/determine motoneurons, interneurons, or sensory neurons in vertebrate spinal cord (Fig. 5A) (Bhati et al., 2008; Ensini et al., 1994). In medial MMC (median motor column) neurons, Lhx3/4 LIM domains bind to the LID of Isl1/2 to form a complex with Ldb1 (Fig. 3E). Interestingly, another LIM-HD factor Lmx1b is involved in motor neuron projection along the dorsoventral axis in murine developing limbs. Lmx1b is expressed on the dorsal sides of limbs (Fig. 5A), and loss of Lmx1b expression leads to ventralization of limbs, resulting in incorrect projection of motor neurons

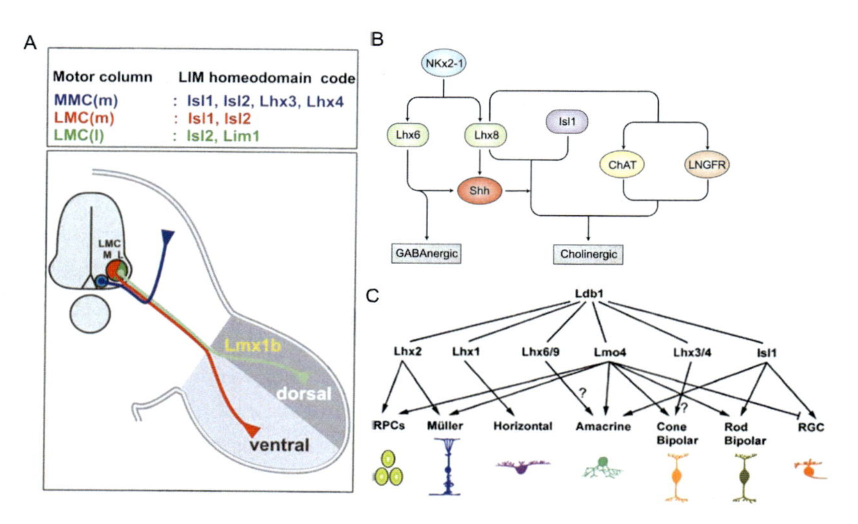

Fig. 5 LIM code regulates neuron cell-type specification/differentiation. (A) Combinatorial LIM-HD expression regulates motor neuron specification in the spinal cord, while Lmx1b regulates regionalization of limbs. MMC(m), medial median motor column; LMC(m), medial lateral motor column; LMC(l), ateral motor column. The diagram is reproduced from Kania, Johnson, and Jessell (2000). (B) Lhx6, Lhx8, and Isl1 regulates neuron cell-type specification in the forebrain. The diagram is reproduced from Zhou et al. (2015). (C) Different LIM-HD regulates terminal differentiation and maintenance of retinal neuron cell-types. The diagram is reproduced from Xiao, Jin, and Xiang (2018).

from the spinal cord (Kania et al., 2000). LIM code-like neuron cell-fate determination was also reported in mouse developing forebrain. Differentiation of GABAergic neurons and cholinergic neurons is regulated by combinations of Lhx6, Lhx8, and Isl1 (reviewed by Zhou et al., 2015) (Fig. 5B). Furthermore, final determination and maintenance of various retinal neuron cell types are regulated by LIM-HDs, LMO4, and Ldb1, in which each cell type is determined by a single LIM-HD, except for Isl1 (Fig. 5C) (Xiao et al., 2018). As described above, Lhx2 and Ebf bind to enhancers named "Greek islands" and recruit Ldb1, forming multi-chromosomal enhancer hubs for singular olfactory receptor gene expression in mature olfactory sensory neurons (Fig. 3G). Considering LIM-HD/LIM-HD interactions as exemplified by Isl1/2 and Lhx3 for motor neuron specification (Fig. 3E), cell type-specific LIM-HD/Ldb/Ssbp complexes must function as cores to form enhanceosomes. Those enhanceosomes bridge cell type-specific enhancers to promoters, in cooperation with partner TFs, which are not necessarily expressed specifically in the cell as long as they interact with the enhanceosome. Partner TFs may determine the specificity of enhancer sequences responding to LIM-HD-containing enhanceosomes. To understand how the LIM code regulates different target genes in each cell type, single-cell transcriptomic/epigenomic analysis would be helpful.

4.2 *Xlim-1/lhx1* functions as a Spemann-Mangold organizer gene

The gastrula organizer, also called the Spemann-Mangold organizer in amphibians, executes the basic body plan by dorsoventral and anteroposterior patterning along the body axes. The dorsal-anterior mesendoderm of gastrulae has been referred to as the head organizer in amphibians, which induces anterior neural tissues and is spatially separated from the trunk organizer (Heasman, 2006; Niehrs, 2004). The head organizer region differentiates into the prechordal plate mesoderm and anterior endoderm in the neurula, whereas the trunk organizer region differentiates into the notochord. As mentioned above, *Xlim-1* was first discovered as a Spemann-Mangold organizer-specific gene, generally called an "organizer gene" (Taira et al., 1992) (Fig. 6A). In the 1990s, many organizer genes were identified one after another at almost the same time as *Xlim-1/lhx1* (Taira et al., 1992). Those are TF-encoding genes, *goosecoid* (Blumberg, Wright, De Robertis, & Cho, 1991; Cho et al., 1991), *otx2* (Blitz & Cho, 1995; Pannese et al., 1995), and *siamois* (*sia1*) (Lemaire, Garrett, & Gurdon, 1995), and signaling molecule-encoding genes, *noggin* (Smith & Harland, 1992), *chordin* (Sasai et al., 1994), and *cerberus* (Bouwmeester, Kim, Sasai, Lu, & De Robertis, 1996). These genes exerted axis duplication activity when their mRNAs were injected individually into the ventral equatorial region of *Xenopus* embryos. This axis-inducing activity by mRNA injection has been postulated as "organizer activity." Organizer genes with organizer activity have been thought to serve essential functions in the Spemann-Mangold organizer. However, a complete secondary axis was only formed by *siamois* mRNA injection, but incomplete axes by the others (e.g., trunk structures by *lhx1, gsc, noggin, and chordin;* a head structure by *cerberus*; and anterior structures by *otx2*).

Among them, one of the most specific genes for the head organizer is *goosecoid* (*gsc*), which is also expressed in homologous tissues in other vertebrates (i.e., a deep region of the shield in zebrafish, the anterior visceral endoderm in mice, and the anterior hypoblast in chickens). These tissues exhibit head organizer functions and express *gsc*. However, this type of tissue, harboring inductive functions and persistently expressing *gsc*, has never

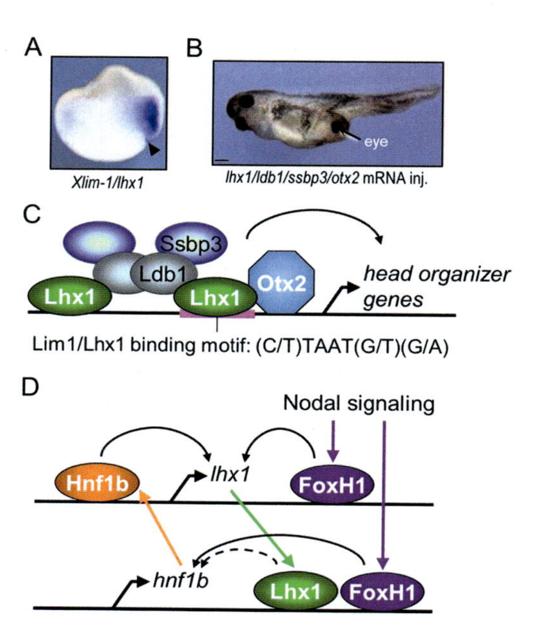

Fig. 6 Xlim-1/Lhx1 serve as a head organizer gene. (A) Xlim-1/lhx1 is expressed in the Spemann-Mangold organizer of *Xenopus tropicalis* embryos. Whole-mount in situ hybridization image for *Xlim-1/lhx1* in early gastrula is reproduced from Yasuoka et al. (2014). A dorsoventral hemisection is shown with dorsal to the right and animal to the top (an arrowhead indicates blastopore). (B) Nearly complete head with an eye is induced by ventral injection of mRNAs for Lhx1/Ldb1/Ssbp3/Otx2. An image of axis-duplicated embryo is reproduced from Yasuoka et al. (2014). (C) Head organizer genes are activated by Lhx1/Ldb1/Ssbp3/Otx2 via *cis*-regulatory modules containing Lim1/Lhx1-binding motifs. The diagram is modified from Yasuoka et al. (2014). (D) A feed-forward loop regulates lim1/lhx1 expression. Nodal/FoxH1 signaling upregulates both *lhx1* and *hnf1*, whereas Lhx1 and Hnf1 activates each other. See text for more details.

been clearly identified in invertebrates, suggesting that the head organizer is a vertebrate innovation (Yasuoka, 2020; Yasuoka, Tando, Kubokawa, & Taira, 2019). Although *gsc* is a good marker for the head organizer, Gsc functions as a transcriptional repressor (Ferreiro, Artinger, Cho, & Niehrs, 1998), implying that head organizer genes are not directly regulated by Gsc.

In contrast to Gsc, Lim1/Lhx1 is a transcriptional activator, which is expressed in the head organizer region (Hiratani et al., 2001; Taira, Otani, Saint-Jeannet, & Dawid, 1994; Yasuoka et al., 2014) (Fig. 6A), and directly activates other head organizer genes such as *gsc*, *otx2*, *chordin*, and *cerberus* (Collart, Verschueren, Rana, Smith, & Huylebroeck, 2005; Hiratani et al., 2001; Mochizuki et al., 2000; Yamamoto et al., 2003).

Importantly, when a mixture of mRNAs for *lhx1*, *ldb1*, *ssbp3*, and *otx2* was injected, a nearly complete head with an eye was induced, suggesting that activated Lhx1 and Otx2 can generate the head organizer (Fig. 6B) (Yasuoka et al., 2014). Although head organizer functions should involve various gene regulatory cascades and networks, as well as morphogenetic mechanisms, Lhx1/Ldb1/Ssbp3/Otx2 can operate them integratively.

To understand more comprehensive gene regulation in the head organizer, we performed a genome-wide study for functions of the head and trunk organizer gene, Lim1/Lhx1, as well as the head organizer genes, Otx2 and Gsc, using *Xenopus tropicalis* gastrulae (Yasuoka et al., 2014). This study revealed that a combination of Lhx1 and Otx2 activates head organizer genes via CRMs containing Lim1-binding motifs (Yasuoka et al., 2014) (Fig. 6C). Intriguingly, a combination of Gsc and Otx2 represses non-head organizer genes, such as trunk organizer genes and ventral genes via Gsc/Otx2-binding motifs. These data suggest that Otx2 functions as an anterior positional cue, and that its combination with the transcriptional activator, Lim1, or the transcriptional repressor, Gsc, activates or represses target genes to form the head organizer. Furthermore, RNA-seq analyses with dissected head organizer regions and remaining regions comprehensively identified 87 head organizer genes (Yasuoka et al., 2014), which included all head organizer genes known at that time. Later, some of them were reportedly confirmed to be enriched in the head organizer (Yasuoka & Taira, 2018); see the references therein). This suggests that almost all head organizer genes are now available for functional analysis. Lim1 also interacts with Mix1 and Sia1 to activate *cerberus* (Fig. 3C) (Sudou et al., 2012; Yamamoto et al., 2003). Our ChIP-seq data for Mix1 and Sia1 in *Xenopus* gastrulae have been published (Jansen et al., 2020). These data provide a platform to analyze interactions between Lhx1, Otx2, Gsc, Mix1, and Sia1 in gene regulatory networks for the head organizer.

lim*1*/*lhx1* is also expressed in the trunk organizer region and its descendants (notochord) (Fig. 8A) (Taira, Otani, Jamrich, & Dawid, 1994). Compared with Lim1 functions in the head organizer, those in the trunk organizer and notochord have been little examined. Some genes expressed in both head and trunk organizer regions, such as *chordin*, can be target genes of Lim1 in the trunk organizer and notochord, but partner TFs for gene activation may be different from those in the head organizer. Candidate partners of Lim1 are trunk-organizer/notochord TFs, such as HNF3β/FoxA2. In fact, HNF3β/FoxA2 genetically interacts with Lim1 to form the anterior primitive streak in mice (Perea-Gómez, Shawlot, Sasaki, Behringer, &

Ang, 1999). Their molecular interactions for notochord development should be analyzed on the basis of genomic data.

Regulatory mechanisms of lim*1*/*lhx1* expression in the gastrula organizer have been analyzed using various animals. In *Xenopus*, *Xlim-1*/*lhx1* promoter assays using a luciferase reporter have shown that *Xlim-1*/*lhx1* is directly activated by Nodal signaling through FoxH1-binding sites in the first intron (Rebbert & Dawid, 1997; Watanabe et al., 2002). Hnf1b also activates *lhx1* in *Xenopus* animal caps through an Hnf1b-binding site just upstream of the *lhx1* promoter (Drews, Senkel, & Ryffel, 2011). Our genomic studies have shown that *hnf1b* mRNA is enriched in the head organizer region (Yasuoka et al., 2014), that Lim1 and Otx2 bind to a putative enhancer containing Lim1-binding motifs in the *hnf1b* intron (Yasuoka et al., 2014), and that the *hnf1b* gene is directly activated by Nodal/Foxh1 signaling (Charney et al., 2017), forming a feed-forward regulatory loop for *lhx1* and *hnf1b* expression (Fig. 6D). In zebrafish, a transgenic reporter construct containing the first intron of *lhx1a* recapitulated organizer expression, which was activated by Nodal signaling (Swanhart et al., 2010). Although induction of *lhx1* expression in the organizer is governed by Nodal/FoxH1 signaling, it is still unclear what kinds of TFs are involved in maintenance of *lhx1* expression in organizer-derived tissues, the prechordal plate, and the notochord. One candidate is FoxA TFs (FoxA1, FoxA2, and FoxA4 in *Xenopus*), because FoxA replaces FoxH1 to maintain CRMs (Charney et al., 2017). Another candidate is Lim1 itself, because a putative enhancer of *lim1*/*lhx1* was activated by a TF cocktail of *lim1*/*ldb1*/*ssbp3*/*otx2*/*gsc*/*tle1* in reporter assays (Yasuoka et al., 2014). We will examine those possibilities in the future.

4.3 Lhx1 is essential for kidney development

Another important expression domain of vertebrate *lim1*/*lhx1* is the developing kidney (see Fig. 8A). *Xlim-1*/*lhx1* expression starts in the middle of lateral mesoderm at the late gastrula stage, and continues in the intermediate mesoderm, and then in the pronephros during *Xenopus* development (Taira, Otani, Jamrich, & Dawid, 1994). Importantly, RA (retinoic acid) treatment of *Xenopus* embryos expands *Xlim-1* expression in the lateral mesoderm, and enlarges the pronephric region (Taira, Otani, Jamrich, & Dawid, 1994). Thus, Lim1/Lhx1 has been recognized as one of the earliest nephrogenic TFs. Gain-of- and loss-of-function analyses of *Xlim-1*/*lhx1* in the pronephric region of *Xenopus* embryos demonstrated that *Xlim-1*/*lhx1* is required for cell fate specification of renal progenitor cells (Carroll & Vize, 1999; Cirio et al., 2011). More recently, *lhx1* knockout with CRISPR-Cas9 in *Xenopus* embryos disrupted pronephric development, resulting in lack of kidney tubules (Fig. 7A) (Delay et al., 2018). Knockout of *lhx1* finally led to edema formation, a phenotype of absent or dysfunctional kidneys, in tadpoles (Fig. 7B). A knockout mouse study also demonstrated that *Lhx1* is essential for proper metanephros development in multiple steps (Kobayashi et al., 2005). Thus, *Lhx1* is indispensable in vertebrate kidney development.

Interestingly, some head organizer genes such as *crescent/frzb2*, *hnf1b*, and *follistatin*, are also specifically expressed in the pronephric kidney of *Xenopus* embryos (Demartis, Maffei, Vignali, Barsacchi, & De Simone, 1994; Hemmati-Brivanlou, Kelly, & Melton, 1994; Shibata, Ono, Hikasa, Shinga, & Taira, 2000). Coexpression of *lhx1* and other head organizer genes in the pronephros implies that Lhx1 also regulates expression of those genes in the pronephric kidney. In fact, *follistatin* expression was reduced in the pronephros when *lhx1* was depleted (Cirio et al., 2011). Given the ancient origin of the excretory organ in bilaterians and that of the gastrula organizer in cnidarians and bilaterians, discussed below, cooption of GRNs from the organizer to the kidney and vice versa may have occurred during animal evolution.

Synergistic activation of *lhx1* by Activin A and retinoic acid (RA) was first reported in *Xenopus* animal explants (Taira et al., 1992). Furthermore, RA treatment of embryos expands the pronephros as mentioned above (Taira, Otani, Jamrich, & Dawid, 1994). Later, it was reported that treatment with both Activin A and RA generates renal progenitor cells from *Xenopus* animal caps and mammalian stem cells (Krneta-Stankic, DeLay, & Miller, 2017; Moriya, Uchiyama, & Asashima, 1993; Osafune, 2010). A time-course study of pronephric gene expression in animal caps treated with Activin A and RA showed that *osr1*, *osr2*, *hnf1b*, and *lhx1* are activated after just 1.5 h of treatment (Fig. 7C) (Drews et al., 2011). However, RA-response elements (RAREs) have not been identified in the *lhx1* genomic locus, which need to be examined. Furthermore, misexpression of *hnf1b* in animal caps induces early nephrogenic genes including *lhx1*. Thus, Lhx1 and its positive regulator Hnf1b are major players in gene regulatory networks of early nephrogenesis.

After nephrectomy, *lhx1* is immediately induced in regenerating nephric tubules in *Xenopus* embryos (Suzuki et al., 2019). Reporter analysis screening coupled with nephrectomy for evolutionarily conserved noncoding sequences in the *lhx1* gene locus identified regeneration signal-response

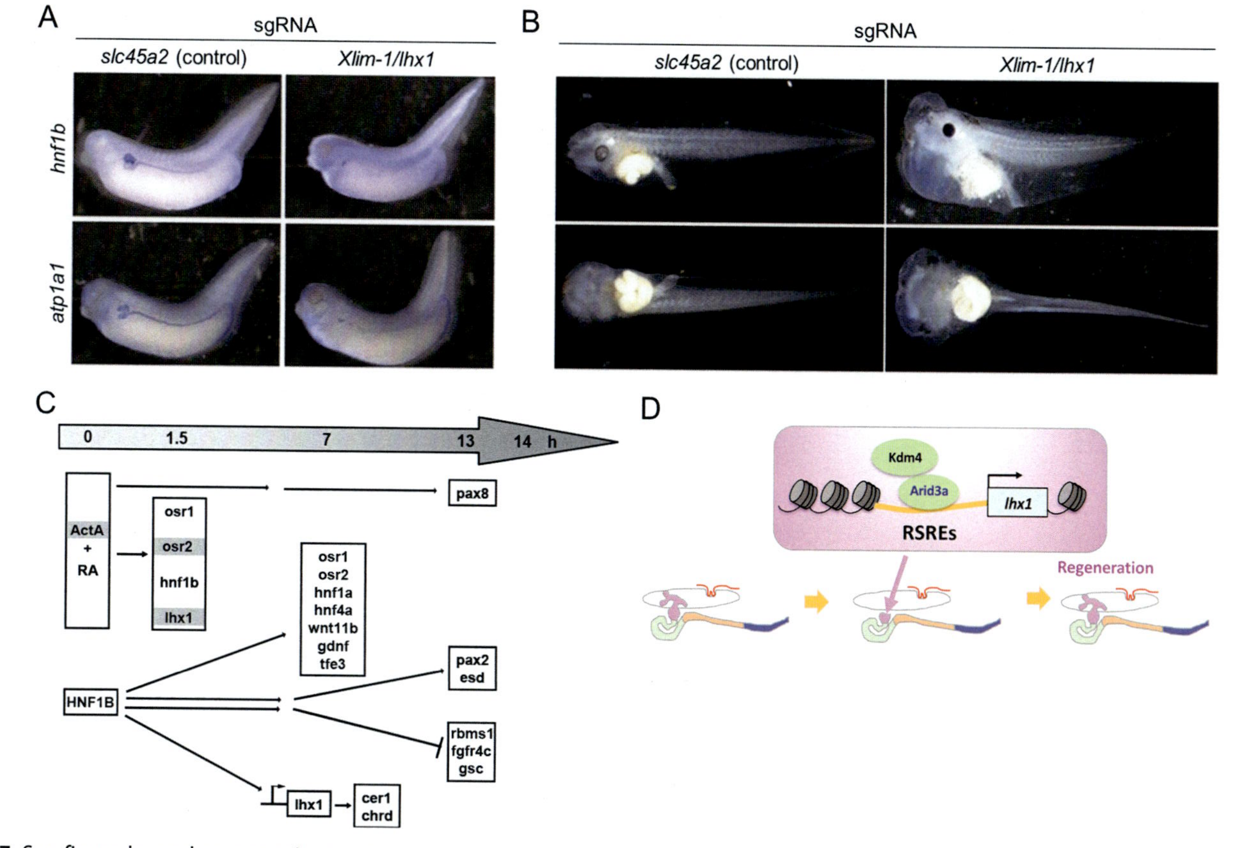

Fig. 7 See figure legend on opposite page.

enhancers (RSREs), in which Arid3a-binding sites were found. Gain-of- and loss-of-function assays showed that Arid3a activates *lhx1* via RSREs (Fig. 7D) (Suzuki et al., 2019). Thus, *lhx1* also plays a key role in pronephric regeneration, though the initial cue to activate *lhx1* through Arid3a remains unclear.

Post-transcriptional regulation of *lhx1* in the pronephros is also important for pronephros formation. *miR-30a-5p* represses *lhx1* expression in the pronephros by binding to its 3′ UTR (Agrawal, Tran, & Wessely, 2009). Microinjection of antisense morpholino oligos against miR-30a-5p impaired terminal differentiation of pronephric cells. These data indicate that controlling *lhx1* expression levels in the nephrogenic field is crucial for appropriate cell specification and differentiation during development.

4.4 Evolutionary aspects of LIM-HD functions

LIM-HD functions have been extensively investigated in invertebrates, showing their conservatism and variability among metazoans. For example, the LIM code for motor neuron specification also applies to *Drosophila*, in which a combination of *Lim3* and *Islet* function as they do in vertebrates (Gill, 2003; Thor, Andersson, Tomlinson, & Thomas, 1999) (see Figs. 3E, 5A). In addition, *apterous/lhx2* is expressed on the dorsal sides of wing discs and confers dorsal identity (Cohen, McGuffin, Pfeifle, Segal, & Cohen, 1992; Gill, 2003). This resembles the *Lmx1b* function in dorsal regionalization of vertebrate limbs (Fig. 5A) (Kania et al., 2000). *apterous/lhx2* and *Lmx1b* belong to different subclasses of LIM-HDs, and *Drosophila* wings and vertebrate limbs are not homologous, indicating that distinct genes were independently co-opted to the dorsal side of appendages and execute similar functions. Therefore, there should be some critical reason to co-opt LIM-HD, instead of other types of TFs, to form a dorsal-specific GRN.

Fig. 7 *Xlim-1/Lhx1* has an indispensable role in the kidney development and regeneration. (A) Knock-out of *Xlim-1/lhx1* but not *slc45a2* (control) by CRISPR-Cas9 depletes expression of *hnf1b* and *atp1a1* in nephric tubules of *Xenopus* embryos. Whole-mount in situ hybridization images are reproduced from Delay et al. (2018). (B) Knock-out of *Xlim-1/lhx1* but not *slc45a2* (control) by CRISPR-Cas9 leads to edema formation in *Xenopus* tadpoles. Images of tadpoles are reproduced from Delay et al. (2018). (C) Nephrogenic genes are sequentially activated after treatment with Activin A and retinoic acid (RA) in animal explants of *Xenopus* embryos. The diagram is reproduced from Drews et al. (2011). (D) *Xlim-1/lhx1* is activated in the regenerating kidney after nephrectomy. This process is mediated by Arid3a, which binds RSREs (regeneration signal-response enhancers) of *lhx1*. The diagram is reproduced from Suzuki, Hirano, Ogino, and Ochi (2019).

4.4.1 Neurons

LIM-HD expression in the nervous system is widely observed across metazoans. Strikingly, in the sea anemone (cnidarian) *Nematostella,* LIM-HD genes are solely or combinatorially expressed in different neuron cell types (Srivastava et al., 2010). The oral nerve ring marked by DOPA-beta-monooxygenase and RFamide (Arg-Phe-NH2), the pharyngeal nerve ring marked by GABA, and the apical tuft marked by GABA express *lmx, lhx1/5,* and *lhx6/8,* respectively. In the pharyngeal endoderm and the directive mesenteries, which express DOPA-beta-monooxygenase, *lhx2/9, isl,* and *lmx* are coexpressed. Coexpression of DOPA-beta-monooxygenase and *lmx* is seen throughout *Nematostella* embryogenesis. In sponge (*Amphimedon*) embryos, both *lhx1/5* and *lhx3/4* are expressed around the pigment spot where photosensory cells occur (Srivastava et al., 2010). Furthermore, in the ctenophore, *Mnemiopsis,* all four LIM-HD genes, *lhx1/5, lhx3/4, isl,* and *lmx,* are expressed in the nervous system (Simmons et al., 2012). Overlapping expression of *lhx1/5, isl,* and *lmx* was observed in putative photosensory cells. Unlike bilaterian motor neurons (see Fig. 5A), *lhx3/4* and *isl* are exclusively expressed in the apical organ, which is a highly innervated sensory structure (see references in Simmons et al., 2012), forming distinct boundaries. Note that *ldb* and *ssbp* have never been found in ctenophore genomes (Fig. 4A), implying that ctenophores use other partner proteins that bind LIM domains by domain–motif interaction for regulatory complex formation. These observations of LIM-HD expression in basal metazoans suggest that the LIM code, like the neuronal cell type specification system, has evolved independently in each metazoan lineage.

4.4.2 Organizer

Does Lim1/Lhx1 rather than other LIM-HDs function solely in the organizer GRN? To address this question, we examined organizer activity of various LIM-HD proteins in *Xenopus* embryos (Yasuoka et al., 2009). When organizer activity of active-form LIM domain mutants ("3m" in Fig. 3B) of LIM-HDs was compared, bilaterian Lhx1 and deuterostome Lhx3 showed strong organizer activity, whereas *Nematostella* Lhx1, *Drosophila* Lhx3, and other LIM-HD proteins from vertebrates exhibited substantially lower organizer activity. Using chimeric constructs and deletion constructs, we further identified the C-terminal region and the linker region between the second LIM domain and homeodomain of Lhx1 and Lhx3 as regions responsible for evolution of organizer activity. Interestingly, ascidian Lhx3 exerted organizer activity as strong as that of

Xlim-1, and similar protein motifs containing aromatic amino acids (phenylalanine and tyrosine) are necessary for transactivation activity in ascidian Lhx3 and Xlim-1 (Hiratani et al., 2001; Yasuoka et al., 2009). These findings suggest that ascidian Lhx3 and bilaterian Lhx1 evolved independently to acquire an interacting surface for transactivation required for organizer activity, but binding proteins to their transactivation motifs have not been identified yet. Why did the ascidian Lhx3 evolve to have "organizer activity"? One possible answer is that ascidians lost *lhx1* expression in the gastrula, concomitant with loss of the organizer. Instead, *lhx3* is expressed in presumptive endoderm at the 32-cell to gastrula stages (Satou, Imai, & Satoh, 2001; Wada, Katsuyama, Yasugi, & Saiga, 1995; Yasuoka et al., 2009). Because *lhx1* expression in the gastrula organizer is widely conserved among cnidarians and bilaterians, including *Nematostella* (Yasuoka et al., 2009), ascidians are a special exception because early ascidian embryogenesis progresses with mosaic maternal factors in the egg, leading to degenerative evolution of the organizer. Endodermal expression of *lhx3* in the gastrula was also observed in amphioxus (Wang, Zhang, Yasui, & Saiga, 2002), but not in other taxa, implying that Lhx3 participated in gastrula organizer gene regulatory networks in the chordate ancestor, and that later, vertebrates lost *lhx3* expression and tunicates lost *lhx1* expression from gastrulae (Yasuoka et al., 2009).

4.4.3 Excretory organs

Regarding kidney evolution, the *lhx1* ortholog in the basal chordate, amphioxus (*Amphi* lim*1/5*), is expressed in the lineage of Hatschek's nephridium and its precursor, suggesting a conservative role of Lhx1 for kidney development in chordates (Langeland, Holland, Chastain, & Holland, 2006). Furthermore, a hypothesis that animal excretory organs share a single origin was recently presented by examining gene expression profiles of TFs for kidney development in various protostomes and deuterostomes (Gąsiorowski et al., 2020). The authors proposed that *lhx1* constitutes the conserved nephridial TF set, together with *eya*, *six1/2*, *pou3*, *sall* and *osr*. To examine indispensable roles of this conserved gene set for nephrogenesis, molecular interactions between these factors on *cis*-regulatory modules and regulatory principles for nephridial genes should be addressed.

4.4.4 Conserved enhancers for LIM-HD

How widely is the activation mechanism of *lhx1* by Nodal/Foxh1 signaling conserved (see Fig. 6D)? To answer this question, enhancer activity of the first intron of amphioxus *lhx1/5*, responding to Nodal/Foxh1 signaling, was demonstrated by luciferase reporter assays using *Xenopus* embryos (Yasuoka et al., 2019). The enhancer sequence in the first intron of *lhx1* is highly variable between amphioxus and vertebrates, whereas activation by Nodal/Foxh1 signaling through Foxh1-binding motifs in the enhancer is conserved. Even in cnidarians, *lhx1* was activated by Nodal signaling in the *Hydra* organizer (Watanabe et al., 2002), suggesting a deeply conserved regulatory axis from Nodal to *lhx1* in cnidarians and bilaterians (reviewed by Yasuoka & Taira, 2018). In addition, RA and its antagonist BMS009 treatment in amphioxus embryos suggested that *lhx1* expression in the embryonic kidney is activated by RA, indicating conservation of regulatory mechanisms of *lhx1* expression in the chordate kidney. Importantly, RA also activates *lhx1* expression in the spinal cord (Sockanathan & Jessell, 1998), proposing possibilities that the same enhancer with RARE(s), if any, regulate *lhx1* expression in both kidney and spinal cord, or that different enhancers with RARE(s) regulate *lhx1* expression in different tissues. More comprehensive and comparative studies on kidney and spinal cord GRNs for *lhx1* would provide further expandability in aspects of evolutionary biology, genome biology, and systems biology (Charney et al., 2017; Yasuoka, 2020; Yasuoka & Taira, 2018).

Another example of the ultra-conserved enhancer for LIM-HD gene expression is the enhancer for *isl*, in the intronic region of a neighboring gene, *scaper*. A recent study demonstrated that both human and sponge (*Amphimedon*) genomes harbor the microsyntenic unit (*isl-scaper*) with *cis*-regulatory sequences (Wong et al., 2020). Surprisingly, although the composition, alignment, and frequency of known TF-binding motifs are highly variable in enhancers shared by sponges, humans, mice, and zebrafish, the sponge sequence and those of vertebrates showed similar enhancer activity in reporter assays using zebrafish embryos. Because the situation of the *lhx1* enhancer in the first intron of chordates is similar to that of *isl-scaper*, it seems that conserved enhancer activity is commonly maintained without strong sequence constraints. If so, it is difficult to predict conserved enhancer activity from comparative genomics, but similar cases for other LIM-HD genes might be discovered by focusing on microsyntenic units, epigenetic landscapes, and TF-binding motif compositions.

4.4.5 Original roles of LIM-HD

As mentioned above, Lhx1 has multiple functions conserved among animals. What is the common feature of neurons, organizers, and kidneys? One key might be "secretion." That is, neurons secrete neuropeptides. The organizer secretes signaling molecules, and tubular secretion of ions and wastes occurs in the kidney. Other LIM-HD genes are also involved in development of endocrine organs, such as Lhx3 in the pituitary and Isl1/2 in the pancreas. When the metazoan ancestor arose, secretory cells should have been an important innovation for multicellularity, by contributing to cell–cell communications and homeostasis through secreted molecules. To achieve those functions of secretory cells, characteristics of LIM-HD proteins (autorepression, various protein–protein interactions, and DNA-looping via Ldb) must have been beneficial in elaborate gene regulatory networks.

5. Conclusion

More than 30 years have passed since identification of LIM-HD genes. In this review, we first summarized literature historically from discovery of the LIM motif and domain and LIM-HD proteins to the subsequent finding of the LIM domain-binding protein, Ldb1/NLI (Fig. 1). Since then, molecular natures and in vivo functions of LIM-HD, LMO, Ldb, and other associated proteins have been enormously clarified. Based on biochemical and 3D analyses, interactions between the two tandem LIM domains of LIM-HDs, the LIDs of Ldb1 and Isl1, and other partner proteins have been revealed as domain–motif or domain–domain interactions (Figs. 2 and 3), which clearly show versatile roles of LIM-HDs and Ldb1 in gene regulation as hub proteins. This review also discusses gene repertoire evolution (Fig. 4), the role of Ldb1 in a long-distance enhancer–promoter communication in *cis* and *trans*, and functions of LIM-HDs during embryonic development (Figs. 5–7). Upstream regulatory factors, downstream target genes, and partner proteins for Lim1/Lhx1 in three aspects of development (neuron, organizer, and kidney) are summarized (Fig. 8).

The original LIM-HD protein was innovated in the common ancestor of metazoans by combining two tandem LIM domains and a homeodomain. During the same era, ancestral Ldb acquired the LID (LIM domain-binding domain) to become a partner of LIM-HD proteins. During gene evolution, six types of LIM-HD proteins (Lhx1/5, Lhx2/9, Lhx3/4, Lhx6/8, Isl1/2, Lmx1a/b) arose by gene duplication in bilaterians, and 12 members in vertebrates (Fig. 1A and Table 1). LIM-HDs and Ldb1 have been utilized to create various GRNs for cell differentiation, which have probably become more and more intricate during metazoan evolution, finally creating humans. Notably, development of the hippocampus in the brain, which

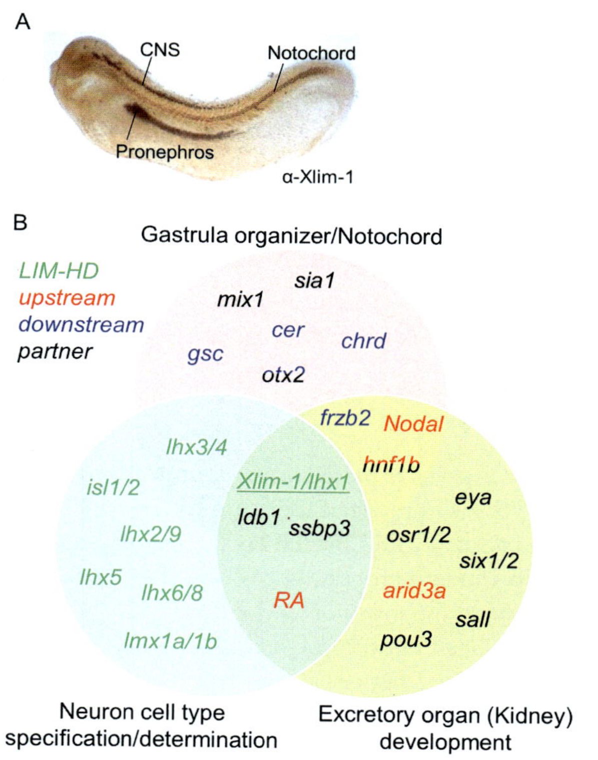

Fig. 8 Versatile functions of Xlim-1/Lhx1 in development. (A) Endogenous Xlim-1/Lhx1 proteins are accumulated in the central nervous system (CNS), notochord, and pronephros in *Xenopus* tailbud embryo. A specific antibody against Xlim-1/Lhx1 proteins were used for whole-mount immunostaining. An image of embryo is reproduced from Sudou, Yamamoto, Ogino, and Taira (2012). (B) A Venn diagram shows upstream genes/signaling, downstream genes, partner proteins, and LIM-HD genes, involved in Lim1/Lhx1 functions during vertebrate development.

is important for long memory in humans, requires Lhx5, the vertebrate paralog (the ohnolog) of Lim1/Lhx1 as a key TF.

One of the remarkable activities of Ldb1 is to mediate remote enhancer functions for long-distance enhancer–promoter communications through interactions of an Ldb1-LIM-HD or -LMO complex with dimerization of Ldb1 and interactions with other TFs and the boundary protein, CTCF. In the case of OR (olfactory receptor) gene clusters on different chromosomes, Ldb1 and Lhx2 mediate *trans*-communications of enhancers near the OR clusters. Another feature of Ldb1 is to form a core complex with Ssbp for assembling enhanceosomes, such as the Wnt enhanceosome.

Thus, combinations of LIM-HDs, Ldb, Ssbp, various TFs and their associated proteins enable a variety of gene regulatory strategies to form elaborate GRNs and to create metazoan diversity.

Regarding developmental roles of LIM-HDs, we introduced LIM codes for neuronal differentiation, Lmx1b and Apterous/Lhx2 for dorsal regionalization of appendages, and Lim1/Lhx1 for gastrula organizer functions and kidney development (Fig. 8). However, our understanding of those functions is still developing. For example, identification of more regulatory sequences, target genes, and binding partners in various cell types using comprehensive "omics" approaches will open the next door to LIM-HD and Ldb functions in development and diseases. Furthermore, comparisons between different groups of cnidarians and bilaterians will offer evolutionary insights into LIM-HD history.

Acknowledgements

We thank Steven D. Aird for technical editing of the manuscript. This work was supported in part by a Grants-in-Aid for Scientific Research from the Japan Society for the Promotion of Science (JSPS) (KAKENHI Grant Nos. 18K14745 and 20H04875 to Y.Y.; and Nos. 25251026 and 18H02447 to M.T.).

References

Acemel, R. D., Maeso, I., & Gómez-Skarmeta, J. L. (2017). Topologically associated domains: A successful scaffold for the evolution of gene regulation in animals. *Wiley Interdisciplinary Reviews: Developmental Biology*, *6*(3), 1–19. https://doi.org/10.1002/wdev.265.

Adl, S. M., Simpson, A. G. B., Lane, C. E., Lukeš, J., Bass, D., Bowser, S. S., et al. (2012). The revised classification of eukaryotes. *Journal of Eukaryotic Microbiology*, *59*(5), 429–514. https://doi.org/10.1111/j.1550-7408.2012.00644.x.

Agrawal, R., Tran, U., & Wessely, O. (2009). The miR-30 miRNA family regulates Xenopus pronephros development and targets the transcription factor Xlim1/Lhx1. *Development*, *136*(23), 3927–3936. https://doi.org/10.1242/dev.037432.

Agulnick, A. D., Taira, M., Breen, J. J., Tanaka, T., Dawid, I. B., & Westphal, H. (1996). Interactions of the LIM-domain-binding factor Ldb1 with LIM homeodomain proteins. *Nature*, *384*(6606), 270–272. https://doi.org/10.1038/384270a0.

Ahlgren, U., Pfaff, S. L., Jessell, T. M., Edlund, T., & Edlund, H. (1997). Independent requirement for ISL1 in formation of pancreatic mesenchyme and islet cells. *Nature*, *385*(6613), 257–260. https://doi.org/10.1038/385257a0.

Bach, I. (2000). The LIM domain: Regulation by association. *Mechanisms of Development*, *91*(1–2), 5–17. https://doi.org/10.1016/S0925-4773(99)00314-7.

Bach, I., Carrière, C., Ostendorff, H. P., Andersen, B., & Rosenfeld, M. G. (1997). A family of LIM domain-associated cofactors confer transcriptional synergism between LIM and Otx homeodomain proteins. *Genes and Development*, *11*(11), 1370–1380. https://doi.org/10.1101/gad.11.11.1370.

Bach, I., Rodriguez-Esteban, C., Carrière, C., Bhushan, A., Krones, A., Rose, D. W., et al. (1999). RLIM inhibits functional activity of LIM homeodomain transcription factors via

recruitment of the histone deacetylase complex. *Nature Genetics*, *22*(4), 394–399. https://doi.org/10.1038/11970.
Bao, F., Azhakanandam, S., & Franks, R. G. (2010). SEUSS and SEUSS-LIKE transcriptional adaptors regulate floral and embryonic development in arabidopsis. *Plant Physiology*, *152*(2), 821–836. https://doi.org/10.1104/pp.109.146183.
Bell, A. C., West, A. G., & Felsenfeld, G. (1999). The protein CTCF is required for the enhancer blocking activity of vertebrate insulators. *Cell*, *98*(3), 387–396. https://doi.org/10.1016/S0092-8674(00)81967-4.
Berlow, R. B., Dyson, H. J., & Wright, P. E. (2015). Functional advantages of dynamic protein disorder. *FEBS Letters*, *589*(19), 2433–2440. https://doi.org/10.1016/j.febslet.2015.06.003.
Bertuzzi, S., Porter, F. D., Pitts, A., Kumar, M., Agulnick, A., Wassif, C., et al. (1999). Characterization of Lhx9, a novel LIM/homeobox gene expressed by the pioneer neurons in the mouse cerebral cortex. *Mechanisms of Development*, *81*(1–2), 193–198. https://doi.org/10.1016/S0925-4773(98)00233-0.
Bhati, M., Lee, C., Nancarrow, A. L., Lee, M., Craig, V. J., Bach, I., et al. (2008). Implementing the LIM code: The structural basis for cell type-specific assembly of LIM-homeodomain complexes. *EMBO Journal*, *27*(14), 2018–2029. https://doi.org/10.1038/emboj.2008.123.
Biryukova, I., & Heitzler, P. (2005). The Drosophila LIM-homeodomain protein Islet antagonizes proneural cell specification in the peripheral nervous system. *Developmental Biology*, *288*(2), 559–570. https://doi.org/10.1016/j.ydbio.2005.09.033.
Blitz, I. L., & Cho, K. W. Y. (1995). Anterior neurectoderm is progressively induced during gastrulation: The role of the Xenopus homeobox gene orthodenticle. *Development*, *121*(4), 993–1004.
Blumberg, B., Wright, C. V., De Robertis, E. M., & Cho, K. W. (1991). Organizer-specific homeobox genes in Xenopus laevis embryos. *Science*, *253*(5016), 194–196. https://doi.org/10.1126/science.1677215.
Boehm, T., Foroni, L., Kaneko, Y., Perutz, M. F., & Rabbitts, T. H. (1991). The rhombotin family of cysteine-rich LIM-domain oncogenes: Distinct members are involved in T-cell translocations to human chromosomes 11p15 and 11p13. *Proceedings of the National Academy of Sciences of the United States of America*, *88*(10), 4367–4371. https://doi.org/10.1073/pnas.88.10.4367.
Bouwmeester, T., Kim, S. H., Sasai, Y., Lu, B., & De Robertis, E. M. (1996). Cerberus is a head-inducing secreted factor expressed in the anterior endoderm of Spemann's organizer. *Nature*. https://doi.org/10.1038/382595a0.
Breen, J. J., Agulnick, A. D., Westphal, H., & Dawid, I. B. (1998). Interactions between LIM domains and the LIM domain-binding protein Ldb1. *The Journal of Biological Chemistry*, *273*(8), 4712–4717.
Capili, A. D., Schultz, D. C., Rauscher, F. J., & Borden, K. L. B. (2001). Solution structure of the PHD domain from the KAP-1 corepressor: Structural determinants for PHD, RING and LIM zinc-binding domains. *EMBO Journal*, *20*(1–2), 165–177. https://doi.org/10.1093/emboj/20.1.165.
Carey, M. (1998). The enhanceosome and transcriptional synergy. *Cell*, *92*(1), 5–8. https://doi.org/10.1016/S0092-8674(00)80893-4.
Carroll, T. J., & Vize, P. D. (1999). Synergism between Pax-8 and lim-1 in embryonic kidney development. *Developmental Biology*, *214*(1), 46–59. https://doi.org/10.1006/dbio.1999.9414.
Charney, R. M., Forouzmand, E., Cho, J. S., Cheung, J., Paraiso, K. D., Yasuoka, Y., et al. (2017). Foxh1 occupies cis-regulatory modules prior to dynamic transcription factor interactions controlling the mesendoderm gene program. *Developmental Cell*, *40*(6), 559–607. https://doi.org/10.1016/j.devcel.2017.02.017.

Chen, H., Lun, Y., Ovchinnikov, D., Kokubo, H., Oberg, K. C., Pepicelli, C. V., et al. (1998). Limb and kidney defects in Lmx1b mutant mice suggest an involvement of LMX1B in human nail patella syndrome. *Nature Genetics*, *19*(1), 51–55. https://doi.org/10.1038/ng0598-51.

Chen, L., Segal, D., Hukriede, N. A., Podtelejnikov, A. V., Bayarsaihan, D., Kennison, J. A., et al. (2002). Ssdp proteins interact with the LIM-domain-binding protein Ldb1 to regulate development. *Proceedings. National Academy of Sciences. United States of America*, *99*(22), 14320–14325. 10.1073pnas.212532399.

Cho, K. W. Y., Blumberg, B., Steinbeisser, H., & De Robertis, E. M. (1991). Molecular nature of Spemann's organizer: The role of the Xenopus homeobox gene goosecoid. *Cell*, *67*(6), 1111–1120. https://doi.org/10.1016/0092-8674(91)90288-A.

Cirio, M. C., Hui, Z., Haldin, C. E., Cosentino, C. C., Stuckenholz, C., Chen, X., et al. (2011). Lhx1 is required for specification of the renal progenitor cell field. *PLoS One*, *6*(4). https://doi.org/10.1371/journal.pone.0018858.

Clevers, H. (2006). Wnt/β-catenin signaling in development and disease. *Cell*, *127*(3), 469–480. https://doi.org/10.1016/j.cell.2006.10.018.

Cohen, B., McGuffin, M. E., Pfeifle, C., Segal, D., & Cohen, S. M. (1992). Apterous, a gene required for imaginal disc development in Drosophila encodes a member of the LIM family of developmental regulatory proteins. *Genes and Development*, *6*(5), 715–729. https://doi.org/10.1101/gad.6.5.715.

Collart, C., Verschueren, K., Rana, A., Smith, J. C., & Huylebroeck, D. (2005). The novel Smad-interacting protein Smicl regulates Chordin expression in the Xenopus embryo. *Development*, *132*(20), 4575–4586. https://doi.org/10.1242/dev.02043.

Dawid, I. B., Breen, J. J., & Toyama, R. (1998). LIM domains: Multiple roles as adapters and functional modifiers in protein interactions. *Trends in Genetics*, *14*(4), 156–162. https://doi.org/10.1016/S0168-9525(98)01424-3.

Deane, J. E., Mackay, J. P., Kwan, A. H. Y., Sum, E. Y. M., Visvader, J. E., & Matthews, J. M. (2003). Structural basis for the recognition of ldb1 by the N-terminal LIM domains of LMO2 and LMO4. *EMBO Journal*, *22*(9), 2224–2233. https://doi.org/10.1093/emboj/cdg196.

Deane, J. E., Ryan, D. P., Sunde, M., Maher, M. J., Guss, J. M., Visvader, J. E., et al. (2004). Tandem LIM domains provide synergistic binding in the LMO4:Ldb1 complex. *EMBO Journal*, *23*(18), 3589–3598. https://doi.org/10.1038/sj.emboj.7600376.

Delay, B. D., Corkins, M. E., Hanania, H. L., Salanga, M., Deng, J. M., Sudou, N., et al. (2018). Tissue-specific gene inactivation in Xenopus laevis: Knockout of lhx1 in the kidney with CRISPR/Cas9. *Genetics*, *208*(2), 673–686. https://doi.org/10.1534/genetics.117.300468.

Demartis, A., Maffei, M., Vignali, R., Barsacchi, G., & De Simone, V. (1994). Cloning and developmental expression of LFB3/HNF1β transcription factor in *Xenopus laevis*. *Mechanisms of Development*, *47*(1), 19–28. https://doi.org/10.1016/0925-4773(94)90092-2.

Deng, W., Lee, J., Wang, H., Miller, J., Reik, A., Gregory, P. D., et al. (2012). Controlling long-range genomic interactions at a native locus by targeted tethering of a looping factor. *Cell*, *149*(6), 1233–1244. https://doi.org/10.1016/j.cell.2012.03.051.

Drews, C., Senkel, S., & Ryffel, G. U. (2011). The nephrogenic potential of the transcription factors osr1, osr2, hnf1b, lhx1 and pax8 assessed in Xenopus animal caps. *BMC Developmental Biology*, *11*(1), 5. https://doi.org/10.1186/1471-213X-11-5.

Emes, R. D., & Ponting, C. P. (2001). A new sequence motif linking lissencephaly, Treacher Collins and oral-facial-digital type 1 syndromes, microtubule dynamics and cell migration. *Human Molecular Genetics*, *10*(24), 2813–2820. https://doi.org/10.1093/hmg/10.24.2813.

Enkhmandakh, B., Makeyev, A. V., & Bayarsaihan, D. (2006). The role of the proline-rich domain of Ssdp1 in the modular architecture of the vertebrate head organizer. *Proceedings of the National Academy of Sciences of the United States of America, 103*(31), 11631–11636. 10.1073pnas.0605209103.

Ensini, M., Morton, S. B., Baldassare, M., Edlund, T., Jessell, T. M., & Pfaff, S. L. (1994). Topographic organization embryonic motor neurons defined by expression of LIM homeobox gene. *Cell, 79*, 957–970.

Faulkner, G., Pallavicini, A., Formentin, E., Comelli, A., Ievolella, C., Trevisan, S., et al. (1999). ZASP: A new Z-band alternatively spliced PDZ-motif protein. *The Journal of Cell Biology, 146*(2), 465–475. https://doi.org/10.1083/jcb.146.2.465.

Faure, A. J., Schmidt, D., Watt, S., Schwalie, P. C., Wilson, M. D., Xu, H., et al. (2012). Cohesin regulates tissue-specific expression by stabilizing highly occupied cis-regulatory modules. *Genome Research, 22*(11), 2163–2175. https://doi.org/10.1101/gr.136507.111.

Fernández-Fúnez, P., Lu, C. H., Rincón-Limas, D. E., García-Bellido, A., & Botas, J. (1998). The relative expression amounts of apterous and its co-factor dLdb/Chip are critical for dorso-ventral compartmentalization in the Drosophila wing. *EMBO Journal, 17*(23), 6846–6853. https://doi.org/10.1093/emboj/17.23.6846.

Ferreiro, B., Artinger, M., Cho, K. W. Y., & Niehrs, C. (1998). Antimorphic goosecoids. *Development, 125*(8), 1347–1359.

Fiedler, M., Graeb, M., Mieszczanek, J., Rutherford, T. J., Johnson, C. M., & Bienz, M. (2015). An ancient Pygo-dependent Wnt enhanceosome integrated by chip/LDB-SSDP. *ELife, 4*, 1–22. https://doi.org/10.7554/eLife.09073.

Franks, R. G., Wang, C., Levin, J. Z., & Liu, Z. (2002). SEUSS, a member of a novel family of plant regulatory proteins, represses floral homeotic gene expression with LEUNIG. *Development, 129*(1), 253–263.

Freyd, G., Kim, S. K., & Horvitz, H. R. (1990). Novel cysteine-rich motif and homeodomain in the product of the *Caenorhabditis elegans* cell lineage gene lin-II. *Nature, 344*(6269), 876–879. https://doi.org/10.1038/344876a0.

Gąsiorowski, L., Andrikou, C., Janssen, R., Bump, P., Budd, G. E., Lowe, C. J., et al. (2020). A single origin of animal excretory organs. *BioRxiv*. https://doi.org/10.1101/2020.11.15.378034.

Gadd, M. S., Bhati, M., Jeffries, C. M., Langley, D. B., Trewhella, J., Guss, J. M., et al. (2011). Structural basis for partial redundancy in a class of transcription factors, the LIM homeodomain proteins, in neural cell type specification. *Journal of Biological Chemistry, 286*(50), 42971–42980. https://doi.org/10.1074/jbc.M111.248559.

Gadd, M. S., Jacques, D. A., Nisevic, I., Craig, V. J., Kwan, A. H., Guss, J. M., et al. (2013). A structural basis for the regulation of the LIM-homeodomain protein islet 1 (Isl1) by intra- and intermolecular interactions. *Journal of Biological Chemistry, 288*(30), 21924–21935. https://doi.org/10.1074/jbc.M113.478586.

Garamszegi, S., Franzosa, E. A., & Xia, Y. (2013). Signatures of pleiotropy, economy and convergent evolution in a domain-resolved map of human-virus protein-protein interaction networks. *PLoS Pathogens, 9*(12), 1–9. https://doi.org/10.1371/journal.ppat.1003778.

German, M. S., Wang, J., Chadwick, R. B., & Rutter, W. J. (1992). Synergistic activation of the insulin gene by a LIM-homeo domain protein and a basic helix-loop-helix protein: Building a functional insulin minienhancer complex. *Genes and Development, 6*(11), 2165–2176. https://doi.org/10.1101/gad.6.11.2165.

Gill, G. N. (2003). Decoding the LIM development code. *Transactions of the American Clinical and Climatological Association, 114*, 179–189.

Gonzalez, D., Bowen, A. J., Carroll, T. S., & Conlan, R. S. (2007). The transcription corepressor LEUNIG interacts with the histone deacetylase HDA19 and mediator components MED14 (SWP) and CDK8 (HEN3) to repress transcription. *Molecular and Cellular Biology, 27*(15), 5306–5315. https://doi.org/10.1128/mcb.01912-06.

Grigoriou, M., Tucker, A. S., Sharpe, P. T., & Pachnis, V. (1998). Expression and regulation of Lhx6 and Lhx7, a novel subfamily of LIM homeodomain encoding genes, suggests a role in mammalian head development. *Development*, *125*(11), 2063–2074.

Heasman, J. (2006). Patterning the early Xenopus embryo. *Development*, *133*(7), 1205–1217. https://doi.org/10.1242/dev.02304.

Heger, P., Marin, B., Bartkuhn, M., Schierenberg, E., & Wiehe, T. (2012). The chromatin insulator CTCF and the emergence of metazoan diversity. *Proceedings of the National Academy of Sciences of the United States of America*, *109*(43), 17507–17512. https://doi.org/10.1073/pnas.1111941109.

Heitzler, P., Vanolst, L., Biryukova, I., & Ramain, P. (2003). Enhancer-promoter communication mediated by Chip during Pannier-driven proneural patterning is regulated by Osa. *Genes and Development*, *17*(5), 591–596. https://doi.org/10.1101/gad.255703.

Hemmati-Brivanlou, A., Kelly, O. G., & Melton, D. A. (1994). Follistatin, an antagonist of activin, is expressed in the Spemann organizer and displays direct neuralizing activity. *Cell*, 77(2), 283–295. https://doi.org/10.1016/0092-8674(94)90320-4.

Hemmati-Brivanlou, A., Frank, D., Bolce, M. E., Brown, B. D., Sive, H. L., & Harland, R. M. (1990). Localization of specific mRNAs in Xenopus embryos by whole-mount in situ hybridization. *Development*, *110*(2), 325–330.

Hiratani, I., Mochizuki, T., Tochimoto, N., & Taira, M. (2001). Functional domains of the LIM homeodomain protein Xlim-1 involved in negative regulation, transactivation, and axis formation in Xenopus embryos. *Developmental Biology*, *229*(2), 456–467. https://doi.org/10.1006/dbio.2000.9986.

Hiratani, I., Yamamoto, N., Mochizuki, T., Ohmori, S.-Y., & Taira, M. (2003). Selective degradation of excess Ldb1 by Rnf12/RLIM confers proper Ldb1 expression levels and Xlim-1/Ldb1 stoichiometry in Xenopus organizer functions. *Development*, *130*(17), 4161–4175. https://doi.org/10.1242/dev.00621.

Hunter, C. S., & Rhodes, S. J. (2005). LIM-homeodomain genes in mammalian development and human disease. *Molecular Biology Reports*, *32*(2), 67–77. https://doi.org/10.1007/s11033-004-7657-z.

Jansen, C., Paraiso, K. D., Zhou, J. J., Blitz, I. L., Fish, M. B., Charney, R. M., et al. (2020). Uncovering the mesendoderm gene regulatory network through multi-omic data integration. *BioRxiv*. https://doi.org/10.1101/2020.11.01.362053.

Jurata, L. W., & Gill, G. N. (1997). Functional analysis of the nuclear LIM domain interactor NLI. *Molecular and Cellular Biology*, *17*(10), 5688–5698. https://doi.org/10.1128/mcb.17.10.5688.

Jurata, L. W., Kenny, D. A., & Gill, G. N. (1996). Nuclear LIM interactor, a rhombotin and LIM homeodomain interacting protein, is expressed early in neuronal development. *Proceedings of the National Academy of Sciences of the United States of America*, *93*(21), 11693–11698. https://doi.org/10.1073/pnas.93.21.11693.

Jurata, L. W., Pfaff, S. L., & Gill, G. N. (1998). The nuclear LIM domain interactor NLI mediates homo- and heterodimerization of LIM domain transcription factors. *Journal of Biological Chemistry*, *273*(6), 3152–3157. https://doi.org/10.1074/jbc.273.6.3152.

Kadrmas, J. L., & Beckerle, M. C. (2004). The lim domain: From the cytoskeleton to the nucleus. *Nature Reviews Molecular Cell Biology*, *5*, 920–931. https://doi.org/10.1038/nrm1499.

Kagey, M. H., Newman, J. J., Bilodeau, S., Zhan, Y., Orlando, D. A., Van Berkum, N. L., et al. (2010). Mediator and cohesin connect gene expression and chromatin architecture. *Nature*, *467*(7314), 430–435. https://doi.org/10.1038/nature09380.

Kania, A., Johnson, R. L., & Jessell, T. M. (2000). Coordinate roles for LIM homeobox genes in directing the dorsoventral trajectory of motor axons in the vertebrate limb. *Cell*, *102*(2), 161–173. https://doi.org/10.1016/S0092-8674(00)00022-2.

Karlsson, O., Thor, S., Norberg, T., Ohlsson, H., & Edlund, T. (1990). Insulin gene enhancer binding protein Isl-1 is a member of a novel class of proteins containing both a homeo- and a Cys-His domain. *Nature*, *344*(6269), 879–882. https://doi.org/10.1038/344879a0.

Kim, M. H., Cooper, D. R., Oleksy, A., Devedjiev, Y., Derewenda, U., Reiner, O., et al. (2004). The structure of the N-terminal domain of the product of the lissencephaly gene Lis1 and its functional implications. *Structure*, *12*(6), 987–998. https://doi.org/10.1016/j.str.2004.03.024.

Kobayashi, A., Kwan, K. M., Carroll, T. J., McMahon, A. P., Mendelsohn, C. L., & Behringer, R. R. (2005). Distinct and sequential tissue-specific activities of the LIM-class homeobox gene Lim1 for tubular morphogenesis during kidney development. *Development*, *132*(12), 2809–2823. https://doi.org/10.1242/dev.01858.

Koch, B. J., Ryan, J. F., & Baxevanis, A. D. (2012). The diversification of the LIM superclass at the base of the Metazoa increased subcellular complexity and promoted multicellular specialization. *PLoS One*, 7(3), e33261. https://doi.org/10.1371/journal.pone.0033261.

Krishna, S. S., Majumdar, I., & Grishin, N. V. (2003). Structural classification of zinc fingers. *Nucleic Acids Research*, *31*(2), 532–550. https://doi.org/10.1093/nar/gkg161.

Krivega, I., Dale, R. K., & Dean, A. (2014). Role of LDB1 in the transition from chromatin looping to transcription activation. *Genes and Development*, *28*(12), 1278–1290. https://doi.org/10.1101/gad.239749.114.

Krneta-Stankic, V., DeLay, B. D., & Miller, R. K. (2017). Xenopus: Leaping forward in kidney organogenesis. *Pediatric Nephrology*, *32*(4), 547–555. https://doi.org/10.1007/s00467-016-3372-y.

Laitaoja, M., Valjakka, J., & Jänis, J. (2013). Zinc coordination spheres in protein structures. *Inorganic Chemistry*, *52*(19), 10983–10991. https://doi.org/10.1021/ic401072d.

Langeland, J. A., Holland, L. Z., Chastain, R. A., & Holland, N. D. (2006). Invaginating organizer region and later in differentiating cells of the kidney and central nervous system. *International Journal of Biological Sciences*, *2*(3), 110–116.

Larroux, C., Luke, G. N., Koopman, P., Rokhsar, D. S., Shimeld, S. M., & Degnan, B. M. (2008). Genesis and expansion of metazoan transcription factor gene classes. *Molecular Biology and Evolution*, *25*(5), 980–996. https://doi.org/10.1093/molbev/msn047.

Layer, J. H., Christy, M., Placek, L., Unutmaz, D., Guo, Y., & Davé, U. P. (2020). LDB1 enforces stability on direct and indirect oncoprotein partners in leukemia. *Molecular and Cellular Biology*, *40*(12), 1–28. https://doi.org/10.1128/mcb.00652-19.

Lee, S. K., & Pfaff, S. L. (2003). Synchronization of neurogenesis and motor neuron specification by direct coupling of bHLH and homeodomain transcription factors. *Neuron*, *38*(5), 731–745. https://doi.org/10.1016/S0896-6273(03)00296-4.

Lee, J., Krivega, I., Dale, R. K., & Dean, A. (2017). The LDB1 complex co-opts CTCF for erythroid lineage-specific long-range enhancer interactions. *Cell Reports*, *19*(12), 2490–2502. https://doi.org/10.1016/j.celrep.2017.05.072.

Lemaire, P., Garrett, N., & Gurdon, J. B. (1995). Expression cloning of Siamois, a xenopus homeobox gene expressed in dorsal-vegetal cells of blastulae and able to induce a complete secondary axis. *Cell*, *81*(1), 85–94. https://doi.org/10.1016/0092-8674(95)90373-9.

Li, H., Witte, D. P., Branford, W. W., Aronow, B. J., Weinstein, M., Kaur, S., et al. (1994). Gsh-4 encodes a LIM-type homeodomain, is expressed in the developing central nervous system and is required for early postnatal survival. *EMBO Journal*, *13*(12), 2876–2885. https://doi.org/10.1002/j.1460-2075.1994.tb06582.x.

Lomvardas, S., Barnea, G., Pisapia, D. J., Mendelsohn, M., Kirkland, J., & Axel, R. (2006). Interchromosomal interactions and olfactory receptor choice. *Cell*, *126*(2), 403–413. https://doi.org/10.1016/j.cell.2006.06.035.

Love, P. E., Warzecha, C., & Li, L. Q. (2014). Ldb1 complexes: The new master regulators of erythroid gene transcription. *Trends in Genetics*, *30*(1), 1–9. https://doi.org/10.1016/j.tig.2013.10.001.

Matsumoto, K., Tanaka, T., Furuyama, T., Kashihara, Y., Mori, T., Ishii, N., et al. (1996). L3, a novel murine LIM-homeodomain transcription factor expressed in the ventral telencephalon and the mesenchyme surrounding the oral cavity. *Neuroscience Letters*, *204*(1), 113–116. https://doi.org/10.1016/0304-3940(96)12341-7.

McGuire, E. A., Davis, A. R., & Korsmeyer, S. J. (1991). T-cell translocation gene 1 (Ttg-1) encodes a nuclear protein normally expressed in neural lineage cells. *Blood*, 77(3), 599–606. https://doi.org/10.1182/blood.v77.3.599.599.

Misas, E., Escandón, P., McEwen, J. G., & Clay, O. K. (2019). The LUFS domain, its transcriptional regulator proteins, and drug resistance in the fungal pathogen Candida auris. *Protein Science*, *28*(11), 2024–2029. https://doi.org/10.1002/pro.3727.

Mochizuki, T., Karavanov, A. A., Curtiss, P. E., Ault, K. T., Sugimoto, N., Watabe, T., et al. (2000). Xlim-1 and LIM domain binding protein 1 cooperate with various transcription factors in the regulation of the goosecoid promoter. *Developmental Biology*, *224*(2), 470–485. https://doi.org/10.1006/dbio.2000.9778.

Monahan, K., Schieren, I., Cheung, J., Mumbey-Wafula, A., Monuki, E. S., & Lomvardas, S. (2017). Cooperative interactions enable singular olfactory receptor expression. *eLife*, *6*, e28620. https://doi.org/10.1101/142489.

Monahan, K., Horta, A., Mumbay-Wafula, A. M., Li, L., Zhao, Y., Love, P. E., et al. (2018). Ldb1 mediates trans enhancement in mammals. *BioRxiv*. https://doi.org/10.1101/287524.

Monahan, K., Horta, A., & Lomvardas, S. (2019). LHX2- and LDB1-mediated trans interactions regulate olfactory receptor choice. *Nature*, *565*(7740), 448–453. https://doi.org/10.1038/s41586-018-0845-0.

Moriya, N., Uchiyama, H., & Asashima, M. (1993). Induction of pronephric tubules by activin and retinoic acid in presumptive ectoderm of *Xenopus laevis* (RA/kidney/mesoderm induction/*Xenopus laevis*). *Development, Growth & Differentiation*, *35*(2), 123–128. https://doi.org/10.1111/j.1440-169X.1993.00123.x.

Nakano, T., Murata, T., Matsuo, I., & Aizawa, S. (2000). OTX2 directly interacts with LIM1 and HNF-3β. *Biochemical and Biophysical Research Communications*, *267*(1), 64–70. https://doi.org/10.1006/bbrc.1999.1872.

Neduva, V., & Russell, R. B. (2006). Peptides mediating interaction networks: New leads at last. *Current Opinion in Biotechnology*, *17*(5), 465–471. https://doi.org/10.1016/j.copbio.2006.08.002.

Niehrs, C. (2004). Regionally specific induction by the Spemann-Mangold organizer. *Nature Reviews Genetics*, *5*(6), 425–434. https://doi.org/10.1038/nrg1347.

Nishioka, N., Nagano, S., Nakayama, R., Kiyonari, H., Ijiri, T., Taniguchi, K., et al. (2005). Ssdp1 regulates head morphogenesis of mouse embryos by activating the Lim1-Ldb1 complex. *Development*, *132*(11), 2535–2546. https://doi.org/10.1242/dev.01844.

Osafune, K. (2010). In vitro regeneration of kidney from pluripotent stem cells. *Experimental Cell Research*, *316*(16), 2571–2577. https://doi.org/10.1016/j.yexcr.2010.04.034.

Ostendorff, H. P., Peirano, R. I., Peters, M. A., Schlüter, A., Bossenz, M., Scheffner, M., et al. (2002). Ubiquitination-dependent cofactor exchange on LIM homeodomain transcription factors. *Nature*, *416*(6876), 99–103. https://doi.org/10.1038/416099a.

Pannese, M., Polo, C., Andreazzoli, M., Vignali, R., Kablar, B., Barsacchi, G., et al. (1995). The Xenopus homologue of Otx2 is a maternal homeobox gene that demarcates and specifies anterior body regions. *Development*, *121*(3), 707–720.

Passier, R., Richardson, J. A., & Olson, E. N. (2000). Oracle, a novel PDZ-LIM domain protein expressed in heart and skeletal muscle. *Mechanisms of Development*, *92*(2), 277–284. https://doi.org/10.1016/S0925-4773(99)00330-5.

Perea-Gómez, A., Shawlot, W., Sasaki, H., Behringer, R. R., & Ang, S. L. (1999). HNF3β and Lim1 interact in the visceral endoderm to regulate primitive streak formation and anterior-posterior polarity in the mouse embryo. *Development*, *126*(20), 4499–4511.

Pérez-Alvarado, G. C., Miles, C., Michelsen, J. W., Louis, H. A., Winge, D. R., Beckerle, M. C., et al. (1994). Structure of the C-terminal LIM domain from the cysteine rich protein, CRP. *Nature Structural & Molecular Biology, 1*, 388–398.
Pérez-Alvarado, G. C., Kosa, J. L., Louis, H. A., Beckerle, M. C., Winge, D. R., & Summers, M. F. (1996). Structure of the cysteine-rich intestinal protein, CRIP. *Journal of Molecular Biology, 257*, 153–174.
Pfaff, S. L., Mendelsohn, M., Stewart, C. L., Edlund, T., & Jessell, T. M. (1996). Requirement for LIM homeobox gene Isl1 in motor neuron generation reveals a motor neuron-dependent step in interneuron differentiation. *Cell, 84*(2), 309–320. https://doi.org/10.1016/S0092-8674(00)80985-X.
Poss, Z. C., Ebmeier, C. C., & Taatjes, D. J. (2013). The Mediator complex and transcription regulation. *Critical Reviews in Biochemistry and Molecular Biology, 48*(6), 575–608. https://doi.org/10.3109/10409238.2013.840259.
Putnam, N. H., Srivastava, M., Hellsten, U., Dirks, B., Chapman, J., Salamov, A., et al. (2007). Sea anemone genome reveals ancestral eumetazoan gene repertoire and genomic organization. *Science, 317*(5834), 86–94. https://doi.org/10.1126/science.1139158.
Rebbert, M. L., & Dawid, I. B. (1997). Transcriptional regulation of the Xlim-1 gene by activin is mediated by an element in intron I. *Proceedings of the National Academy of Sciences of the United States of America, 94*(18), 9717–9722. https://doi.org/10.1073/pnas.94.18.9717.
Ren, G., Jin, W., Cui, K., Rodrigez, J., Hu, G., Zhang, Z., et al. (2017). CTCF-mediated enhancer-promoter interaction is a critical regulator of cell-to-cell variation of gene expression. *Molecular Cell, 67*(6). https://doi.org/10.1016/j.molcel.2017.08.026. 1049-1058.e6.
Renko, M., Fiedler, M., Rutherford, T. J., Schaefer, J. V., Plückthun, A., & Bienz, M. (2019). Rotational symmetry of the structured Chip/LDB-SSDP core module of the Wnt enhanceosome. *Proceedings of the National Academy of Sciences of the United States of America, 116*(42), 20977–20983. https://doi.org/10.1073/pnas.1912705116.
Robertson, N. O., Smith, N. C., Manakas, A., Mahjoub, M., McDonald, G., Kwan, A. H., et al. (2018). Disparate binding kinetics by an intrinsically disordered domain enables temporal regulation of transcriptional complex formation. *Proceedings of the National Academy of Sciences of the United States of America, 115*(18), 4643–4648. https://doi.org/10.1073/pnas.1714646115.
Robustelli, P., Piana, S., & Shaw, D. E. (2020). Mechanism of coupled folding-upon-binding of an intrinsically disordered protein. *Journal of the American Chemical Society, 142*(25), 11092–11101. https://doi.org/10.1021/jacs.0c03217.
Sala, S., & Ampe, C. (2018). An emerging link between LIM domain proteins and nuclear receptors. *Cellular and Molecular Life Sciences, 75*(11), 1959–1971. https://doi.org/10.1007/s00018-018-2774-3.
Sánchez-García, I., Osada, H., Forster, A., & Rabbitts, T. H. (1993). The cysteine-rich LIM domains inhibit DNA binding by the associated homeodomain in Isl-1. *EMBO Journal, 12*(11), 4243–4250. https://doi.org/10.1002/j.1460-2075.1993.tb06108.x.
Sasai, Y., Lu, B., Steinbeisser, H., Geissert, D., Gont, L. K., & De Robertis, E. M. (1994). Xenopus chordin: A novel dorsalizing factor activated by organizer-specific homeobox genes. *Cell, 79*(5), 779–790. https://doi.org/10.1016/0092-8674(94)90068-X.
Satou, Y., Imai, K. S., & Satoh, N. (2001). Early embryonic expression of a LIM-homeobox gene Cs-lhx3 is downstream of beta-catenin and responsible for the endoderm differentiation in Ciona savignyi embryos. *Development (Cambridge, England), 128*(18), 3559–3570.
Schmidt, D., Schwalie, P. C., Ross-Innes, C. S., Hurtado, A., Brown, G. D., Carroll, J. S., et al. (2010). A CTCF-independent role for cohesin in tissue-specific transcription. *Genome Research, 20*(5), 578–588. https://doi.org/10.1101/gr.100479.109.

Schwaiger, M., Schönauer, A., Rendeiro, A. F., Pribitzer, C., Schauer, A., Gilles, A. F., et al. (2014). Evolutionary conservation of the eumetazoan gene regulatory landscape. *Genome Research*, *24*(4), 639–650. https://doi.org/10.1101/gr.162529.113.

Shawlot, W., & Behringer, R. R. (1995). Requirement for Lim1 in head-organizer function. *Nature*, *374*, 425–430.

Sheng, H. Z., Zhadanov, A. B., Mosinger, B., Fujii, T., Bertuzzi, S., Grinberg, A., et al. (1996). Specification of pituitary cell lineages by the LIM homeobox gene Lhx3. *Science*, *272*(5264), 1004–1007. https://doi.org/10.1126/science.272.5264.1004.

Shibata, M., Ono, H., Hikasa, H., Shinga, J., & Taira, M. (2000). Xenopus crescent encoding a Frizzled-like domain is expressed in the Spemann organizer and pronephros. *Mechanisms of Development*, *96*(2), 243–246. https://doi.org/10.1016/S0925-4773(00)00399-3.

Simmons, D. K., Pang, K., & Martindale, M. Q. (2012). Lim homeobox genes in the ctenophore *Mnemiopsis leidyi*: The evolution of neural cell type specification. *EvoDevo*, *3*(2), 1–11.

Smith, W. C., & Harland, R. M. (1992). Expression cloning of noggin, a new dorsalizing factor localized to the Spemann organizer in Xenopus embryos. *Cell*, *70*(5), 829–840. https://doi.org/10.1016/0092-8674(92)90316-5.

Sockanathan, S., & Jessell, T. M. (1998). Motor neuron-derived retinoid signaling specifies the subtype identity of spinal motor neurons. *Cell*, *94*(4), 503–514. https://doi.org/10.1016/S0092-8674(00)81591-3.

Spemann, H., & Mangold, H. (1924). über Induktion von Embryonalanlagen durch Implantation artfremder Organisatoren. *Archiv Für Mikroskopische Anatomie Und Entwicklungsmechanik*, *100*(3), 599–638. https://doi.org/10.1007/BF02108133.

Splinter, E., Heath, H., Kooren, J., Palstra, R. J., Klous, P., Grosveld, F., et al. (2006). CTCF mediates long-range chromatin looping and local histone modification in the β-globin locus. In *Vol. 20. Genes and development* (pp. 2349–2354). https://doi.org/10.1101/gad.399506.

Sridhar, V. V., Surendrarao, A., Gonzalez, D., Steven Conlan, R., & Liu, Z. (2004). Transcriptional repression of target genes by LEUNIG and SEUSS, two interacting regulatory proteins for Arabidopsis flower development. *Proceedings of the National Academy of Sciences of the United States of America*, *101*(31), 11494–11499. https://doi.org/10.1073/pnas.0403055101.

Srivastava, V., & Verma, P. K. (2017). The plant LIM proteins: Unlocking the hidden attractions. *Planta*, *246*(3), 365–375. https://doi.org/10.1007/s00425-017-2715-7.

Srivastava, M., Larroux, C., Lu, D. R., Mohanty, K., Chapman, J., Degnan, B. M., et al. (2010). Early evolution of the LIM homeobox gene family. *BMC Biology*, *8*, 4. https://doi.org/10.1186/1741-7007-8-4.

Su, C., Li, Y., Lu, Y., & Chen, J. (2009). Mss11, a transcriptional activator, is required for hyphal development in Candida albicans. *Eukaryotic Cell*, *8*(11), 1780–1791. https://doi.org/10.1128/EC.00190-09.

Sudou, N., Yamamoto, S., Ogino, H., & Taira, M. (2012). Dynamic in vivo binding of transcription factors to cis-regulatory modules of cer and gsc in the stepwise formation of the Spemann-Mangold organizer. *Development*, *139*(9), 1651–1661. https://doi.org/10.1242/dev.068395.

Suzuki, N., Hirano, K., Ogino, H., & Ochi, H. (2019). Arid3a regulates nephric tubule regeneration via evolutionarily conserved regeneration signal-response enhancers. *eLife*, *8*, 1–28. https://doi.org/10.7554/ELIFE.43186.

Swanhart, L. M., Takahashi, N., Jackson, R. L., Gibson, G. A., Watkins, S. C., Dawid, I. B., et al. (2010). Characterization of an lhx1a transgenic reporter in zebrafish. *International Journal of Developmental Biology*, *54*(4), 731–736. https://doi.org/10.1387/ijdb.092969ls.

Taira, M., Jamrich, M., Good, P. J., & Dawid, I. B. (1992). The LIM domain-containing homeo box gene Xlim-1 is expressed specifically in the organizer region of Xenopus gastrula embryos. *Genes and Development*, *6*(3), 356–366.

Taira, M., Hayes, W. P., Otani, H., & Dawid, I. B. (1993). Expression of LIM class homeobox gene Xlim-3 in Xenopus development is limited to neural and neuroendocrine tissues. *Developmental Biology*, *159*(1), 245–256.

Taira, M., Otani, H., Jamrich, M., & Dawid, I. B. (1994). Expression of the LIM class homeobox gene *Xlim-1* in pronephros and CNS cell lineages of *Xenopus* embryos is affected by retinoic acid and exogastrulation. *Development*, *120*(6), 1525–1536.

Taira, M., Otani, H., Saint-Jeannet, J.-P., & Dawid, I. B. (1994). Role of the LIM class homeodomain protein Xlim-1 in neural and muscle induction by the Spemann organizer in Xenopus. *Nature*, *372*(6507).

Taira, M., Evrard, J. L., Steinmetz, A., & Dawid, I. B. (1995). Classification of LIM proteins. *Trends in Genetics*, *11*(11), 431–432. https://doi.org/10.1016/S0168-9525(00)89139-8.

Thaler, J. P., Lee, S.-K., Jurata, L. W., Gill, G. N., & Pfaff, S. L. (2002). LIM factor Lhx3 contributes to the specification of motor neuron and interneuron identity through cell-type-specific protein-protein interactions. *Cell*, *110*. https://doi.org/10.1016/s0092-8674(02)00823-1.

Thor, S., Andersson, S. G. E., Tomlinson, A., & Thomas, J. B. (1999). A LIM-homeodomain combinatorial code for motor neuron pathway selection. *Nature*, *397*, 76–80.

Toyama, R., Curtiss, P. E., Otani, H., Kimura, M., Dawid, I. B., & Taira, M. (1995). The LIM class homeobox gene lim5: Implied role in CNS patterning in Xenopus and Zebrafish. *Developmental Biology*, *170*(2). https://doi.org/10.1006/dbio.1995.1238.

Toyama, R., Kobayashi, M., Tomita, T., & Dawid, I. B. (1998). Expression of LIM-domain binding protein (Ldb) genes during zebrafish embryogenesis. *Mechanisms of Development*, *71*(1–2), 197–200. https://doi.org/10.1016/S0925-4773(97)00202-5.

Tran, Y. H., Xu, Z., Kato, A., Mistry, A. C., Goya, Y., Taira, M., et al. (2006). Spliced isoforms of LIM-domain-binding protein (CLIM/NLI/Ldb) lacking the LIM-interaction domain. *Journal of Biochemistry*, *140*(1), 105–119. https://doi.org/10.1093/jb/mvj134.

Tran, M. K., Kurakula, K., Koenis, D. S., & de Vries, C. J. M. (2016). Protein–protein interactions of the LIM-only protein FHL2 and functional implication of the interactions relevant in cardiovascular disease. *Biochimica et Biophysica Acta (BBA)—Molecular Cell Research*, *1863*(2), 219–228. https://doi.org/10.1016/j.bbamcr.2015.11.002.

Tsuchida, T., Ensini, M., Morton, S. B., Baldassare, M., Edlund, T., Jessell, T. M., et al. (1994). Topographic organization of embryonic motor neurons defined by expression of LIM homeobox genes. *Cell*, *79*(6), 957–970. https://doi.org/10.1016/0092-8674(94)90027-2.

van Meyel, D. J. (2003). Ssdp proteins bind to LIM-interacting co-factors and regulate the activity of LIM-homeodomain protein complexes in vivo. *Development*, *130*(9), 1915–1925. https://doi.org/10.1242/dev.00389.

Wada, S., Katsuyama, Y., Yasugi, S., & Saiga, H. (1995). Spatially and temporally regulated expression of the LIM class homeobox gene Hrlim suggests multiple distinct functions in development of the ascidian, Halocynthia roretzi. *Mechanisms of Development*, *51*(1), 115–126. https://doi.org/10.1016/0925-4773(95)00359-9.

Wadman, I. A., Osada, H., Grütz, G. G., Agulnick, A. D., Westphal, H., Forster, A., et al. (1997). The LIM-only protein Lmo2 is a bridging molecule assembling an erythroid, DNA-binding complex which includes the TAL1, E47, GATA-1 and Ldb1/NLI proteins. *EMBO Journal*, *16*(11), 3145–3157. https://doi.org/10.1093/emboj/16.11.3145.

Wang, Y., Zhang, P. J., Yasui, K., & Saiga, H. (2002). Expression of Bblhx3, a LIM-homeobox gene, in the development of amphioxus *Branchiostoma belcheri tsingtauense*. *Mechanisms of Development*, *117*(1–2), 315–319. https://doi.org/10.1016/S0925-4773(02)00197-1.

Wang, H., Kim, J., Wang, Z., Yan, X. X., Dean, A., & Xu, W. (2020). Crystal structure of human LDB1 in complex with SSBP2. *Proceedings of the National Academy of Sciences of the United States of America*, *117*(2), 1042–1048. https://doi.org/10.1073/pnas.1914181117.

Watanabe, M., Rebbert, M. L., Andreazzoli, M., Takahashi, N., Toyama, R., Zimmerman, S., et al. (2002). Regulation of the Lim-1 gene is mediated through conserved FAST-1/FoxH1 sites in the first intron. *Developmental Dynamics*, *225*(4), 448–456. https://doi.org/10.1002/dvdy.10176.

Way, J. C., & Chalfie, M. (1988). mec.3, a homeobox-containing gene that specifies differentiation of the touch receptor neurons in *C. elegans*. *Cell*, *54*.

Wilkinson-White, L., Gamsjaeger, R., Dastmalchi, S., Wienert, B., Stokes, P. H., Crossley, M., et al. (2011). Structural basis of simultaneous recruitment of the transcriptional regulators LMO2 and FOG1/ZFPM1 by the transcription factor GATA1. *Proceedings of the National Academy of Sciences of the United States of America*, *108*(35), 14443–14448. https://doi.org/10.1073/pnas.1105898108.

Wong, E. S., Zheng, D., Tan, S. Z., Bower, N. I., Garside, V., Vanwalleghem, G., et al. (2020). Deep conservation of the enhancer regulatory code in animals. *Science*, *370*-(6517). https://doi.org/10.1126/science.aax8137.

Xiao, D., Jin, K., & Xiang, M. (2018). Necessity and sufficiency of Ldb1 in the generation, differentiation and maintenance of non-photoreceptor cell types during retinal development. *Frontiers in Molecular Neuroscience*, *11*, 1–16. https://doi.org/10.3389/fnmol.2018.00271.

Xu, Y., Baldassare, M., Fisher, P., Rathbun, G., Oltz, E. M., Yancopoulos, G. D., et al. (1993). LH-2: A LIM/homeodomain gene expressed in developing lymphocytes and neural cells. *Proceedings of the National Academy of Sciences of the United States of America*, *90*(1), 227–231. https://doi.org/10.1073/pnas.90.1.227.

Xu, Z., Meng, X., Cai, Y., Liang, H., Nagarajan, L., & Brandt, S. J. (2007). Single-stranded DNA-binding proteins regulate the abundance of LIM domain and LIM domain-binding proteins. *Genes and Development*, *21*(8), 942–955. https://doi.org/10.1101/gad.1528507.

Xue, D., Tu, Y., & Chalfie, M. (1993). Cooperative interactions between the *Caenorhabditis elegans* homeoproteins UNC-86 and MEC-3. *Science*, *261*(5126), 1324–1328. https://doi.org/10.1126/science.8103239.

Yamamoto, S., Hikasa, H., Ono, H., & Taira, M. (2003). Molecular link in the sequential induction of the Spemann organizer: Direct activation of the cerberus gene by Xlim-1, Xotx2, Mix.1, and Siamois, immediately downstream from Nodal and Wnt signaling. *Developmental Biology*, *257*(1), 190–204. https://doi.org/10.1016/S0012-1606(03)00034-4.

Yasuoka, Y. (2020). Enhancer evolution in chordates: Lessons from functional analyses of cephalochordate cis-regulatory modules. *Development Growth and Differentiation*, *62*(5), 279–300. https://doi.org/10.1111/dgd.12684.

Yasuoka, Y., & Taira, M. (2018). *The molecular basis of the gastrula organizer in amphibians and cnidarians* (pp. 667–708). https://doi.org/10.1007/978-4-431-56609-0_31.

Yasuoka, Y., Kobayashi, M., Kurokawa, D., Akasaka, K., Saiga, H., & Taira, M. (2009). Evolutionary origins of blastoporal expression and organizer activity of the vertebrate gastrula organizer gene lhx1 and its ancient metazoan paralog lhx3. *Development*, *136*(12), 2005–2014. https://doi.org/10.1242/dev.028530.

Yasuoka, Y., Suzuki, Y., Takahashi, S., Someya, H., Sudou, N., Haramoto, Y., et al. (2014). Occupancy of tissue-specific cis-regulatory modules by Otx2 and TLE/Groucho for embryonic head specification. *Nature Communications*, *5*, 4322. https://doi.org/10.1038/ncomms5322.

Yasuoka, Y., Tando, Y., Kubokawa, K., & Taira, M. (2019). Evolution of cis-regulatory modules for the head organizer gene goosecoid in chordates: Comparisons between Branchiostoma and Xenopus. *Zoological Letters*, *5*(1), 1–17. https://doi.org/10.1186/s40851-019-0143-1.

Zhang, H., Wang, W.-P., Guo, T., Yang, J.-C., Chen, P., Ma, K.-T., et al. (2009). The LIM-homeodomain protein ISL1 activates insulin gene promoter directly through synergy with BETA2. *Journal of Molecular Biology*, *392*(3), 566–577. https://doi.org/10.1016/j.jmb.2009.07.036.

Zhao, Y., Sheng, H. Z., Amini, R., Grinberg, A., Lee, E., Huang, S., et al. (1999). Control of hippocampal morphogenesis and neuronal differentiation by the LIM homeobox gene Lhx5. *Science*, *284*(5417), 1155–1158. https://doi.org/10.1126/science.284.5417.1155.

Zhou, Q., Ruiz-Lozano, P., Martone, M. E., & Chen, J. (1999). Cypher, a striated muscle-restricted PDZ and LIM domain-containing protein, binds to α-actinin-2 and protein kinase C. *Journal of Biological Chemistry*, *274*(28), 19807–19813. https://doi.org/10.1074/jbc.274.28.19807.

Zhou, C., Yang, G., Chen, M., He, L., Xiang, L., Ricupero, C., et al. (2015). Lhx6 and Lhx8: Cell fate regulators and beyond. *The FASEB Journal*, *29*(10), 4083–4091. https://doi.org/10.1096/fj.14-267500.

Zimmermann, B., Robb, S. M. C., Genikhovich, G., Fropf, W. J., Weilguny, L., He, S., et al. (2020). Sea anemone genomes reveal ancestral metazoan chromosomal macrosynteny. *BioRxiv*. https://doi.org/10.1101/2020.10.30.359448.

CHAPTER SIX

Control of zygotic genome activation in *Xenopus*

Ira L. Blitz* and Ken W.Y. Cho*
Department of Developmental and Cell Biology, University of California, Irvine, CA, United States
*Corresponding authors: e-mail address: ilblitz@uci.edu; kwcho@uci.edu

Contents

Abstract

The fertilized frog egg contains all the materials needed to initiate development of a new organism, including stored RNAs and proteins deposited during oogenesis, thus the earliest stages of development do not require transcription. The onset of transcription from the zygotic genome marks the first genetic switch activating the gene regulatory network that programs embryonic development. Zygotic genome activation occurs after an initial phase of transcriptional quiescence that continues until the midblastula stage, a period called the midblastula transition, which was first identified in *Xenopus*. Activation of transcription is programmed by maternally supplied factors and is regulated at multiple levels. A similar switch exists in most animals and is of great interest both to developmental biologists and to those interested in understanding nuclear reprogramming. Here we review in detail our knowledge on this major switch in transcription in *Xenopus* and place recent discoveries in the context of a decades old problem.

Current Topics in Developmental Biology, Volume 145
ISSN 0070-2153
https://doi.org/10.1016/bs.ctdb.2021.03.003

1. Introduction

Zygotic genome activation (ZGA), the first burst of transcription in an embryo, marks the initiation of the new organism's gene regulatory network (GRN) hierarchy and thus is the first translation of the embryonic genotype into phenotype. In amphibians, like in most animals, much of early development is initially controlled by maternally supplied factors that program the events of ZGA. Eggs contain the raw materials to "unpack" and reprogram the incoming sperm nucleus, transforming it from a genome bound by protamines to one that, like the egg genome, is wrapped in nucleosomes. Following fertilization, the *Xenopus* zygote undergoes a regular series of embryonic cleavage divisions, without growth in volume, creating a multicellular blastula composed of pluripotent cells. These cells quickly differentiate to form the primary germ layers, which occurs simultaneous with the early specification of the embryonic dorsal-ventral and anterior-posterior axes, and then subdivision of these axes into the organ primordia needed to create the adult organism. For more than 60 years, embryos of the allo-tetraploid frog *Xenopus laevis* have been a model system to study fundamental questions in biology, uncovering the molecular, cellular and developmental biology of the early vertebrate embryo. More recently, the diploid frog *Xenopus tropicalis* has also entered into wide use (Harland & Grainger, 2011). Both frogs have contributed to studies of the maternal-to-zygotic transition (MZT), the mother-to-embryo hand off in genetic control, and provide insights impacting biological questions in many other systems.

During the cleavage stages, embryonic genomes are initially transcriptionally inactive, with a major burst of zygotic transcription beginning during blastula stage. New transcription begins shortly before germ layer specification and is required for initiation of the massive tissue rearrangements of gastrulation (Newport & Kirschner, 1982a). The major transcriptional burst of ZGA coincides with other changes in the early embryo including a switch from synchronous to asynchronous cell divisions, a slowing of the cell cycle, and an acquisition of cell motility (Newport & Kirschner, 1982a). This collection of changes was recognized as an important moment in development and named the midblastula transition (MBT) (Gerhart, 1980; Newport & Kirschner, 1982a; Signoret & Lefresne, 1971). Thus, ZGA and MBT are part of the larger MZT, the broader series of events comprising the handoff of maternal-to-zygotic control. The transcriptional switch, ZGA, is not only an interesting phenomenon in early embryogenesis, but also provides an attractive model to study genetic and epigenomic transcriptional regulatory mechanisms that have relevance across all of biology. We recommend a number of excellent reviews that provide a multi-species overview of the MZT and ZGA (Jukam, Shariati, & Skotheim, 2017; Lee, Bonneau, & Giraldez, 2014; Wu & Vastenhouw, 2020; to name a few). Here we specifically review the regulation of transcription at ZGA in *Xenopus*, permitting a deeper dive into the literature within this one system, including exciting new findings that enhance our understanding of this important genetic switch.

2. On the biology of zygotic genome activation

2.1 Characteristics of early *Xenopus* embryos

Two different *Xenopus* species are in common use today. Their early morphology, developmental/molecular mechanisms, and transcriptome dynamics are very similar (Harland & Grainger, 2011; Yanai, Peshkin, Jorgensen, & Kirschner, 2011), and we consider the early embryology of these species to be relatively interchangeable. A comparison of their features can be found in Table 1. Following fertilization, the first cell cycle, demarcated by initiation of first cleavage, ends at ~1.5 h postfertilization (hpf) in *X. laevis* and ~1.25 hpf in *X. tropicalis*. The subsequent cleavages are metasynchronous—cleavage of larger vegetal blastomeres slightly lags behind the smaller animal blastomeres (Chen, Einstein, et al., 2019; Satoh, 1977). The early cleavages are noteworthy for lacking G_1 and G_2 phases, and thus rapidly cycle between S and M (Gerhart, 1980). Time lapse videomicroscopy of the animal hemisphere reveals that, following the 12th cleavage (~4000 cell stage) when MBT is reached, *X. laevis* cell cycles slow and become asynchronous. Embryos begin to gastrulate ~25 min after the 14th cleavage (Satoh, 1977), probably containing ~16,000 cells. The cell cycle times get progressively longer over several divisions and cycles acquire G1 and G2 phases (Gerhart, 1980; Newport & Kirschner, 1982a; Satoh, 1977).

2.2 ZGA: Revealing the phenomenon

The first successful measurements of RNA synthesis in vertebrate embryos were performed in *Xenopus laevis*. Injection of adult females with radiolabeled sodium phosphate resulted in incorporation into newly synthesized RNA in developing oocytes, and then embryos derived from these (Brown & Littna, 1964a, 1964b). During development, labeled 28S rRNA was first detected at early gastrula stage, albeit weakly, and increased

Table 1 Comparisons between *Xenopus tropicalis* and *Xenopus laevis*.

	Diameter egg/embryo	Volume[a]	Temperature range	Ploidy, genome size[b]	Cell cycle time[c]	Time to first cleavage[d]	Time to ZGA/stage 8[d]	Time to gastrulation[d]
X. tropicalis	~0.8 mm	~0.27 μL	22–27 °C	Diploid, ~1.5 Gbp	~21–23 min	~1.25 h	~4 h	~6.5 h
X. laevis	~1.2 mm	~0.9 μL	16–22 °C	Allotetraploid, ~3.1 Gbp	~30–35 min	~1.5 h	~8 h	~10 h

[a]Calculated. This ~3.4-fold difference is borne out by differences in polyA+ mRNA levels in both species (Owens et al., 2016).
[b]See citations within genome papers (Hellsten et al., 2010; Mitros et al., 2019; Session et al., 2016). ~72% of the *X. laevis* genes are tetraploid, being represented by the presence of homeologous pairs, while ~28% of genes are present in a diploid state (Session et al., 2016).
[c]For *X. laevis* see Satoh (1977), Gerhart (1980), Newport and Kirschner (1982a), Kimelman, Kirschner, and Scherson (1987), Chen, Einstein, Little, and Good (2019). For *X. tropicalis* see Khokha et al. (2002) and Owens et al. (2016).
[d]From Nieuwkoop and Faber (1994) for *X. laevis* and Owens et al. (2016) for *X. tropicalis*.

thereafter. Newly synthesized tRNA (4S RNA) and a high molecular weight RNA class was also detected in the period of midblastula to early gastrula (Brown & Littna, 1966a, 1966b). By bisecting de-jellied embryos to expose the interior to ^{3}H-labeled uridine, transcription was shown to be lacking during early cleavage stages and newly synthesized transcripts were first detected at stage 8.5 (Bacharova & Davidson, 1966; Bachvarova, Davidson, Allfrey, & Mirsky, 1966). These studies showed that synthesis of both high molecular weight RNAs (mRNA) and tRNA begins in blastula stage embryos and confirmed that ZGA begins around the 12th cleavage. A two-hour pulse labeling demonstrated that mRNA and tRNA are synthesized in both animal and vegetal halves (Woodland & Gurdon, 1969), while rRNA incorporated radiolabel at a very low rate and apparently only in the animal pole. (Knowland, 1970) confirmed the timing of tRNA expression and found that rRNA synthesis, transcribed by RNA polymerase I (RNAPI), is insignificant until late gastrula stage 12. Since mRNA and tRNA are synthesized by RNA polymerases II (RNAPII) and III (RNAPIII), respectively, these observation suggests that the timing of transcriptional onset by both of these polymerases might be affected by a common mechanism.

The spatial distribution of zygotic RNA synthesis was examined by autoradiographic analysis of sectioned embryos (Bacharova & Davidson, 1966). Counting silver grains per nucleus (a measure of radioisotopic decay) in each germ layer revealed more isotopic incorporation in endodermal and mesodermal nuclei than in ectoderm, and their first appearance occurred around midblastula stage 8.5 but not before. By microinjecting ^{32}P-labeled rUTP, the timing of incorporation was confirmed to occur after 12th cleavage (stage 8.5). Gel autoradiography resolved labeled 4S, 5S, 7S, Ul and U2 snRNAs (Newport & Kirschner, 1982a), but no incorporation into RNAPI-synthesized 18S or 28S rRNAs was detected until late gastrula/ early neurula stages. Autoradiography also revealed that transcription begins in all cells relatively synchronously. Importantly, low-level incorporation of labeled rUTP occurs *before* MBT into high molecular weight RNA, and this transcription is sensitivity to low doses of α-amanitin, suggesting this to be RNAPII-dependent mRNA synthesis (Bacharova & Davidson, 1966; Kimelman et al., 1987; Nakakura, Miura, Yamana, Ito, & Shiokawa, 1987; Shiokawa et al., 1989). Since the level of pre-MBT labeling was extremely low compared to that at MBT, it became clear that the *major* burst of transcription at stage 8.5 marks an important biological switch—a major transition in the RNA biosynthetic machinery. It is important to note that ZGA does not occur at MBT in the small population of primordial germ cells, which are intermixed within the vegetally localized endoderm of the blastula, and transcription is repressed in these specialized cells until early neurula stage (Venkatarama et al., 2010).

The spatial onset of ZGA has recently been revisited using new techniques, which suggest that the view presented above needs revision. Metabolic labeling employing 5-ethynyl uridine (5-EU), an alkyne-modified ribonucleotide, permits use of "click" chemistry to visualize spatial expression of newly synthesized RNA. 5-EU labeled "bulk RNA" was visualized with confocal microscopy to show that nascent RNA synthesis in *X. laevis* embryos begins in cells located at the extreme animal pole (ectoderm) by cell cycle 12 (Fig. 1), and gradually progresses vegetally until reaching the vegetal pole cells (endoderm) during cell cycle 14 (Chen, Einstein, et al., 2019). Cell numbers were monitored during this period to show that it takes ~80 additional minutes for vegetal cells to undergo the 13th division and ~8000-cell embryos contain populations of cells that have completed 12th, 13th, and 14th divisions. Because 5-EU signal was sensitive to low dose α-amanitin, most or all of this labeling represents synthesis by RNAPII. While this result, on its face, seems contradictory to the silver grain counting data discussed above (Bacharova & Davidson, 1966; Newport & Kirschner, 1982a), these studies can be reconciled by assuming that measuring radioisotope incorporation is more sensitive to very low levels of incorporation (see also Section 2.3) than confocal fluorescence measurements. Thus, pre-MBT transcription was not detected in the recent study by Chen, Einstein, et al. (2019), and the bulk of ZGA at MBT

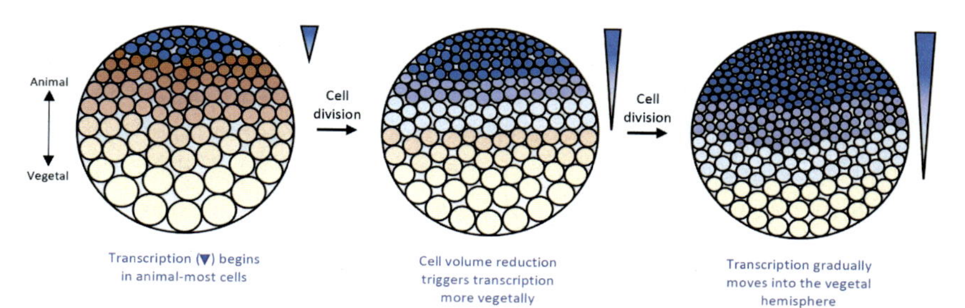

Fig. 1 DNA-to-cytoplasmic ratio is a major determinant of the timing of ZGA. As cell divisions subdivide the volume of embryonic cytoplasm, there is an increase in the number of genomes relative to cytoplasmic volume. Since the cells at the animal hemisphere are smaller, transcription (indicated by shades of blue fill) begins in these before the rest of the embryo. As cells along the animal-vegetal axis continue dividing and achieve smaller volumes, a wave (indicated by inverted blue triangles) of transcription travels vegetally. See text in Section 2.4 for details. Cells along the animal-vegetal axis and cell numbers are not drawn to scale.

shows an animal-to-vegetal spatial progression (Fig. 1). These recent findings add a new perspective on ZGA regulation that are discussed in Section 2.4.

2.3 Transcribing to the beat of a different drummer: Pre-MBT gene expression

While low level metabolic labeling was observed before the burst at MBT, conclusive demonstration that some genes are expressed before MBT required injecting very high doses of ^{32}P-rUTP, combined with long exposure gel autoradiography (Kimelman et al., 1987). Incorporation into high molecular weight RNA was shown to occur in *X. laevis* as early as the 128-cell stage (~4 hpf, 7th cleavage) and these conclusions were confirmed and extended to one cycle earlier, by the 64-cell stage (Nakakura et al., 1987; Shiokawa et al., 1989). Pre-MBT incorporation was subsequently shown to represent polyadenylated (polyA+) RNA (Yang, Tan, Darken, Wilson, & Klein, 2002).

Pre-MBT transcription came into sharper focus when *nodal5* and *nodal6*, encoding ligands of the Tgfβ superfamily, became the first genes identified to be transcribed before MBT (Blythe, Cha, Tadjuidje, Heasman, & Klein, 2010; Skirkanich, Luxardi, Yang, Kodjabachian, & Klein, 2011; Yang et al., 2002). *nodal5/6* mRNAs are detected by reverse transcriptase-polymerase chain reaction (RT-PCR) as early as 128–256 cell stage (cell cycles 7–8, 5–6 cleavage divisions before the major onset of ZGA during the MBT), but not before (Yang et al., 2002). These genes are dorsovegetally expressed and regulated by the maternal Wnt signaling cascade. Blockade of Wnt signaling effectors during the 32–64-cell stage inhibited the expression of both *nodal5* and *nodal6* and also disrupted normal dorsal specification, while LiCl hyperactivated their expression (Blythe et al., 2010; Skirkanich et al., 2011; Yang et al., 2002). Similarly, various manipulations of the Nodal signaling pathway also showed that pre-MBT activation of Nodal signaling is necessary for transcriptional activation of pathway targets at these early stages (Skirkanich et al., 2011). These studies taken in combination provided strong evidence that pre-MBT transcription is functionally important for normal embryogenesis.

More recently, RNA-seq has enabled a more expansive view into the spatiotemporal dynamics of the *Xenopus* transcriptome spanning ZGA in both *X. tropicalis* and *X. laevis* (Collart et al., 2014; Owens et al., 2016; Paranjpe, Jacobi, van Heeringen, & Veenstra, 2013; Peshkin et al., 2015; Session et al., 2016; Tan et al., 2013; Yanai et al., 2011). These studies verified that the bulk of new RNA synthesis, including both messenger and long noncoding (Forouzmand et al., 2017; Paranjpe et al., 2013;

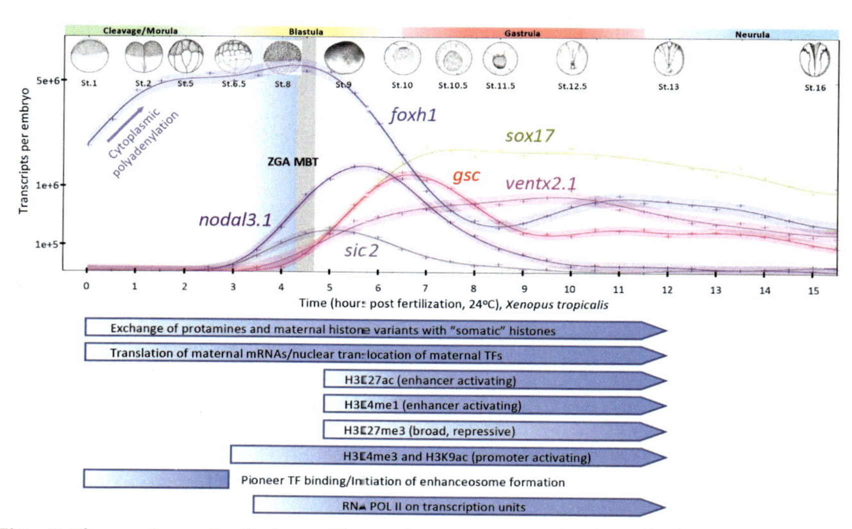

Fig. 2 The embryonic timing of zygotic genome activation. Embryonic stages of *Xenopus tropicalis* development (Nieuwkoop & Faber, 1994) are superimposed on a graph of mRNA expression profiles of several genes. Time in hours post fertilization (hpf) is presented on the X-axis. *foxh1* mRNA is shown as an example of a maternally deposited message that increases its levels by cytoplasmic polyadenylation in the period before ZGA. *nodal3.1* and *sia2* are two examples of genes that begin their pre-MBT transcription between 3 and 3.5 hpf, which corresponds to the period between 64 cell and stage 7. *gsc*, *ventx* and *sox17* are examples of genes that begin transcription at MBT. Data and 95% confidence intervals are derived from Owens et al. (2016) and the Y-axis values use a square root scaling.

Tan et al., 2013) RNAs, in *X. tropicalis*, begins around 4–4.5 hpf (stages 8–9), in good agreement with the timing of the of MBT at stage 8.5 in *X. laevis*. From a high temporal resolution RNA-seq dataset incorporating spike-in RNAs, we have a quantitative view of the kinetics of this process (Fig. 2, top) measured in numbers of transcripts per embryo (Owens et al., 2016). In addition to *nodal5* and *nodal6*, transcripts of *nodal3* and the *sia1* and *sia2* homeobox genes are detectable pre-MBT and as early as the 32- to 128-cell stages (~2.5–3.5 hpf, cell cycles 5–7; see also figure S7 of Owens et al., 2016).

A drawback to standard methods of bulk RNA-seq is that it is difficult to distinguish between pre-existing maternal RNAs and newly synthesized message. Such discrimination can be achieved by metabolic labeling and subsequent purification of the labeled nascent RNA followed by RNA-seq, which also offers higher sensitivity. By microinjecting 5-EU into *X. laevis* embryos and performing RNA-seq on labeled RNA at late blastula (stage 9), 1315 newly transcribed genes were identified (a ~fourfold increase in sensitivity) (Chen, Einstein, et al., 2019). A similar approach in *X. tropicalis* using 4-thiouracil (4-TU) labeling identified mRNA representing 27 zygotically expressed genes as early as 32-cell stage, 144 genes by 128-cell stage, and 1044 genes by 1024-cell stage (Gentsch, Owens, & Smith, 2019). Interestingly, the genes transcribed pre-ZGA were shorter at 32 cell, with an average length of ~1 kb, than the ~6 kb genes transcribed at 1024 cell. Early expressed zygotic genes in zebrafish are also shorter than those expressed later (Heyn et al., 2014).

Newly identified genes by 4-TU labeling play roles in nucleosome assembly, nucleic acid synthesis and translation, and also encode a number of zinc finger transcription factors (TFs) and pri-mir427 (Gentsch, Owens, & Smith, 2019; Owens et al., 2016). Similarly, transcription of the orthologous mir430, and also numerous zinc finger TFs occurs very early in zebrafish ZGA (Heyn et al., 2014; White et al., 2017). Interestingly, genes known to be responsive to Nodal (e.g., *sox17a/b*, *lefty*, *mix1*) and Bmp signaling (e.g., *ventx2.2*) were transcribed prior to MBT (Gentsch, Owens, & Smith, 2019). Smads are intracellular signal transducers for Nodal and Bmp signaling cascades and are phosphorylated upon ligand binding to receptors. Previous findings indicated that phosphorylation of Smads are not detected until ZGA (Faure, Lee, Keller, ten Dijke, & Whitman, 2000; Lee, Heasman, & Whitman, 2001; Saka, Hagemann, Piepenburg, & Smith, 2007; Schohl & Fagotto, 2002). The discrepancy of the timing of Smad phosphorylation by Nodal and Bmp signaling with target gene activation suggests that these genes are initially weakly activated pre-MBT, perhaps independent of Smad function, but become further activated in response to these ligands at MBT. Alternatively, the antibody-based in situ methods (histochemical staining and fluorescence) previously used to demonstrate Smad phosphorylation were not sensitive enough to detect activation at earlier stages. Support for the latter notion has been provided by Skirkanich et al. (2011), who showed that Nodal-Smad2/3 signaling is active pre-MBT and that immunoprecipitation followed by western blotting can detect phospho-Smad2/3 at these earlier stages, consistent with the nascent RNA-seq results. These new data from 4-TU labeling suggest that further investigation is needed to determine the onset of Smad1/9 phosphorylation and Bmp signaling activity before MBT.

These observations provide strong evidence that while most zygotic transcription does occur at MBT, there is a smaller set of genes that are transcribed and functionally relevant before the major burst of ZGA. Pre-MBT transcription suggests that there must be something special about regulation of these genes, perhaps serving as a key regulatory node controlling the earliest zygotic gene regulation.

2.4 Models for regulation of ZGA timing: Repression and nuclear (DNA)-to-cytoplasmic ratio

Early evidence that the pre-MBT embryo is in a transcriptionally repressed state came from nuclear transplantation experiments. When nuclei from neurula stage embryos are transplanted into enucleated eggs, transcription off genomes from the incoming nuclei was rapidly silenced (Gurdon & Woodland, 1969). Expression of mRNA and tRNA was then reactivated in accordance with the normal timing of MBT in these embryos. These observations suggest that a maternally expressed repressor(s) maintains the genome in an inactive state, and transcriptional activation during MBT is likely to involve a developmentally programmed alleviation from negative regulation.

Newport and Kirschner (1982a) argued that because dissociated blastomeres begin transcription at the same time as intact sibling controls, the timing mechanism must be cell autonomous. Inhibitors of cytokinesis still permitted continued DNA replication on schedule, with the usual slowing at MBT, and concomitant with on-time transcription. Furthermore, by creating a constriction in the embryo during early cleavage stages they could asymmetrically partition nuclei into daughter cells, creating two half-sized embryos with different numbers of nuclei, and each with roughly half the cytoplasm. Half embryos containing more nuclei underwent ZGA earlier than normal, while those with fewer nuclei had delayed ZGA. From these observations, they concluded that ZGA is not likely controlled by counting rounds of DNA replication or through the use of a developmental clock/timer, but mostly likely utilized a measuring of nuclear-to-cytoplasmic (N:C) ratio. Consistent with this model, ZGA occurs prematurely in polyspermic embryos (Newport & Kirschner, 1982a), which also show premature hyperphosphorylation on the C-terminal domain of the large subunit of RNAPII (Palancade, Bellier, Almouzni, & Bensaude, 2001). Polyspermic embryos created from ~7 sperm showed initiation of ZGA one cell cycle earlier than normal, which is the number predicted if embryos used a mechanism measuring N:C ratio (Newport & Kirschner, 1982a). Thus, during normal embryogenesis the increase in N:C ratio achieved by midblastula stage overcomes the inhibition of ZGA present during cleavage and early blastula stages, when fewer nuclei are present in a larger cytoplasmic volume.

Sensing of N:C ratio might occur by stoichiometric titration of a maternally supplied repressor in the cytoplasm, and the increasing DNA content at each cell cycle was proposed as a titration mechanism (Newport & Kirschner, 1982b). Microinjection of 1 ng per embryo of plasmid containing a yeast leucine tRNA gene results in its immediate transcription that is rapidly silenced (within 1–2 h) (Newport & Kirschner, 1982b). However, when the embryos reached MBT, transcription from the dormant plasmid was reactivated alongside the endogenous genome. Injected DNA remains extrachromosomal and therefore control of the plasmid's reactivation is not occurring in *cis*, but instead *trans* acting factors must be involved in both early activation/repression. Derepression of transcription from this plasmid could be reversed by coinjecting 24 ng of competitor plasmid to titrate repressor. This corresponds to the amount of DNA equivalent to the number of nuclear genomes present after 12 embryonic cleavages. Since tRNA genes are transcribed by RNAPIII, a plasmid reporter containing an RNAPII-regulated gene was also examined. Use of the *Xenopus myc* (c-myc) promoter, and others dependent on RNAPII, confirmed the findings from the RNAPIII-driven leucine tRNA gene (Almouzni & Wolffe, 1995; Prioleau, Huet, Sentenac, & Méchali, 1994; and references therein). Taken together, these results suggest that transcription by both RNAPII and III is repressed during the pre-MBT period, and that a titratable repressor present in a fixed cytoplasmic volume is depleted by increasing numbers of genomes as embryos undergo rounds of DNA replication. More recent experiments employing precise measurements of nuclear volumes across early cleavage stages, in conjunction with overexpression of factors that alter nuclear volume, have supported the interpretation that embryos measure the N:C, and/or DNA-to-cytoplasmic (DNA:C) ratio (Jevtić & Levy, 2015, 2017). Further discussion of DNA:C ratio can be found in Section 3.6.

Due to the gradation of yolk distribution in mesolecithal *Xenopus* embryos, N:C ratios are not uniform throughout the embryo. Cell volumes differ along the animal-vegetal axis of the embryo and this raises questions about the role of N:C ratio in regulation of ZGA timing. As the embryo enters MBT there is ~50- to 100-fold difference in volume between the largest vegetal cells and the smallest animal cells (Fig. 1) (Chen, Einstein, et al., 2019; Newport & Kirschner, 1982a). To address this issue of ZGA regulation, *X. laevis* nascent RNAs were metabolically labeled with 5-EU to allow for biotinylation and imaging by confocal microscopy (Chen, Einstein, et al., 2019). This permitted direct bulk visualization of nascent transcripts at the single cell level in whole, unmanipulated, fixed embryos at various stages of early development. This approach revealed that animal-most cells begin transcription during the 12th cleavage cycle (~4000 cells), but not in more vegetal cells (Fig. 1). As development continues, cell labeling gradually moves toward the vegetal pole, with vegetal cells beginning transcription late in the 13th cycle (~8000 cells). By ~16,000 cell stage (log2 cell number = 14), ~80% of all cells were labeled with 5-EU. Low dose α-amanitin blocked most incorporation, supporting the notion that the signal is due to RNAPII dependent transcription. Analysis of cell sizes during cell cycle 13 showed that when diameters drop below ~40 μm, transcription dramatically increases, whereas cells more vegetally with sizes ~50 μm fail to incorporate 5-EU. Cells with an ~6-fold difference in volume showed 20-fold difference in 5-EU incorporation. This translates to an ~4.5 cell cycle difference from the onset of ZGA in the animal-most cells until vegetal-most cells decrease to a size permitting transcription. Interestingly, measurements of cell size at the corresponding onset of ZGA in vegetal cells (~14K+ cells) shows that they begin transcription at ~62 μm. The relative concentrations of histones between the animal and vegetal poles were also examined. If the quantity of histones was lower vegetally than animally, then their repressive effects (see Section 3.6) would be titrated at a lower DNA concentration and ZGA would commence when the cell volume was larger than animally. Quantitative western blots for H2b, H3 and H4 showed all three are present at higher levels (averaging ~1.6-fold) animally than vegetally during this period. A combination of careful measurements and computational modeling showed that achieving the appropriate threshold cell size fits the observed spatiotemporal pattern of 5-EU incorporation the best. These researchers argued that their data, which also involved repeating the Newport and Kirschner constriction experiment with 5-EU incorporation, and modeling, rules out multiple other scenarios including models based on a global timer, spatial position, a mitotic gradient, and a cell cycle counter. Since cleavage of cytoplasm into smaller volumes occurs without growth of the embryo, such a mechanism allows for a cell size dependent sensor.

Finally, none of these models provide an explanation for pre-MBT transcription occurring at the 32–128 cell stages, discussed in Section 2.3, nor the transcription of pri-mir427, also an RNAPII-dependent process, which begins as early as the 8-cell stage (Lund, Liu, Hartley, Sheets, & Dahlberg, 2009; Owens et al., 2016; N. Owens et al., unpublished observations). Therefore, a subset of genes are likely to be operating by different rules than those we have discussed so far.

3. Regulation of RNA polymerase II-mediated transcription

3.1 Regulation of RNA polymerase activity

In the large volumes of mature *X. laevis* oocytes, RNAPII is present at 10^4–10^5 the levels of normal somatic cells, and RNAPII from both oocytes and cleavage stage embryos possesses in vitro transcriptional activity on various promoter DNA templates (Roeder, 1974; Toyoda & Wolffe, 1992). Therefore, RNAPII is both present, and the bulk of the polymerase is in a transcriptionally capable state, before MBT. Is RNAPII positioned on promoters before ZGA and waiting for a trigger to begin transcription? A timecourse of ChIP-seq covering blastula stages 7, 8, 9 and early gastrula stage 10.5 found that RNAPII covers "gene bodies" (transcription units) starting around stage 8 (Fig. 2), with signal detected at stage 7 only on genes expressed pre-MBT (Charney, Forouzmand, et al., 2017; Charney, Paraiso, Blitz, & Cho, 2017; Gentsch, Owens, & Smith, 2019; our unpublished observations). A small number of genes show RNAPII engaged as early as stage 6 (Gentsch, Owens, & Smith, 2019). Thus far, no clear evidence has been found supporting the notion that RNAPII is pre-loaded on the promoter regions of genes at any stage examined. This observation appears to rule out models whereby either RNAPII is pre-bound at promoters in a poised state, or is paused downstream of cap sites.

Phosphorylation of the large subunit of RNAPII, Polr2a, occurs on serines 2 and 5 of its C-terminal domain (CTD) repeats and westerns using antibodies specific to the phospho-forms suggest very low levels before MBT, but abundant levels of both phosphoserine 2 and 5 appear at MBT (Palancade et al., 2001; Veenstra, Destrée, & Wolffe, 1999; Veenstra, Mathu, & Destrée, 1999; Toyoda & Wolffe, 1992). This suggests that while RNAPII is capable of functioning in transcription, it is not activated by CTD kinases to any large extent until ZGA has begun. However, a small fraction of total RNAPII is engaged in transcription before MBT and therefore might be missed in western blots. Overall, these data are consistent with observations described above that RNAPII is not present on most genes to any significant extent before ZGA (Charney, Forouzmand, et al., 2017; Charney, Paraiso, et al., 2017; Gentsch, Owens, & Smith, 2019). Therefore, it seems that fully functional RNA polymerases and their associated cofactors are available in advance of ZGA and that there is a failure to recruit polymerase pre-initiation complexes to gene promoters. Since there are a small number of genes that are expressed before MBT, these must rely on a different set of rules to overcome genome-wide repression of transcription.

3.2 Regulation of TF nuclear localization

Many proteins with nuclear functions that are present in the egg and cleavage stages are cytoplasmically localized and enter the nucleus in the late blastula stage (Dreyer, 1987). This raises the possibility that cytoplasmic anchoring and timed nuclear translocation of transcriptional machinery controls the onset of ZGA at MBT. The TFs Mier1 and Pou2f1 are cytoplasmically anchored until MBT (Luchman, Paterno, Kao, & Gillespie, 1991; Post, Luchman, Mercer, Paterno, & Gillespie, 2005; Veenstra, Mathu, et al., 1999), as is Cbtf122 (Ilf3), a subunit of maternal CCAAT-binding factor (Brewer, Guille, Fear, Partington, & Patient, 1995). Cbtf122 binds RNA in the cytoplasm, anchoring it until late blastula stage 9, and by early gastrula stage 10 it is released and translocates to the nucleus (Brzostowski et al., 2000). Xnf7, which may function in transcription, is cytoplasmically anchored and released upon phosphorylation (Li, Shou, Kloc, Reddy, & Etkin, 1994; Miller et al., 1991). In the case of Mybl2, its nuclear localization signal is actively masked, preventing transport across the nuclear pore complex (Humbert-Lan & Pieler, 1999). TF release from anchoring complexes is likely to be an important step in regulating the timing of target gene expression and therefore ZGA, but this hypothesis has yet to be formally tested.

Importantly, not all TFs are prevented from nuclear entry before the onset of ZGA. For example, all three *Xenopus* RNA polymerases are localized in the nucleus before MBT (Roeder, 1974). Ctnnb1 (β-catenin), Foxh1, Otx1, Sox3, and Vegt, have all been shown to be nuclear (either directly or have been demonstrated to be bound to enhancers) between the 16 and 64 cell stages (Charney, Forouzmand, et al., 2017; Charney, Paraiso, et al., 2017; Gentsch, Spruce, Owens, & Smith, 2019; Larabell et al., 1997; Paraiso et al., 2019; Stennard, Zorn, Ryan, Garrett, & Gurdon, 1999). These examples suggest that many TFs might have access to the nuclear genome shortly after fertilization (Fig. 2). Some TFs require signaling to break free of their cytoplasmic anchors and enter the nucleus and thus the signal controls this timing. Smad2 is C-terminally phosphorylated prior to ZGA in response to overexpressed Nodal5 and Activins, but not Nodal1, and immunoprecipitation followed by western blotting detected phospho-Smad2 prior to ZGA (Faure, Lee, Keller, ten Dijke, & Whitman, 2000; Skirkanich et al., 2011). While Smad2 can be efficiently activated before MBT upon overexpression of ligands, Smad1 fails to be phosphorylated until MBT, even when Bmps are overexpressed (Faure et al., 2000). This suggests that Smad1's cytoplasmic localization is regulated by a factor that only permits activation at or after ZGA. In summary, while clearly many TFs do access nuclear chromatin before ZGA, the timing of translocation of some may be a regulated step. For most, the role of such regulation of nuclear translocation in controlling the timing of transcriptional activation remains to be established. A time course using quantitative proteomics would be valuable to explore the temporal dynamics of TF protein expression levels, and proteomics on nuclei (Amin et al., 2014; Peshkin et al., 2015, 2019; Wühr et al., 2014, 2015) would be critical to providing a global view of nuclear localization of the entire TF repertoire in the buildup to ZGA.

3.3 Regulation of TF translation

RNA-seq shows that all of the machinery critical for transcription appears to be expressed in the oocyte, from subunits of the RNAPII, general transcription factors (Gtfs) involved in formation of the pre-initiation complex (TATA binding protein, Tbp; and Tbp-associated factors, TAFs), to enhancer binding TFs and recruited chromatin modifiers. A major point of control in the embryo for many of these is the regulation of their expression at the protein level. Many RNAs present in the oocyte are either not polyadenylated or contain relatively short polyA tracts, and thus they are not readily translated immediately following fertilization (reviewed in Woodland, 1982). Comparisons between RNA-seq data from polyadenylated (polyA+) mRNA and ribosomally depleted total RNA identified differentially adenylated RNAs (Collart et al., 2014), revealing that more than 2100 (~10%) genes have a sustained increase in polyA+ levels during early cleavage (e.g., *foxh1* mRNA in Fig. 2, top). Cytoplasmic polyadenylation provides a mechanism for controlled deployment of transcripts during a period when very few genomes are available for rapid transcription (Collart et al., 2014; Woodland, 1982). A developmental timer mechanism to control initiation of ZGA is suggested to rely on the gradual buildup of sufficient protein concentrations by increasing translation of these maternal mRNAs following cytoplasmic polyadenylation. Such a mechanism for regulating the timing of availability of transcriptional machinery is likely to be an important contributor to the timing of ZGA, but how such events would be tightly coordinated with DNA:C ratio measuring system are yet to be determined.

Detailed expression dynamics at the protein level for most of the transcriptional machinery is not known. All three RNA polymerases are present before ZGA, and show in vitro transcriptional activity across early development (Roeder, 1974). Western blot analysis using an antibody to the largest RNAPII subunit, Polr2a showed little signal before fertilization, but by 2-cell stage Polr2a was already detectable (Jallow, Jacobi, Weeks, Dawid, & Veenstra, 2004; Veenstra, Destrée, et al., 1999). Other "core" transcriptional components such as Gtf2b (aka TFIIB) and the Gtf2f1 (aka Rap74) subunit of TFIIF are expressed at uniform levels from oocytes and unfertilized eggs through all of early embryogenesis (Veenstra, Destrée, et al., 1999). While the above observations suggest that many components of the core machinery for transcription are present in the developing embryo before ZGA, a slightly different pattern is observed for TATA-box binding protein (Tbp). *tbp* mRNA undergoes cytoplasmic polyadenylation and protein is not detectable on western blotting in 2-cell stage embryos (Veenstra, Destrée, et al., 1999), through very low levels are detectable after enrichment by immunoprecipitation from 1000 eggs (Bell & Scheer, 1999). Tbp protein levels rise by the 64-cell stage, gradually increasing ~sevenfold (estimated here from Veenstra, Destrée, et al., 1999) by blastula stage 9, whereupon levels rise sharply to a peak between stage 9 and late gastrula stage 12. This translational increase led to the hypothesis that Tbp availability may be an important regulated step in controlling transcription at ZGA. Tbp knockdown embryos do initiate gastrulation but die before its completion (Veenstra, Weeks, & Wolffe, 2000). Interestingly, two other Tbp paralogs exist, Tbpl1 and Tbpl2 (aka Tbp2). Maternal polyA+ *tbpl1* mRNA is present at high levels in eggs. Tbpl1 knockdown embryos do not appear to progress past stage 8 (Veenstra et al., 2000). Maternal Tbpl2 protein levels are high in oocytes but taper off after fertilization and Tbpl2 has been shown to be critical for gastrulation (Jallow et al., 2004). Presumably this combination of TATA binding factors drives differential gene expression in the early embryo, but their interplay is complex as revealed by a triple knockdown (Gazdag, Jacobi, van Kruijsbergen, Weeks, & Veenstra, 2016; Jacobi et al., 2007). The roles of Tbps in regulating the timing of transcription at ZGA will require further investigation.

The protein level expression of several TFs that interact with enhancers has also been studied, albeit to limited extents. Maternal *foxh1* mRNA undergoes rapid cytoplasmic polyadenylation (Fig. 2) and by the 8-cell stage its protein level has already reached a peak that is maintained roughly constant until it declines during early gastrulation (Charney, Forouzmand, et al., 2017; Charney, Paraiso, et al., 2017). The maternal isoform of Vegt protein is found at low levels in unfertilized eggs and increases by stage 6 (32-cells) (Stennard et al., 1999). Similarly, maternal Vegt, Otx1 and Sox3 have also been shown to be detectable at the protein level using various assays by at least the 32- to 64-cell stage (Gentsch, Spruce, et al., 2019; Paraiso et al., 2019). Thus, some enhancer-binding TFs are available far in advance of ZGA. It is clear that more sensitive proteomic timecourse data (e.g., Peshkin et al., 2015) is needed to determine whether levels of key machinery might be rate limiting before ZGA.

3.4 Regulation of chromatin accessibility by histone variants

Accessibility of regulatory DNA sequences to TFs and polymerase might be an important step for regulation of transcription at ZGA. The egg is endowed with large maternal stores of both histone mRNAs and protein, and metabolic labeling studies reveal massive additional translation of histones during early cleavage stages (Adamson & Woodland, 1974; Woodland & Adamson, 1977). Histone variants have been shown to play a role in gene regulation in a variety of developmental contexts and in cancer (Loppin & Berger, 2020; Martire & Banaszynski, 2020). Several variant histones are maternally supplied and are exchanged with alternative histones as *Xenopus* embryonic development proceeds (Fig. 2). One example is histone H3.3, which is translated off maternal mRNA, and protein levels climb between 2- and 16-cell stage (Szenker, Lacoste, & Almouzni, 2012). H3.3 is a "replacement" histone, providing a continuous source of histone H3 throughout the cell cycle. In various systems H3.3 is found in transcriptionally active regions of genomes, and opposes the recruitment of histone H1 to promote a more open chromatin state (Jin & Felsenfeld, 2007; Loppin & Berger, 2020; Szenker et al., 2012). H3.3 is phosphorylated on Ser31, which leads to an up-regulation of Ep300 activity and H3K27 acetylation in both *Xenopus* and embryonic stem cells (Martire et al., 2019; Sitbon, Boyarchuk, Dingli, Loew, & Almouzni, 2020). Morpholino knockdown of H3.3 in *X. laevis* causes severe gastrulation defects, but does not completely abrogate ZGA (Sitbon et al., 2020; Szenker et al., 2012). While expression of some zygotically activated genes were inhibited (i.e., *fgf4, myf5, myod1, tbxt, wnt11*), many other genes were unaffected (i.e., *a2m, eomes, gdf3, mixer, nodal2, sox17*). Since a transcriptome-wide analysis was not performed, and especially shortly after onset of ZGA, it is difficult to assess the full extent of the impact of loss of H3.3 on transcription at MBT. A transcriptome-wide view at earlier timepoints is needed to better understand H3.3's impact on ZGA.

H1-8 (also known as B4, H1M, and H1foo) is a maternal histone present in cleavage stage embryos that replaces the somatic linker histone H1, which is absent from oocytes, sperm and eggs (see Dimitrov, Almouzni, Dasso, & Wolffe, 1993; Ura, Nightingale, & Wolffe, 1996; and references cited therein). H1-8 is 30% identical to H1 at the amino acid level and has lower affinity for chromatin that H1 (Ura et al., 1996). It is believed that H1-8 contributes to a generally less compact chromatin state during cleavage and blastula stages and this likely permits more ready access of TFs to the chromatin. Though H1 protein is absent maternally, its RNA is present in eggs. Translation during cleavage and blastula leads to a ~1:1 molar ratio between H1-8 and H1 by MBT, and H1-8 declines to less than 5% of total linker histone by the beginning of neurulation. Molecular evidence is needed to determine whether loss of function would impact expression of all genes at ZGA. A better understanding of the genomic distribution of H1-8 is also needed.

Lastly, the genome in mature *Xenopus* sperm is packaged in a specialized chromatin that contains protamines, but also retains histones. Entry into the egg leads to rapidly remodeling of the incoming sperm chromatin to initiate early embryogenesis by exchanging the protamines with its vast pool of nucleosomes. Recent evidence suggests that sperm chromatin not only contains classical nucleosome octamers but, its genome also contains nonclassical "variant" nucleosomes—subnucleosomal particles that deviate from the octameric stoichiometry of standard nucleosomes (Oikawa et al., 2020). The potential roles of nucleosomes and subnucleosomes on the sperm genome is discussed in detail in the next section.

Histone variants, by creating an open chromatin environment, likely function in early embryogenesis to maintain a permissive state for transcription, but more work is needed in this area to determine how they impact expression of individual genes at ZGA.

3.5 Regulation at the level of epigenetic chromatin modifications

3.5.1 DNA methylation studies

Epigenetic control of gene expression has been intensively studied in many biological systems. Two modifications to the epigenome that influence gene expression are DNA cytosine methylation (5-methylcytosine) and posttranslational modification of histone tails (histone marks). DNA methylation is known to influence gene transcription by recruiting repressive chromatin modifications to methylated regions (see citations in Bogdanovic et al., 2011). Recently it has also been shown that methylation impacts many TF-DNA interactions by reducing TF binding affinities when 5-methylcytosine residues are present in their sequence motifs (Yin et al., 2017). Regulatory regions of genes containing nucleosomes with certain histone marks are either more or less accessible to TFs. And likewise, TFs can influence histone mark deposition and nucleosome behavior: some TFs recruit the enzymatic "writers" of these epigenomic marks. A number of studies have examined the behavior of DNA methylation and histone marking during early *Xenopus* development.

In *Xenopus* early embryos, unlike mammals, widespread demethylation of cytosines is not observed and the genome shows sustained levels of 5-methylcytosine content throughout early embryogenesis (Veenstra & Wolffe, 2001). Genome-wide analysis of the DNA methylome in *X. tropicalis* showed hypermethylation on promoters and transcription units, but transcription start sites (TSSs) containing the active H3K4me3 mark are cytosine hypomethylated (Bogdanovic et al., 2011). However, no correlation was found to support a role for repression of transcription by DNA methylation during blastula and gastrula stages. The repressive H3K27me3 histone mark is found on regions of the genome lacking DNA methylation, and is minimally present until stage 9, after MBT onset, but gradually strengthens by late gastrula stage 12 (Fig. 2) (Akkers et al., 2009; Gupta, Wills, Ucar, & Baker, 2014; van Heeringen et al., 2014). Acquisition of H3K27me3 marks in zebrafish similarly begins after the onset of ZGA (Vastenhouw et al., 2010). As in embryonic stem cells, *Xenopus* promoters were identified that appeared to be bivalently marked with both H3K4me3 (active) and H3K27me3 (repressive) modifications, but sequential ChIP followed by qPCR (ChIP-qPCR) showed that these marks are not present on DNA from the same cells, and thus the most parsimonious explanation is that they occur in different cell populations (Akkers et al., 2009).

3.5.2 Histone modifications studies

Spatial analysis examining dissected tissue fragments of early gastrula stage embryos showed that, while H3K4me3 is present in both animal and vegetal hemispheres, H3K27me3 is enriched animally on genes preferentially expressed in the vegetal pole (e.g., H3K27me3 was found preferentially decorating *vegt* in animal pole cells) (Akkers et al., 2009). This suggests that

H3K27me3 deposition occurs in the animal pole to repress vegetal gene expression. H3K27me3 is deposited largely by polycomb repressive complex 2 (PRC2), which is composed of various protein subunits including Ezh2 and Jarid2. ChIP-seq on Ezh2 and Jarid2 showed PRC2 binding to the genome is detectable at blastula stage 9 (van Heeringen et al., 2014). Since PRC2 deposition of H3K27me3 occurs in DNA methylation-free regions of the genome, DNA methylation may function to establish genomic regions for future H3K27me3 marking, but only if genes within these regions are not first transcriptionally activated. This mechanism can ensure specific genomic regions to be regionally marked for repression.

Temporally, the active enhancer marks H3K4me1 and H3K27ac appear around the onset of ZGA (Fig. 2) (Gupta et al., 2014; van Heeringen et al., 2014), with H3K27ac signal being relatively weaker. Furthermore, the major histone acetyltransferase responsible for deposition of the H3K27ac mark, Ep300, is not bound to enhancers until MBT (Hontelez et al., 2015). Ep300 lacks a direct DNA binding domain and therefore is recruited to enhancers by binding to TFs occupying these elements. Ep300 has also been suggested to be a rate limiting factor driving ZGA in zebrafish (Chan et al., 2019). The timing of recruitment of writers for the H3K4me1 mark, Kmt2c (Mll3) and Kmt2d (Mll4), has not been reported. Active enhancers usually contain flanking nucleosomes bearing both H3K4me1 and H3K27ac marks, and are said to be in a "primed" state before they are active, when they contain H3K4me1 but lack H3K27ac (reviewed in Calo & Wysocka, 2013). However, the timing of deposition of these marks in *Xenopus* suggests that cis regions regulating gene expression at ZGA may not require these activating marks in order to initiate enhancer influence on transcription, but acquire these marks shortly after ZGA commences (see also Section 3.6).

While epigenetic histone marks associated with active enhancers or polycomb repression are not found before ZGA, a large fraction of epigenetic modifications is under direct maternal control. In the presence of α-amanitin, an inhibitor of RNAPII elongation, only a minority, 15%, of the regions bound by Ep300 are capable of recruiting this factor and hence are under maternal control, while 85% of Ep300 recruitment appears to require zygotic TF expression (Hontelez et al., 2015). This is consistent with the observation that H3K27ac signal, which is deposited by Ep300, is weak in late blastulae and only increases during early gastrula and beyond (Gupta et al., 2014; van Heeringen et al., 2014). Unlike Ep300 recruitment, a higher percentage of H3K4me1 deposition on enhancers is under maternal control, with 48%, 36%, and 30% of H3K4me1 in dissected ectoderm, mesoderm and endoderm, respectively, being α-amanitin resistant (assayed at stage 10.5) (Paraiso et al., manuscript in prep). Both 90% of polycomb repressive mark H3K27me3 and 85% of active promoter mark H3K4me3 are still present even in the presence of α-amanitin (Hontelez et al., 2015), suggesting that most regions of deposition of these marks are under maternal control.

H3K4me3 is a mark of active promoters that has been shown to recruit the TAF3 component of TFIID (Lauberth et al., 2013; Vermeulen et al., 2007). Both H3K4me3 and H3K9ac are found on promoters before ZGA, appearing by at least stage 6.5–7 (Fig. 2) (Akkers et al., 2009; Blythe et al., 2010; Hontelez et al., 2015). However, these active promoter marks do not correlate well with transcriptional activity at ZGA (Hontelez et al., 2015). Maternally regulated H3K4me3 is associated with promoters in regions of DNA hypomethylation, whereas promoter regions with high DNA methylation were dependent on zygotic transcription for H3K4me3 deposition. Several pre-MBT expressed dorsal Wnt target genes are decorated with H3K4me3, and deposition of this mark is dependent on Ctnnb1 and RNAPII recruitment (Blythe et al., 2010). The promoters of these genes are also decorated with H3K9ac and H3K14ac during the pre-MBT stages, and H3K9ac was significantly lower on *sia1* and *nodal3.1* by late blastula, suggesting that H3K9 acetylation is dynamic and coincides with the timing of competence for Wnt signaling during the pre-MBT period (Blythe et al., 2010; Esmaeili et al., 2020). Consistent with this notion, inhibition of histone deacetylases (HDACs) with trichostatin A beginning at stage 6.5 extended the period of H3K9 acetylation into late blastula stage 9, and also Wnt responsiveness (Esmaeili et al., 2020). HDAC activity is therefore involved in "decertifying" these regulatory regions during blastula stages. Ctnnb1 is also required to recruit the arginine methyltransferase Prmt2, which is responsible for deposition of H3R8me2 on Wnt target gene promoters (Blythe et al., 2010). These observations suggest that maternally expressed Ctnnb1 is required for recruitment of histone marks in the early embryo.

Super enhancers (SEs) are a subset of enhancers that are clustered, contain high densities of active enhancer marks, and are associated with transcriptionally active developmental "master" genes (Hnisz et al., 2013; Parker et al., 2013; Whyte et al., 2013). Using various metrics, SEs have been identified in *X. tropicalis* early embryogenesis (Hontelez et al., 2015; Paraiso et al., 2019; Gentsch, Spruce, et al., 2019; Paraiso et al., manuscript in prep). SEs acquire dense enhancer histone marking after the onset of ZGA and on the same temporal schedule as "regular" enhancers (those outside of enhancer clusters). Ep300 recruitment to SEs (identified at stage 11) in the presence of α-amanitin suggests that <35% of are under maternal control, while ~50% of SEs require zygotic transcription for Ep300 interactions (Hontelez et al., 2015). SEs also acquire germ layer specific H3K4me1 by early gastrula stage (Paraiso et al., 2019; Paraiso et al., manuscript in prep). Endoderm-specific SE marking is sequentially dependent on both maternal and zygotic TFs. Binding of maternal endodermal TFs Otx1, Vegt and ubiquitously expressed Foxh1 TF is enriched in a population of 441 identified endodermal SEs (Paraiso et al., 2019), and knockdown of these TFs results in a loss of H3K4me1 marks on the overwhelming majority of the endodermal SEs (Paraiso et al., 2019). These findings reinforce the notion that maternal TFs playing an essential role in establishment of epigenetic marks shortly after ZGA.

3.5.3 DNA accessibility and looping studies

Assay for Transposase Accessible Chromatin using sequencing (ATAC-seq) and DNAse-seq have been used to assess open chromatin in early embryos (Bright et al., 2021; Esmaeili et al., 2020; Gentsch, Spruce, et al., 2019; Jansen et al., 2020). ATAC-seq on *X. laevis* gastrula stage 10 animal caps (ACs) revealed approximately 70,000 regions of open chromatin with ~8% containing TSSs and ~16% located in intergenic regions (Esmaeili et al., 2020). DNAse hypersensitive regions identified by DNAse-seq found numerous open chromatin regions that frequently correlated with RNAPII binding and H3K4me1 marking on putative cis-regulatory DNA (Gentsch, Spruce, et al., 2019). ATAC-seq peaks mapped in *X. tropicalis* embryos across multiple stages showed that those exhibiting dynamic changes between stages 9–16 were found to correlate well with Ep300-associated enhancers (Bright et al., 2021). Increasing accessibility was also correlated with increasing expression of genes associated with ATAC-seq peaks. ATAC-seq performed on stage 10.5 *X. tropicalis* animal cap ectodermal cells (which have declining pluripotency by this stage) shows open chromatin on ectodermally expressed genes including *grhl3* and *tfap2a*, as well as Spemann organizer (dorsal marginal zone) genes including *chrd* and *gsc* (Bright et al., 2021). This is in stark contrast to chromatin in the more lineage restricted dorsal marginal zone cells, which shows open regions on organizer but not ectodermal genes, and therefore displays better correlation between open chromatin and active gene expression. Therefore, chromatin opening alone is not a good predictor of enhancer and promoter activities. Chromatin may be in a more open state across the embryo at earlier stages. It is noteworthy that a timecourse of ATAC-seq spanning the 64-cell to dome stages fails to identify regions of hypersensitivity until ~1000 cell stage (ZGA in zebrafish), though genes expressed before ZGA do show accessibility as early as the 64-cell stage (Liu, Wang, Hu, Wang, & Zhang, 2018). The timing of chromatin opening relative to gene activation needs to be determined in *Xenopus* to better understand the role of DNA accessibility in transcriptional regulation at ZGA.

Long range looping interactions create topologically associated domains (TADs) in chromatin, which compartmentalize genomes and can restrain enhancers from inappropriate interactions with genes/promoters located outside of TADs (Yu & Ren, 2017). To examine the dynamics of TAD formation spanning ZGA, Hi-C was performed during early *X. tropicalis* embryogenesis (Niu et al., 2020). No evidence was found for the presence of TADs before ZGA (stage 8) and TADs arise in two "waves," an initial wave at ZGA and a second at gastrula stage 11. A stable 1300 TAD boundaries were detectable at both stages 9 and 10, which increased to 2662 at stage 11 and these were then maintained throughout the later stages analyzed. Morpholino knockdowns of Ctcf and Rad21 (a component of the Cohesin complex), proteins that complex at the base of TAD loops (Yu & Ren, 2017), showed that single knockdowns of Ctcf or Rad21 weakened TAD boundaries, while knockdown of both abolished TAD formation. Since blocking RNAPII elongation with α-amanitin in other systems had no effect on TADs, but transcription was shown to be necessary for TAD formation in human embryos (Chen, Ke, et al., 2019), Polr2a KD using a morpholino was examined. Loss of Polr2a weakened TAD structures at ZGA, but not at later developmental stages. It is currently unclear whether the requirement for Polr2a is due to a direct role of RNAPII complexes in TAD formation at ZGA, independent of its role in transcription, or whether transcription is the determining factor. The appearance of TADs at ZGA in *Xenopus*, with an increase in TAD numbers as development proceeds, generally aligns with observations in other model organisms (Niu et al., 2020). However, observations in zebrafish suggest a more complex picture, where TADs are reported to be present pre-ZGA, then lost around ZGA, and then re-established during gastrulation (Kaaij, van der Weide, Ketting, & de Wit, 2018).

3.5.4 Nucleosome studies

Do epigenetic marks pass through the germline to influence ZGA? DNA methylation levels do not appear to change significantly from oocyte through early development (Veenstra & Wolffe, 2001). Therefore, regions marked by DNA methylation in the oocyte might be passed to the embryo, however this needs further investigation. In the male germline, higher DNA methylation was found on mature sperm TSSs compared to spermatid TSSs, revealing a transition in DNA methylation on TSSs during spermiogenesis (Teperek et al., 2016). Nucleosomes may also contain epigenetic information conveyed from sperm to the embryo. In *X. laevis*, ~60–70% of histones H2a and H2b are lost from sperm chromatin, but histones H3 and H4 are retained (Oikawa et al., 2020). Furthermore, analysis of nucleosomal positioning, as revealed by MNase-seq, suggests higher nucleosome occupancy on sperm TSSs than spermatids (Teperek et al., 2016). Interestingly, MNase not only liberates the expected ~150 bp DNA fragments typical of histone octamers, but also ~70 and ~110 bp subnucleosome-sized DNA fragments (Oikawa et al., 2020). Quantitative protein mass spectrometry comparisons suggested that the ~70 and ~110 bp chromatin fragments represent protamine-associated subnucleosomal particles consisting of $(H3/H4)_2$ tetramers and $(H3/H4)_2(H2A/H2B)$ hexamers, respectively. Notably, ~0.4% and 6% of the sperm genome was homogenously (i.e., across the sperm population) marked by H3K4me3 or H3K27me3, respectively, suggesting that octameric nucleosomes or subnucleosomes specifically positioned on genes might, upon fertilization, carry epigenetic marks into the egg. High H3K4me3 was found on TSSs while high H3K27me3 was found marking a ~4 kb wide region centered on TSSs. Homogeneous H3 trimethylation and histone particle composition were found to occur on different sets of genes, with homogeneous H3 trimethylation largely occurring on octameric nucleosomes. Several such clusters of genes have developmental functions and also appear to be bivalently marked in all sperm. To examine the role of these marks in sperm, embryos were produced from eggs overexpressing either human histone demethylases KDM5B, which acts on H3K4me2/3, or KDM6B, which demethylates H3K27me3, and RNA-seq was performed at early to midgastrula. Demethylation of H3K27me3 at fertilization resulted in misregulation of gene expression in the resulting embryos that correlated with sites of homogeneous trimethylation in the sperm genome (Oikawa et al., 2020; Teperek et al., 2016). It remains unclear how these sperm histone marks convey instructions that are then transmitted through cleavage stages to ZGA, but these observations suggest that chromatin in the gametes may contain epigenetic information that is utilized in early gene activation.

3.6 Regulation of enhanceosome formation

Substantial evidence suggests that nucleosomal wrapping of the genome competes with TF binding to promoters and enhancers (e.g., see references cited in Almouzni & Wolffe, 1995; Prioleau et al., 1994), and therefore nucleosomes may repress formation of functional enhanceosomes to regulate ZGA. Histones have been implicated as a major inhibitor of ZGA and therefore nucleosomes may be the sought after repressor being titrated (Prioleau et al., 1994). Typical somatic cells contain an ~1:1 mass ratio of histones to DNA, but a combination of maternally stored histones and massive histone translation during early cleavage produces an excess capable of supporting ~30,000 nuclei (in *Xenopus laevis*) (Adamson & Woodland, 1974; Woodland & Adamson, 1977). Coinjection with competitor DNA titrates these nucleosomes, thereby permitting exogenously added reporter genes to be transcribed once the permissive DNA:nucleosome ratio is reached. Interestingly, coinjection of plasmids (containing either the *myc* or cytomegalovirus promoters) together with Tbp protein induced pre-MBT transcription from these templates (Almouzni & Wolffe, 1995; Prioleau et al., 1994). This suggested that Tbp's access to the promoter is limited by the level of endogenous nucleosomes. In the case of the cytomegalovirus promoter, coinjection of plasmid and Tbp together with nonspecific DNA was necessary to observe this effect (Almouzni & Wolffe, 1995), while the use of the *myc* promoter only required coinjected Tbp (Prioleau et al., 1994). Differences in the amounts of these injected plasmids likely accounts for some differences in experimental outcomes (Prioleau, Buckle, & Méchali, 1995). Coinjection of the cytomegalovirus promoter, together with Tbp, nonspecific DNA, and all four core histone proteins, reversed the effects of the nonspecific DNA (Almouzni & Wolffe, 1995). The mass amount of core histones required was equivalent to that of the nonspecific DNA, and thus these experiments supported the notion that nucleosomes are a driver of transcriptional repression before MBT. *Xenopus* egg extracts were used to identify a transcriptional repressor responsible for the nonspecific DNA titration effects observed in vivo (Amodeo, Jukam, Straight, & Skotheim, 2015). Using in vitro transcription with a RNAPIII transcriptional output, histones H3 and H4 were biochemically purified and identified on the basis of their major repressive activities. Further in vivo support was obtained by MO knockdown of H3 expression to ~50%, which shifted transcription one cell cycle earlier.

Experiments described thus far suggest that high histone concentrations in early embryos produces a competition between nucleosomes and TFs for binding DNA motifs, thereby preventing TF interactions with promoters and enhancers, resulting in transcriptional repression. Repression is then alleviated by increasing numbers of genomes that titrate nucleosomes by MBT leading to ZGA (with the caveat that a small minority of genes are able to escape the repressive environment and are transcribed pre-MBT). This implies that a competition between nucleosomes and TFs for regulatory elements on ZGA genes favors nucleosomal occupancy until MBT, whereas TFs "win" this competition to activate transcription before MBT on a small set of genes. Recently studies in the early zebrafish embryo have corroborated this view from *Xenopus* (Joseph et al., 2017; Miao et al., 2020; Pálfy, Schulze, Valen, & Vastenhouw, 2020; Reisser et al., 2018; Veil, Yampolsky, Grüning, & Onichtchouk, 2019; Zhang et al., 2014).

Pioneer factors (Zaret, 1999; Zaret & Carroll, 2011) are TFs that are the first to bind to enhancers in chromatin and either actively open chromatin locally, or passively bind until other ("settler") TF interactions (perhaps with cooperative binding) lead to the assembly of functional enhanceosomes. Do pioneer TFs play a direct role in the timing of ZGA? Among the 1250 genes encoding TFs in *X. tropicalis* (Blitz et al., 2017) several hundred are expressed at levels with TPMs (transcripts per million) of >10 in the early embryo (Blitz et al., 2017; Owens et al., 2016; Paraiso et al., 2019). However, only a small fraction of all these TFs acting during early development has been studied in detail. The maternally expressed forkhead TF, Foxh1, is a major cofactor mediating Nodal-Smad2/3 signaling. Foxh1 binds to the genome dynamically across early stages of *X. tropicalis* embryogenesis (Charney, Forouzmand, et al., 2017; Charney, Paraiso, et al., 2017; Chiu et al., 2014) and appears to provide a paradigm for maternal TFs that act as the earliest pioneer factors. ChIP-seq analysis at multiple closely spaced developmental stages, pre-ZGA blastula stage 8, post-ZGA blastula stage 9 and early gastrula stage 10.5 revealed a total of ~41,000 bound regions (Charney, Forouzmand, et al., 2017; Charney, Paraiso, et al., 2017). The two blastula stages had a large overlap, with ~29–30,000 regions each at the two blastula stages (~55–60%), but only ~1300 sites were identified at early gastrula. This drop off is due to a significant decline in expression of Foxh1 by this stage (Charney, Forouzmand, et al., 2017). 954 regions, associated with 611 genes, were identified as being bound across all 3 stages, which span only ~3.5 h of developmental time. These 954 sites correlate better with the co-binding of Ep300 and a bimodal H3K4me1 distribution than the entire set of sites. Furthermore, Ep300-bound regions identified genome-wide were enriched for Foxh1 and Smad2/3 motifs (Hontelez et al., 2015), suggesting that these factors are abundant among all Ep300-marked enhancers. Furthermore, ChIP-qPCR performed at the 32-cell stage (stage 6, ~1.5–2h before ZGA in *X. tropicalis*) showed that Foxh1 is already bound to its sites several cell divisions before target gene activation. Since these sites were Foxh1 bound before the appearance of enhancer marks (and before the engagement of RNAPII on these genes), this demonstrates that maternal TF pioneering activity occurs before enhancer "activation" by chromatin modifications (Fig. 2). Mining persistent Foxh1-bound regions for enrichment of TF motifs implicated a number of maternal and zygotic TFs such as Sox3 and Pou5f3 that collaborate with Foxh1 to regulate expression of these target genes (Charney, Forouzmand, et al., 2017; Charney, Paraiso, et al., 2017; Chiu et al., 2014; Gentsch, Spruce, et al., 2019; Paraiso et al., 2019).

Other maternal TFs have shown a generally similar pattern of behavior. Otx1 is a vegetally expressed maternal TF that positively co-regulates endodermal genes together with the well-known endodermal TF Vegt, while Otx1 and Vegt interact in repression of mesodermal gene expression (Paraiso et al., 2019). Sequential ChIP followed by qPCR demonstrated that these TFs are indeed co-bound to cis elements. Genome-wide binding patterns elucidated by ChIP-seq at stage 8 identified ~5000 and ~22,000 Otx1- and Vegt-bound regions with 64% and 18% of these representing regions of likely co-binding, respectively. As with Foxh1, Otx1 and Vegt also co-bind to regulatory regions as early as the 32-cell stage, and >70% of Otx1 and >25% of Vegt binding overlap with regions of Foxh1 binding at stage 8. Morpholino knockdown of Foxh1 reduced binding of Otx1 and Vegt to these sites suggesting a cooperative binding interaction between all three ("OVF") TFs to form enhanceosomes involved in endodermal gene activation at ZGA. Consistent with the Foxh1 analysis, Otx1 and Vegt bound regions also contain enrichment for Sox and Pou motifs that may implicate coregulation by Sox3 and Pou5f3. These findings with OVF TFs are consistent with the model in which combinations of maternal TFs bind motifs before ZGA to pre-select specific DNA regulatory regions for gene activation and also to regulate the epigenetic landscape of the embryonic genome (Charney, Forouzmand, et al., 2017; Paraiso et al., 2019).

One characteristic of enhancers is RNAPII association. Approximately 27,000 intergenic regions were identified based on RNAPII binding with

~650 regions bound as early as the 32-cell stage, increasing to more than 10,000 by ~1000 cell stage (Gentsch, Spruce, et al., 2019). Mining these regions for enrichment of TF binding motifs, presumably reflecting pre-MBT TF recruitment, implicated Fox, Pou and Sox family TFs, among others. MO knockdown of Pou5f3 expression had only modest effects on select genes, but RNA-seq revealed that double knockdowns of Pou5f3 and Sox3 affected up to 25% of all zygotic genes expressed at early gastrula (Chiu et al., 2014; Gentsch, Spruce, et al., 2019). Interesting, Pou5f3/Sox3 double knockdown reduced chromatin accessibility to DNAse I (assayed by DNAse-seq) on ~41% of 16,637 putative regulatory elements, concomitant with loss of bimodal H3K4me1 marking and binding by RNAPII, Smad2 and Ctnnb1, supporting the notion that Pou5f3/Sox3 act as pioneer TFs to open chromatin (Gentsch, Spruce, et al., 2019). These observations of widespread effects of Pou5f3 on gene expression are also consistent with recent findings in zebrafish, where Pou5f3, SoxB1-type and Nanog TFs have been suggested to regulate numerous genes and the ZGA timing of their expression (Gao et al., 2020; Lee et al., 2013; Leichsenring, Maes, Mössner, Driever, & Onichtchouk, 2013; Miao et al., 2020; Pálfy et al., 2020; Veil et al., 2019). Despite these findings, there is as yet no evidence that Pou5f3 and/or Sox3 specifically control the timing of ZGA in *Xenopus* (all anuran genomes thus far sequenced lack a *nanog* gene; Blitz et al., 2017; Hellsten et al., 2010; Session et al., 2016; our unpublished observations).

While many maternal TFs interact with enhancers during early cleavage stages, most of the associated genes are not expressed until MBT. Targets of transcriptional activation before MBT are regulated by the maternal Wnt signaling cascade suggesting that Ctnnb1 activation might be a trigger for pre-MBT gene expression (Blythe et al., 2010; Skirkanich et al., 2011; Yang et al., 2002). Since Ctnnb1 lacks a DNA binding domain, its recruitment to target sites occurs through direct association with Tcf/Lef TFs. Ctnnb1 also interacts with members of the Sox family of TFs, including Sox17 (Mukherjee et al., 2020; Zorn et al., 1999). Ctnnb1 genome-wide binding dynamics between stages 7 and early gastrula stage 10.5 yields some surprising behavior in *X. laevis* and *X. tropicalis* (Afouda et al., 2020; Kjolby & Harland, 2017; Mukherjee et al., 2020; Nakamura, de Paiva, Veenstra, & Hoppler, 2016). Since maternal Wnt signaling, acting through Ctnnb1, controls dorsal gene expression both before and at MBT, whereas zygotic Wnt signaling controls ventral gene expression shortly thereafter, Ctnnb1 was expected to shift its association with target genes that function in these two distinct dorsal-ventral specification events. ChIP-seq spanning this period compared with Wnt-regulated transcriptomic data revealed that Ctnnb1 interacts with both dorsal pre-MBT/MBT targets and ventral post-MBT targets early, but the pre-MBT binding is lost by gastrula stage. The dorsal Wnt/Ctnnb1 targets activated at MBT and post-MBT ventral targets retained Ctnnb1 binding into early gastrula. Regulation of these different classes of genes appears to require not only Ctnnb1 binding, but also coordinate activity of other associated TFs (Afouda et al., 2020; Kjolby & Harland, 2017; Mukherjee et al., 2020; Nakamura et al., 2016). Both the pre-MBT and MBT dorsal gene targets of Ctnnb1 are associated with Foxh1 and are co-regulated by Nodal signaling, whereas the later, post-MBT ventral targets are co-regulated by Bmp and Fgf pathways. Therefore, how the small number of targets expressed before MBT escape pre-MBT repression remains unclear but Ctnnb1 does not appear to be the decisive factor. While the pre-MBT transcribed genes *nodal5* and *nodal6* are targets of both maternal Ctnnb1 and Vegt (Takahashi et al., 2006), Vegt also does not appear to be a specific trigger for pre-MBT transcription. This is exemplified by *X. laevis sox17a*, which is a Vegt target (Howard, Rex, Clements, & Woodland, 2007), but is not transcribed until MBT. Interesting, recent findings also show that Ctnnb1 can direct endodermal gene transcription at MBT in the vegetal pole independent of Tcf/Lef association (Mukherjee et al., 2020; Zorn et al., 1999). Therefore, the enhancer-TF logic of pre-MBT expressed genes that allows these to break free of the repressive states of the early embryo remains unknown and requires further investigation.

Since enhancers are platforms for combinatorial docking of TFs to produce functional enhanceosomes, understanding how different enhancers integrate sets of TF inputs to control transcriptional timing (and spatial expression) is expected to provide insights into the timing of ZGA. Multifactorial binding is one marker of functional enhancer elements, in addition to various epigenetic marks, RNAPII binding and production of eRNAs. CRMs identified in multiple studies have shown that genomic regions binding maternal factors during late blastula stage, are later bound by zygotic transcription factors during gastrulation (Charney, Forouzmand, et al., 2017; Charney, Paraiso, et al., 2017; Gentsch, Spruce, et al., 2019; Paraiso et al., 2019). This property of CRMs is consistent with the view that maternal pioneer TFs establish the first interactions with enhancers and gene activation. Pioneer TF binding may simply maintain a local open chromatin landscape that is permissive (Zaret, 1999; Zaret & Carroll, 2011) for subsequent zygotic TF binding. The notion that only a small set of TFs have pioneering activity is brought into question by recent high throughput analyses showing that many TFs are capable of accessing binding sites not only on linker DNA between nucleosomes, but also when their sites are positioned on the surface of nucleosomes (Fernandez Garcia et al., 2019; Iwafuchi et al., 2020; Michael et al., 2020; Soufi et al., 2015; Zhu et al., 2018). Regardless, how enhanceosomes interact with combinations of TFs to comply with a hardwired regulatory logic that determines genes temporal and spatial transcription at, or before, MBT remains an area of great interest.

4. Conclusion

In this review, we examined various mechanisms that regulate *Xenopus* ZGA. The rich history of *Xenopus* research combined with modern genomics have significantly increased the power of gene expression analysis controlling ZGA. Based on the current data, it is clear that control of ZGA is not a one step process, but is a collection of several distinct events regulating transcriptional activation. Future challenges will be to test some of the specific models proposed here. More explicit testing of the nucleosome competition model is needed. Do maternal pioneering TFs that bind the genome during cleavage stages displace nucleosomes from enhancers by, for example, recruiting SWI/SNF complexes that have nucleosome "sliding" activities? Investigation of the 3D architecture of chromatin as embryos develop is needed, including visualizing the changes in chromatin dynamics in vivo using high resolution imaging in real time. When do enhanceosomes recruit mediator and cohesin complexes to facilitate enhancer-promoter looping and is this a regulated step that is critical for polymerase recruitment to pre-initiation complexes on promoters? Since SEs are enriched within nuclear condensates in cell culture systems, do similar condensates appear before transcription at ZGA begins, or contemporaneous with the start of transcription, or after ZGA? Is there a role for genomewide repression by transcriptional corepressors? We believe that the *Xenopus* system will lead discovery in this area and contribute to basic principles not only regulating ZGA, but also in controlling transcription in all cell types.

Acknowledgments

The authors were supported by grants from NIH (R01GM126395, R35GM139617) and NSF (1755214). The authors apologize to all colleagues whose important studies were not cited due to space restrictions.

References

Adamson, E. D., & Woodland, H. R. (1974). Histone synthesis in early amphibian development: Histone and DNA syntheses are not co-ordinated. *Journal of Molecular Biology*, *88*, 263–285.

Afouda, B. A., Nakamura, Y., Shaw, S., Charney, R. M., Paraiso, K. D., Blitz, I. L., et al. (2020). Foxh1/nodal defines context-specific direct maternal Wnt/β-catenin target gene regulation in early development. *iScience*, *23*, 101314.

Akkers, R. C., van Heeringen, S. J., Jacobi, U. G., Janssen-Megens, E. M., Françoijs, K. J., Stunnenberg, H. G., et al. (2009). A hierarchy of H3K4me3 and H3K27me3 acquisition in spatial gene regulation in *Xenopus* embryos. *Developmental Cell*, *17*, 425–434.

Almouzni, G., & Wolffe, A. P. (1995). Constraints on transcriptional activator function contribute to transcriptional quiescence during early Xenopus embryogenesis. *The EMBO Journal*, *14*, 1752–1765.

Amin, N. M., Greco, T. M., Kuchenbrod, L. M., Rigney, M. M., Chung, M. I., Wallingford, J. B., et al. (2014). Proteomic profiling of cardiac tissue by isolation of nuclei tagged in specific cell types (INTACT). *Development*, *141*, 962–973.

Amodeo, A. A., Jukam, D., Straight, A. F., & Skotheim, J. M. (2015). Histone titration against the genome sets the DNA-to-cytoplasm threshold for the *Xenopus* midblastula transition. *Proceedings of the National Academy of Sciences of the United States of America*, *112*, E1086–E1095.

Bacharova, R., & Davidson, E. H. (1966). Nuclear activation at the onset of amphibian gastrulation. *The Journal of Experimental Zoology*, *163*, 285–296.

Bachvarova, R., Davidson, E. H., Allfrey, V. G., & Mirsky, A. E. (1966). Activation of RNA synthesis associated with gastrulation. *Proceedings of the National Academy of Sciences of the United States of America*, *55*, 358–365.

Bell, P., & Scheer, U. (1999). Developmental changes in RNA polymerase I and TATA box-binding protein during early Xenopus embryogenesis. *Experimental Cell Research*, *248*, 122–135.

Blitz, I. L., Paraiso, K. D., Patrushev, I., Chiu, W. T. Y., Cho, K. W. Y., & Gilchrist, M. J. (2017). A catalog of *Xenopus tropicalis* transcription factors and their regional expression in the early gastrula stage embryo. *Developmental Biology*, *426*, 409–417.

Blythe, S. A., Cha, S. W., Tadjuidje, E., Heasman, J., & Klein, P. S. (2010). Beta-catenin primes organizer gene expression by recruiting a histone H3 arginine 8 methyltransferase, Prmt2. *Developmental Cell*, *19*, 220–231.

Bogdanovic, O., Long, S. W., van Heeringen, S. J., Brinkman, A. B., Gómez-Skarmeta, J. L., Stunnenberg, H. G., et al. (2011). Temporal uncoupling of the DNA methylome and transcriptional repression during embryogenesis. *Genome Research*, *21*, 1313–1327.

Brewer, A. C., Guille, M. J., Fear, D. J., Partington, G. A., & Patient, R. K. (1995). Nuclear translocation of a maternal CCAAT factor at the start of gastrulation activates Xenopus GATA-2 transcription. *The EMBO Journal*, *14*, 757–766.

Bright, A. R., van Genesen, S., Li, Q., Grasso, A., Frölich, S., van der Sande, M., et al. (2021). Combinatorial transcription factor activities on open chromatin induce embryonic heterogeneity in vertebrates. *The EMBO Journal*, e104913.

Brown, D. D., & Littna, E. (1964a). RNA synthesis during the development of *Xenopus laevis*, the South African clawed toad. *Journal of Molecular Biology*, *8*, 669–687.

Brown, D. D., & Littna, E. (1964b). Variations in the synthesis of stable RNA's during oogenesis and development of *Xenopus laevis*. *Journal of Molecular Biology*, *8*, 688–695.

Brown, D. D., & Littna, E. (1966a). Synthesis and accumulation of DNA-like RNA during embryogenesis of *Xenopus laevis*. *Journal of Molecular Biology*, *20*, 81–94.

Brown, D. D., & Littna, E. (1966b). Synthesis and accumulation of low molecular weight RNA during embryogenesis of *Xenopus laevis*. *Journal of Molecular Biology*, *20*, 95–112.

Brzostowski, J., Robinson, C., Orford, R., Elgar, S., Scarlett, G., Peterkin, T., et al. (2000). RNA-dependent cytoplasmic anchoring of a transcription factor subunit during *Xenopus* development. *The EMBO Journal, 19*, 3683–3693.

Calo, E., & Wysocka, J. (2013). Modification of enhancer chromatin: What, how, and why? *Molecular Cell, 49*, 825–837.

Chan, S. H., Tang, Y., Miao, L., Darwich-Codore, H., Vejnar, C. E., Beaudoin, J. D., et al. (2019). Brd4 and P300 confer transcriptional competency during zygotic genome activation. *Developmental Cell, 49*, 867–881.

Charney, R. M., Forouzmand, E., Cho, J. S., Cheung, J., Paraiso, K. D., Yasuoka, Y., et al. (2017). Foxh1 occupies cis-regulatory modules prior to dynamic transcription factor interactions controlling the mesendoderm gene program. *Developmental Cell, 40*, 595–607.

Charney, R. M., Paraiso, K. D., Blitz, I. L., & Cho, K. W. Y. (2017). A gene regulatory program controlling early Xenopus mesendoderm formation: Network conservation and motifs. *Seminars in Cell & Developmental Biology, 66*, 12–24.

Chen, X., Ke, Y., Wu, K., Zhao, H., Sun, Y., Gao, L., et al. (2019). Key role for CTCF in establishing chromatin structure in human embryos. *Nature, 576*, 306–310.

Chen, H., Einstein, L. C., Little, S. C., & Good, M. C. (2019). Spatiotemporal patterning of zygotic genome activation in a model vertebrate embryo. *Developmental Cell, 49*, 852–866.

Chiu, W. T., Charney Le, R., Blitz, I. L., Fish, M. B., Li, Y., Biesinger, J., et al. (2014). Genome-wide view of TGFβ/Foxh1 regulation of the early mesendoderm program. *Development. 141*, 4537-4547.

Collart, C., Owens, N. D., Bhaw-Rosun, L., Cooper, B., De Domenico, E., Patrushev, I., et al. (2014). High-resolution analysis of gene activity during the *Xenopus* mid-blastula transition. *Development, 141*, 1927–1939.

Dimitrov, S., Almouzni, G., Dasso, M., & Wolffe, A. P. (1993). Chromatin transitions during early Xenopus embryogenesis: Changes in histone H4 acetylation and in linker histone type. *Developmental Biology, 160*, 214–227.

Dreyer, C. (1987). Differential accumulation of oocyte nuclear proteins by embryonic nuclei of Xenopus. *Development, 101*, 829–846.

Esmaeili, M., Blythe, S. A., Tobias, J. W., Zhang, K., Yang, J., & Klein, P. S. (2020). Chromatin accessibility and histone acetylation in the regulation of competence in early development. *Developmental Biology, 462*, 20–35.

Faure, S., Lee, M. A., Keller, T., ten Dijke, P., & Whitman, M. (2000). Endogenous patterns of TGFbeta superfamily signaling during early *Xenopus* development. *Development, 127*, 2917–2931.

Fernandez Garcia, M., Moore, C. D., Schulz, K. N., Alberto, O., Donague, G., Harrison, M. M., et al. (2019). Structural features of transcription factors associating with nucleosome binding. *Molecular Cell, 75*, 921–932.

Forouzmand, E., Owens, N. D. L., Blitz, I. L., Paraiso, K. D., Khokha, M. K., Gilchrist, M. J., et al. (2017). Developmentally regulated long non-coding RNAs in Xenopus tropicalis. *Developmental Biology, 426*, 401-408.

Gao, M., Veil, M., Rosenblatt, M., Gebhard, A., Hass, H., Buryanova, L., et al. (2020). Pluripotency factors select gene expression repertoire at zygotic genome activation. *bioRxiv*. 2020.02.16.949362.

Gazdag, E., Jacobi, U. G., van Kruijsbergen, I., Weeks, D. L., & Veenstra, G. J. (2016). Activation of a T-box-Otx2-Gsc gene network independent of TBP and TBP-related factors. *Development. 143*, 1340-1350.

Gentsch, G. E., Owens, N. D. L., & Smith, J. C. (2019). The spatiotemporal control of zygotic genome activation. *iScience, 16*, 485–498.

Gentsch, G. E., Spruce, T., Owens, N. D. L., & Smith, J. C. (2019). Maternal pluripotency factors initiate extensive chromatin remodelling to predefine first response to inductive signals. *Nature Communications*, *10*, 4269.
Gerhart, J. G. (1980). Mechanisms regulating pattern formation in the amphibian egg and early embryo. In R. F. Goldberger (Ed.), *Vol. 2. Biological regulation and development* (pp. 133–315). New York: Plenum Press.
Gupta, R., Wills, A., Ucar, D., & Baker, J. (2014). Developmental enhancers are marked independently of zygotic nodal signals in *Xenopus*. *Developmental Biology*, *395*, 38–49.
Gurdon, J. B., & Woodland, H. R. (1969). The influence of the cytoplasm on the nucleus during cell differentiation, with special reference to RNA synthesis during amphibian cleavage. *Proceedings of the Royal Society of London—Series B: Biological Sciences*, *173*, 99–111.
Harland, R. M., & Grainger, R. M. (2011). Xenopus research: Metamorphosed by genetics and genomics. *Trends in Genetics*, *27*, 507–515.
Hellsten, U., Harland, R. M., Gilchrist, M. J., Hendrix, D., Jurka, J., Kapitonov, V., et al. (2010). The genome of the Western clawed frog *Xenopus tropicalis*. *Science*, *328*, 633–636.
Heyn, P., Kircher, M., Dahl, A., Kelso, J., Tomancak, P., Kalinka, A. T., et al. (2014). The earliest transcribed zygotic genes are short, newly evolved, and different across species. *Cell Reports*, *6*, 285–292.
Hnisz, D., Abraham, B. J., Lee, T. I., Lau, A., Saint-André, V., Sigova, A. A., et al. (2013). Super-enhancers in the control of cell identity and disease. *Cell*, *155*, 934–947.
Hontelez, S., van Kruijsbergen, I., Georgiou, G., van Heeringen, S. J., Bogdanovic, O., Lister, R., et al. (2015). Embryonic transcription is controlled by maternally defined chromatin state. *Nature Communications*, *6*, 10148.
Howard, L., Rex, M., Clements, D., & Woodland, H. R. (2007). Regulation of the *Xenopus* Xsox17alpha(1) promoter by co-operating VegT and Sox17 sites. *Developmental Biology*, *310*, 402–415.
Humbert-Lan, G., & Pieler, T. (1999). Regulation of DNA binding activity and nuclear transport of B-Myb in Xenopus oocytes. *Journal of Biological Chemistry*, *274*, 10293–10300.
Iwafuchi, M., Cuesta, I., Donahue, G., Takenaka, N., Osipovich, A. B., Magnuson, M. A., et al. (2020). Gene network transitions in embryos depend upon interactions between a pioneer transcription factor and core histones. *Nature Genetics*, *52*, 418–427.
Jacobi, U. G., Akkers, R. C., Pierson, E. S., Weeks, D. L., Dagle, J. M., & Veenstra, G. J. (2007). TBP paralogs accommodate metazoan- and vertebrate-specific developmental gene regulation. *The EMBO Journal*, *26*, 3900-3909.
Jallow, Z., Jacobi, U. G., Weeks, D. L., Dawid, I. B., & Veenstra, G. J. (2004). Specialized and redundant roles of TBP and a vertebrate-specific TBP paralog in embryonic gene regulation in *Xenopus*. *Proceedings of the National Academy of Sciences of the United States of America*, *101*, 13525–13,530.
Jansen, C., Paraiso, K. D., Zhou, J. J., Blitz, I. L., Fish, M. B., Charney, R. M., et al. (2020). Uncovering the mesendoderm gene regulatory network through multi-omic data integration. *bioRxiv*, 2020.11.01.362053.
Jevtić, P., & Levy, D. L. (2015). Nuclear size scaling during *Xenopus* early development contributes to midblastula transition timing. *Current Biology*, *25*, 45–52.
Jevtić, P., & Levy, D. L. (2017). Both nuclear size and DNA amount contribute to midblastula transition timing in *Xenopus laevis*. *Scientific Reports*, *7*, 7908.
Jin, C., & Felsenfeld, G. (2007). Nucleosome stability mediated by histone variants H3.3 and H2A.Z. *Genes & Development*, *21*, 1519–1529.
Joseph, S. R., Pálfy, M., Hilbert, L., Kumar, M., Karschau, J., Zaburdaev, V., et al. (2017). Competition between histone and transcription factor binding regulates the onset of transcription in zebrafish embryos. *eLife*, *6*, e23326.

Jukam, D., Shariati, S. A. M., & Skotheim, J. M. (2017). Zygotic genome activation in vertebrates. *Developmental Cell, 42*, 316–332.

Kaaij, L. J. T., van der Weide, R. H., Ketting, R. F., & de Wit, E., (2018). Systemic loss and gain of chromatin architecture throughout zebrafish development. *Cell Reports, 24*, 1-10.e4.

Khokha, M. K., Chung, C., Bustamante, E. L., Gaw, L. W., Trott, K. A., Yeh, J., et al. (2002). Techniques and probes for the study of Xenopus tropicalis development. *Developmental Dynamics, 225*, 499–510.

Kimelman, D., Kirschner, M., & Scherson, T. (1987). The events of the midblastula transition in *Xenopus* are regulated by changes in the cell cycle. *Cell, 48*, 399–407.

Kjolby, R. A. S., & Harland, R. M. (2017). Genome-wide identification of Wnt/beta-catenin transcriptional targets during *Xenopus* gastrulation. *Developmental Biology, 426*, 165–175.

Knowland, J. S. (1970). Polyacrylamide gel electrophoresis of nucleic acids synthesised during the early development of Xenopus laevis Daudin. *Biochimica et Biophysica Acta, 204*, 416–429.

Larabell, C. A., Torres, M., Rowning, B. A., Yost, C., Miller, J. R., Wu, M., et al. (1997). Establishment of the dorso-ventral axis in Xenopus embryos is presaged by early asymmetries in beta-catenin that are modulated by the Wnt signaling pathway. *The Journal of Cell Biology, 136*, 1123–1136.

Lauberth, S. M., Nakayama, T., Wu, X., Ferris, A. L., Tang, Z., Hughes, S. H., et al. (2013). H3K4me3 interactions with TAF3 regulate preinitiation complex assembly and selective gene activation. *Cell, 152*, 1021–1036.

Lee, M. A., Heasman, J., & Whitman, M. (2001). Timing of endogenous activin-like signals and regional specification of the *Xenopus* embryo. *Development, 128*, 2939–2952.

Lee, M. T., Bonneau, A. R., Takacs, C. M., Bazzini, A. A., DiVito, K. R., Fleming, E. S., et al. (2013). Nanog, Pou5f1 and SoxB1 activate zygotic gene expression during the maternal-to-zygotic transition. *Nature, 503*, 360–364.

Lee, M. T., Bonneau, A. R., & Giraldez, A. J. (2014). Zygotic genome activation during the maternal-to-zygotic transition. *Annual Review of Cell and Developmental Biology, 30*, 581–613.

Leichsenring, M., Maes, J., Mössner, R., Driever, W., & Onichtchouk, D. (2013). Pou5f1 transcription factor controls zygotic gene activation in vertebrates. *Science, 341*, 1005–1009.

Li, X., Shou, W., Kloc, M., Reddy, B. A., & Etkin, L. D. (1994). Cytoplasmic retention of Xenopus nuclear factor 7 before the mid blastula transition uses a unique anchoring mechanism involving a retention domain and several phosphorylation sites. *The Journal of Cell Biology, 124*, 7–17.

Liu, G., Wang, W., Hu, S., Wang, X., & Zhang, Y. (2018). Inherited DNA methylation primes the establishment of accessible chromatin during genome activation. *Genome Research, 28*, 998–1007.

Loppin, B., & Berger, F. (2020). Histone variants: The nexus of developmental decisions and epigenetic memory. *Annual Review of Genetics, 54*, 121–149.

Luchman, H. A., Paterno, G. D., Kao, K. R., & Gillespie, L. L. (1991). Differential nuclear localization of ER1 protein during embryonic development in *Xenopus laevis*. *Mechanisms of Development, 80*, 111–114.

Lund, E., Liu, M., Hartley, R. S., Sheets, M. D., & Dahlberg, J. E. (2009). Deadenylation of maternal mRNAs mediated by miR-427 in *Xenopus laevis* embryos. *RNA, 15*, 2351–2363.

Martire, S., & Banaszynski, L. A. (2020). The roles of histone variants in fine-tuning chromatin organization and function. *Nature Reviews. Molecular Cell Biology, 21*, 522–541.

Martire, S., Gogate, A. A., Whitmill, A., Tafessu, A., Nguyen, J., Teng, Y. C., et al. (2019). Phosphorylation of histone H3.3 at serine 31 promotes p300 activity and enhancer acetylation. *Nature Genetics*, *51*, 941–946.

Miao, L., Tang, Y., Bonneau, A. R., Chan, S. H., Kojima, L., Pownall, M. E., et al. (2020). Synergistic activity of Nanog, Pou5f3, and Sox19b establishes chromatin accessibility and developmental competence in a context-dependent manner. *bioRxiv*. 2020.09.01. 278796.

Michael, A. K., Grand, R. S., Isbel, L., Cavadini, S., Kozicka, Z., Kempf, G., et al. (2020). Mechanisms of OCT4-SOX2 motif readout on nucleosomes. *Science*, *368*, 1460–1465.

Miller, M., Reddy, B. A., Kloc, M., Li, X. X., Dreyer, C., & Etkin, L. D. (1991). The nuclear-cytoplasmic distribution of the Xenopus nuclear factor, xnf7, coincides with its state of phosphorylation during early development. *Development*, *113*, 569–575.

Mitros, T., Lyons, J. B., Session, A. M., Jenkins, J., Shu, S., Kwon, T., et al. (2019). A chromosome-scale genome assembly and dense genetic map for *Xenopus tropicalis*. *Developmental Biology*, *452*, 8–20.

Mukherjee, S., Chaturvedi, P., Rankin, S. A., Fish, M. B., Wlizla, M., Paraiso, K. D., et al. (2020). Sox17 and β-catenin co-occupy Wnt-responsive enhancers to govern the endoderm gene regulatory network. *eLife*, *9*, e58029.

Nakakura, N., Miura, T., Yamana, K., Ito, A., & Shiokawa, K. (1987). Synthesis of heterogeneous mRNA-like RNA and low-molecular-weight RNA before the midblastula transition in embryos of Xenopus laevis. *Developmental Biology*, *123*, 421–429.

Nakamura, Y., de Paiva, A. E., Veenstra, G. J., & Hoppler, S. (2016). Tissue- and stage-specific Wnt target gene expression is controlled subsequent to beta-catenin recruitment to cis-regulatory modules. *Development*, *143*, 1914–1925.

Newport, J., & Kirschner, M. (1982a). A major developmental transition in early *Xenopus* embryos: I. Characterization and timing of cellular changes at the midblastula stage. *Cell*, *30*, 675–686.

Newport, J., & Kirschner, M. (1982b). A major developmental transition in early *Xenopus* embryos: II. Control of the onset of transcription. *Cell*, *30*, 687–696.

Nieuwkoop, P. D., & Faber, J. (1994). *Normal table of Xenopus laevis (daudin): A systematical and chronological survey of the development from the fertilized egg till the end of metamorphosis*. New York: Garland Publishing.

Niu, L., Shen, W., Shi, Z., He, N., Wan, J., Sun, J., et al. (2020). Systematic chromatin architecture analysis in *Xenopus tropicalis* reveals conserved three-dimensional folding principles of vertebrate genomes. *bioRxiv*, 2020.04.02.021378.

Oikawa, M., Simeone, A., Hormanseder, E., Teperek, M., Gaggioli, V., O'Doherty, A., et al. (2020). Epigenetic homogeneity in histone methylation underlies sperm programming for embryonic transcription. *Nature Communications*, *11*, 3491.

Owens, N. D. L., Blitz, I. L., Lane, M. A., Patrushev, I., Overton, J. D., Gilchrist, M. J., et al. (2016). Measuring absolute RNA copy numbers at high temporal resolution reveals transcriptome Kinetics in development. *Cell Reports*, *14*, 632–647.

Palancade, B., Bellier, S., Almouzni, G., & Bensaude, O. (2001). Incomplete RNA polymerase II phosphorylation in *Xenopus laevis* early embryos. *Journal of Cell Science*, *114*, 2483–2489.

Pálfy, M., Schulze, G., Valen, E., & Vastenhouw, N. L. (2020). Chromatin accessibility established by Pou5f3, Sox19b and Nanog primes genes for activity during zebrafish genome activation. *PLoS Genetics*, *16*, e1008546.

Paraiso, K. D., Blitz, I. L., Coley, M., Cheung, J., Sudou, N., Taira, M., et al. (2019). Endodermal maternal transcription factors establish super-enhancers during zygotic genome activation. *Cell Reports*, *27*, 2962–2977.

Paranjpe, S. S., Jacobi, U. G., van Heeringen, S. J., & Veenstra, G. J. (2013). A genome-wide survey of maternal and embryonic transcripts during *Xenopus tropicalis* development. *BMC Genomics*, *14*, 762.

Parker, S. C., Stitzel, M. L., Taylor, D. L., Orozco, J. M., Erdos, M. R., Akiyama, J. A., et al. (2013). Chromatin stretch enhancer states drive cell-specific gene regulation and harbor human disease risk variants. *Proceedings of the National Academy of Sciences of the United States of America*, *110*, 17921–17926.

Peshkin, L., Wühr, M., Pearl, E., Haas, W., Freeman, R. M., Jr., Gerhart, J. C., et al. (2015). On the relationship of protein and mRNA dynamics in vertebrate embryonic development. *Developmental Cell*, *35*, 383–394.

Peshkin, L., Lukyanov, A., Kalocsay, M., Gage, R. M., Wang, D. Z., Pells, T. J., et al. (2019). The protein repertoire in early vertebrate embryogenesis. *bioRxiv*, 10.1101/571174.

Post, J. N., Luchman, H. A., Mercer, F. C., Paterno, G. D., & Gillespie, L. L. (2005). Developmentally regulated cytoplasmic retention of the transcription factor XMI-ER1 requires sequence in the acidic activation domain. *The International Journal of Biochemistry & Cell Biology*, *37*, 463–477.

Prioleau, M. N., Huet, J., Sentenac, A., & Méchali, M. (1994). Competition between chromatin and transcription complex assembly regulates gene expression during early development. *Cell*, *77*, 439–449.

Prioleau, M. N., Buckle, R. S., & Méchali, M. (1995). Programming of a repressed but committed chromatin structure during early development. *The EMBO Journal*, *14*, 5073–5084.

Reisser, M., Palmer, A., Popp, A. P., Jahn, C., Weidinger, G., & Gebhardt, J. C. M. (2018). Single-molecule imaging correlates decreasing nuclear volume with increasing TF-chromatin associations during zebrafish development. *Nature Communications*, *9*, 5218.

Roeder, R. G. (1974). Multiple forms of deoxyribonucleic acid-dependent ribonucleic acid polymerase in *Xenopus laevis*. Levels of activity during oocyte and embryonic development. *The Journal of Biological Chemistry*, *249*, 249–256.

Saka, Y., Hagemann, A. I., Piepenburg, O., & Smith, J. C. (2007). Nuclear accumulation of Smad complexes occurs only after the midblastula transition in *Xenopus*. *Development*, *134*, 4209–4218.

Satoh, N. (1977). Metachronous cleavage and initiation of gastrulation in amphibian embryos. *Development, Growth and Differentiation*, *19*, 111–117.

Schohl, A., & Fagotto, F. (2002). Beta-catenin, MAPK and Smad signaling during early *Xenopus* development. *Development*, *129*, 37–52.

Session, A. M., Uno, Y., Kwon, T., Chapman, J. A., Toyoda, A., Takahashi, S., et al. (2016). Genome evolution in the allotetraploid frog *Xenopus laevis*. *Nature*, *538*, 336–343.

Shiokawa, K., Misumi, Y., Tashiro, K., Nakakura, N., Yamana, K., & Oh-uchida, M. (1989). Changes in the patterns of RNA synthesis in early embryogenesis of *Xenopus laevis*. *Cell Differentiation and Development*, *28*, 17–25.

Signoret, J., & Lefresne, J. (1971). Contribution a l'etude de la segmentation de l'oef d'axolotl. I. Definition de la transition blastuleenne. *Annals of Embryology and Morphogenesis*, *4*, 113–123.

Sitbon, D., Boyarchuk, E., Dingli, F., Loew, D., & Almouzni, G. (2020). Histone variant H3.3 residue S31 is essential for *Xenopus* gastrulation regardless of the deposition pathway. *Nature Communications*, *11*, 1256.

Skirkanich, J., Luxardi, G., Yang, J., Kodjabachian, L., & Klein, P. S. (2011). An essential role for transcription before the MBT in *Xenopus laevis*. *Developmental Biology*, *357*, 478–491.

Soufi, A., Garcia, M. F., Jaroszewicz, A., Osman, N., Pellegrini, M., & Zaret, K. S. (2015). Pioneer transcription factors target partial DNA motifs on nucleosomes to initiate reprogramming. *Cell*, *161*, 555–568.

Stennard, F., Zorn, A. M., Ryan, K., Garrett, N., & Gurdon, J. B. (1999). Differential expression of VegT and antipodean protein isoforms in Xenopus. *Mechanisms of Development, 86*, 87–98.
Szenker, E., Lacoste, N., & Almouzni, G. (2012). A developmental requirement for HIRA-dependent H3.3 deposition revealed at gastrulation in *Xenopus*. *Cell Reports, 1*, 730–740.
Takahashi, S., Onuma, Y., Yokota, C., Westmoreland, J. J., Asashima, M., & Wright, C. V. (2006). Nodal-related gene Xnr5 is amplified in the Xenopus genome. *Genesis, 44*, 309–321.
Tan, M. H., Au, K. F., Yablonovitch, A. L., Wills, A. E., Chuang, J., Baker, J. C., et al. (2013). RNA sequencing reveals a diverse and dynamic repertoire of the *Xenopus tropicalis* transcriptome over development. *Genome Research, 23*, 201–216.
Teperek, M., Simeone, A., Gaggioli, V., Miyamoto, K., Allen, G. E., Erkek, S., et al. (2016). Sperm is epigenetically programmed to regulate gene transcription in embryos. *Genome Research, 26*, 1034–1046.
Toyoda, T., & Wolffe, A. P. (1992). Characterization of RNA polymerase II-dependent transcription in Xenopus extracts. *Developmental Biology, 153*, 150–157.
Ura, K., Nightingale, K., & Wolffe, A. P. (1996). Differential association of HMG1 and linker histones B4 and H1 with dinucleosomal DNA: Structural transitions and transcriptional repression. *The EMBO Journal, 15*, 4959–4969.
van Heeringen, S. J., Akkers, R. C., van Kruijsbergen, I., Arif, M. A., Hanssen, L. L., Sharifi, N., et al. (2014). Principles of nucleation of H3K27 methylation during embryonic development. *Genome Research, 24*, 401–410.
Vastenhouw, N. L., Zhang, Y., Woods, I. G., Imam, F., Regev, A., Liu, X. S., et al. (2010). Chromatin signature of embryonic pluripotency is established during genome activation. *Nature, 464*, 922–926.
Veenstra, G. J., & Wolffe, A. P. (2001). Constitutive genomic methylation during embryonic development of *Xenopus*. *Biochimica et Biophysica Acta, 1521*, 39–44.
Veenstra, G. J., Mathu, M. T., & Destrée, O. H. (1999). The Oct-1 POU domain directs developmentally regulated nuclear translocation in *Xenopus* embryos. *Biological Chemistry, 380*, 253–257.
Veenstra, G. J., Destrée, O. H., & Wolffe, A. P. (1999). Translation of maternal TATA-binding protein mRNA potentiates basal but not activated transcription in *Xenopus* embryos at the midblastula transition. *Molecular and Cellular Biology, 19*, 7972–7982.
Veenstra, G. J., Weeks, D. L., & Wolffe, A. P. (2000). Distinct roles for TBP and TBP-like factor in early embryonic gene transcription in *Xenopus*. *Science, 290*, 2312–2315.
Veil, M., Yampolsky, L. Y., Grüning, B., & Onichtchouk, D. (2019). Pou5f3, SoxB1, and Nanog remodel chromatin on high nucleosome affinity regions at zygotic genome activation. *Genome Research, 29*, 383–395.
Venkatarama, T., Lai, F., Luo, X., Zhou, Y., Newman, K., & King, M. L. (2010). Repression of zygotic gene expression in the Xenopus germline. *Development, 137*, 651–660.
Vermeulen, M., Mulder, K. W., Denissov, S., Pijnappel, W. W., van Schaik, F. M., Varier, R. A., et al. (2007). Selective anchoring of TFIID to nucleosomes by trimethylation of histone H3 lysine 4. *Cell, 131*, 58–69.
White, R. J., Collins, J. E., Sealy, I. M., Wali, N., Dooley, C. M., Digby, Z., et al. (2017). A high-resolution mRNA expression time course of embryonic development in zebrafish. *eLife, 6*, e30860.
Whyte, W. A., Orlando, D. A., Hnisz, D., Abraham, B. J., Lin, C. Y., Kagey, M. H., et al. (2013). Master transcription factors and mediator establish super-enhancers at key cell identity genes. *Cell, 153*, 307–319.

Woodland, H. (1982). The translational control phase of early development. *Bioscience Reports*, *2*, 471–491.
Woodland, H. R., & Adamson, E. D. (1977). The synthesis and storage of histones during the oogenesis of Xenopus laevis. *Developmental Biology*, *57*, 118–135.
Woodland, H. R., & Gurdon, J. B. (1969). RNA synthesis in an amphibian nuclear-transplant hybrid. *Developmental Biology*, *20*, 89–104.
Wu, E., & Vastenhouw, N. L. (2020). From mother to embryo: A molecular perspective on zygotic genome activation. *Current Topics in Developmental Biology*, *140*, 209–254.
Wühr, M., Freeman, R. M., Jr., Presler, M., Horb, M. E., Peshkin, L., Gygi, S., et al. (2014). Deep proteomics of the *Xenopus laevis* egg using an mRNA-derived reference database. *Current Biology*, *24*, 1467–1475.
Wühr, M., Güttler, T., Peshkin, L., McAlister, G. C., Sonnett, M., Ishihara, K., et al. (2015). The nuclear proteome of a vertebrate. *Current Biology*, *25*, 2663–2671.
Yanai, I., Peshkin, L., Jorgensen, P., & Kirschner, M. W., (2011). Mapping gene expression in two Xenopus species: Evolutionary constraints and developmental flexibility. *Developmental Cell, 20*, 483-496.
Yang, J., Tan, C., Darken, R. S., Wilson, P. A., & Klein, P. S. (2002). Beta-catenin/Tcf-regulated transcription prior to the midblastula transition. *Development*, *129*, 5743–5752.
Yin, Y., Morgunova, E., Jolma, A., Kaasinen, E., Sahu, B., Khund-Sayeed, S., et al. (2017). Impact of cytosine methylation on DNA binding specificities of human transcription factors. *Science*, *356*, eaaj2239.
Yu, M., & Ren, B. (2017). The three-dimensional organization of mammalian genomes. *Annual Review of Cell and Developmental Biology*, *33*, 265–289.
Zaret, K. (1999). Developmental competence of the gut endoderm: Genetic potentiation by GATA and HNF3/fork head proteins. *Developmental Biology*, *209*, 1–10.
Zaret, K. S., & Carroll, J. S. (2011). Pioneer transcription factors: Establishing competence for gene expression. *Genes & Development*, *25*, 2227–2241.
Zhang, Y., Vastenhouw, N. L., Feng, J., Fu, K., Wang, C., Ge, Y., et al. (2014). Canonical nucleosome organization at promoters forms during genome activation. *Genome Research*, *24*, 260–266.
Zhu, F., Farnung, L., Kaasinen, E., Sahu, B., Yin, Y., Wei, B., et al. (2018). The interaction landscape between transcription factors and the nucleosome. *Nature*, *562*, 76–81.
Zorn, A. M., Barish, G. D., Williams, B. O., Lavender, P., Klymkowsky, M. W., & Varmus, H. E. (1999). Regulation of Wnt signaling by Sox proteins: XSox17 alpha/beta and XSox3 physically interact with beta-catenin. *Molecular Cell*, *4*, 487–498.

CHAPTER SEVEN

Mass spectrometry based proteomics for developmental neurobiology in the amphibian *Xenopus laevis*

Aparna B. Baxi[a,b], Leena R. Pade[a], and Peter Nemes[a,b,*]

[a]Department of Chemistry & Biochemistry, University of Maryland, College Park, College Park, MD, United States

[b]Department of Anatomy and Cell Biology, The George Washington University, Washington, DC, United States

*Corresponding author: e-mail address: nemes@umd.edu

Contents

Abstract

The South African clawed frog (*Xenopus laevis*), a prominent vertebrate model in cell and developmental biology, has been instrumental in studying molecular mechanisms of neural development and disease. Recently, high-resolution mass spectrometry (HRMS), a bioanalytical technology, has expanded the molecular toolbox of protein detection and characterization (proteomics). This chapter overviews the characteristics, advantages, and challenges of this biological model and technology. Discussions are offered on their combined use to aid studies on cell differentiation and development of neural tissues. Finally, the emerging integration of proteomics and other 'omic technologies is reflected on to generate new knowledge, drive and test new hypotheses, and ultimately, advance the understanding of neural development during states of health and disease.

Current Topics in Developmental Biology, Volume 145
ISSN 0070-2153
https://doi.org/10.1016/bs.ctdb.2021.04.002

1. Introduction

Understanding molecular mechanisms of neural development and their impact on the brain is central to advancing health sciences. The CDC reports that 1 in ~2700 newborns have life-threatening neural tube defects in the US and their treatment and care amounts to $1.5 billion/year (Arth et al., 2017). The need is high and still unmet to expand our basic understanding of the molecular drivers of normal development and the design of next-generation therapeutics to remedy such disorders. Physical sciences, such as bioanalytical chemistry, fill this gap by innovating tools and technologies to enable the study of neural development using model organisms. The South African clawed frog (*Xenopus laevis*) is one such model that has supported important discoveries in developmental neurobiology (Borodinsky, 2017), including neural induction, differentiation, regeneration, and human genetic diseases. Readers interested in these topics are referred to the recent *Xenopus Community White Papers* (Xenopus Community, 2020) and reviews on these topics (Blum & Ott, 2018; Duncan & Khokha, 2016; Hwang, Marquez, & Khokha, 2019; Pratt & Khakhalin, 2013; Sater & Moody, 2017).

Characterization of all the biomolecules is essential to understanding neural development and homeostasis; these molecules range from transcripts to proteins, peptides, and metabolites. Technological advances over the last four decades have enabled the detection and quantification of these molecules with scalability in time and space. Today, molecular amplification by polymerase chain reaction (PCR) and next-generation sequencing (NGS) are routinely used to analyze genes and transcripts at trace levels. Likewise, antibodies equipped with signal amplifying "molecular handles" offer sufficient sensitivity to measure select proteins in low abundance. More recently, high-resolution mass spectrometry (HRMS) has advanced studies in developmental biology by enabling the detection of hundreds to thousands of proteins from complex biological samples (Hashimoto, Greco, & Cristea, 2019; Lombard-Banek, Portero, Onjiko, & Nemes, 2017). These studies also allow for the identification of genes and gene products for testing function. HRMS may be integrated with tools of cell, developmental, and molecular biology including cell fate tracing, transcription/translation-blocking molecules (e.g., morpholinos, shRNA, miRNA), and gene editing tools (e.g., ZFNs, TALENs, CRISPR-Cas9). Discovery analysis of biomolecules followed by hypothesis driven functional studies bridge the biochemical and functional pillars of systems biology.

This chapter focuses on the advantages of HRMS-based proteomic studies (proteomics) using *Xenopus* to advance developmental neurobiology. We then provide vignettes of biological studies that capitalized on HRMS to answer important questions in neuroscience. Where pertinent, biological insights are highlighted that were made possible due to the integration of *Xenopus* with HRMS-based proteomics. We also discuss how, in return, *Xenopus* embryos have spurred the development of ultrasensitive instruments and methodologies for proteomic and metabolomic studies in increasing resolution in time and space (e.g., tissues to single cells). Although the focus of this chapter is on developmental neurobiology, HRMS and *Xenopus* have provided important insights in a broader context of biology and health research (Federspiel et al., 2019; Peshkin et al., 2015; Presler et al., 2017; Sun et al., 2014; Wuehr et al., 2014). Likewise, HRMS has also supported proteomics in other amphibian species. Most recent topics include organ regeneration using the axolotl and questions pertaining to evolutionary biology using the Chinese giant salamander (Demircan, Sibai, & Altuntas, 2020; Geng et al., 2019). We apologize to anyone whose work could not be discussed in this chapter due to space constraints.

2. Mass spectrometry proteomics for *Xenopus*

2.1 Proteomics by HRMS

The proteome, the full suite of proteins produced in a biological system, can be used as an efficient descriptor of the molecular phenotype. Because proteins are produced downstream of transcription, they are often influenced by intrinsic and extrinsic events to the system. Post-translational modifications (PTMs), such as phosphorylation, further expand the proteomic landscape carrying out biological functions to "proteoforms" (Fig. 1A). As shown for the early *Xenopus* embryo in Fig. 1B, complex dynamics between gene transcription and translation during development challenge the use of mRNA as a general surrogate for quantifying protein expression (Peshkin et al., 2015; Smits et al., 2014). Directly measuring proteins and their PTMs is essential for developing and testing hypotheses for protein-targeted studies.

HRMS is the leading technology for detecting and quantifying proteins and their PTMs. Mass spectrometers enable unbiased, direct, and specific detection and quantification of large numbers of proteins (Aebersold, Burlingame, & Bradshaw, 2013). HRMS attains unparalleled molecular specificity by "weighing" mass (m/z value) to sub-millidalton (<10 ppm)

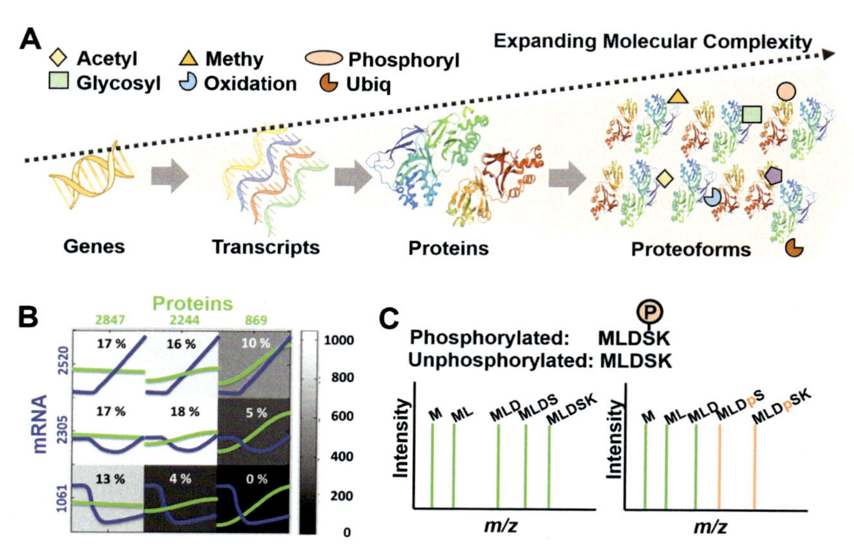

Fig. 1 Measurement of proteins in biological systems. (A) Complex proteomic landscape from molecular systems biology, expanded by post-translational modifications (PTMs). Representative PTMs are illustrated. (B) Complex temporal correlation between mRNA and protein abundances in developing systems call for direct measurement of proteins. The example clusters mRNA and protein expression in the *X. laevis* embryo. Number of genes (gray scale, % genes shown), transcripts (blue), and proteins (green) are marked for each cluster. Each cluster marks the number of transcripts, proteins, and the % of gene products belonging. (C) HRMS provides unmatched specificity and selectivity to identify and quantify proteins, including their PTMs. The example shows the sequencing of a peptide and localization of phosphorylation. *Adapted with permission from Peshkin, L., Wuhr, M., Pearl, E., Haas, W., Freeman, R. M., Gerhart, J. C., et al. (2015). On the relationship of protein and mRNA dynamics in vertebrate embryonic development.* Developmental Cell, *35, 383–394.*

accuracy and obtaining sequence-specific information on protein/peptide molecules (Aebersold et al., 2013; Walther & Mann, 2010; Zhang, Fonslow, Shan, Baek, & Yates, 2013). HRMS also enables the identification and precise localization of PTMs on a protein based on mass measurements. So far ~400 known PTMs (http://www.unimod.org/modifications_list.php) have been explored using HRMS, including more than 200 in biological contexts spanning from signal transduction to cellular localization (Duan & Walther, 2015; Virag et al., 2020). Fig. 1C exemplifies the sequencing of a peptide backbone and accurate localization of phosphorylation (see serine residues) based on gas-phase ion fragmentation during high-resolution tandem MS. For example, HRMS-based quantification revealed dynamic changes in protein phosphorylation as *Xenopus* eggs undergo meiotic exit (Presler et al., 2017). Detection and quantification of broad types of PTMs through HRMS has opened new possibilities in the research of protein function, regulation, and turnover.

HRMS aids scientific rigor in biology by supplying rich and high-fidelity quantitative data. Table 1 compares the instrumental and technical characteristics of HRMS with antibody-based technologies from classical molecular biology, such as western blots, immunofluorescence, and ELISA. A statistical survey from 2015 found that poorly characterized and non-specific antibodies have taxed the economy of research by over $350 million annually in the US alone (Bradbury & Pluckthun, 2015). Proteomics by HRMS addresses many challenges of protein detection using antibodies. Because detection based on mass (m/z values) does not require functioning probes,

Table 1 Comparison of protein detection by high-resolution mass spectrometry (HRMS) and classical antibody-based assays (e.g., Western blot, immunofluorescence, ELISA).

Figure of merit	HRMS	Antibody-based assays
Experimental strategy	Discovery or targeted	Targeted
Mechanism of molecular identification	MS/MS, (accurate) m/z, separation	Antibody detects epitope and (optionally) drives signal amplification reaction
Specificity, accuracy	Exquisite specificity and accuracy (<5 ppm accuracy routine)	Depends on antibody specificity, requires validation
Sensitivity	Fmol–zmol	Down to low fmol levels
Quantification	Yes; absolute and relative quantification with precision and accuracy	Semi-quantitative based on a calibration curve (absolute)
Throughput	High throughput: thousands of proteins identified and quantified in a single experiment in ~3 h. Sample multiplexing common	Low throughput: usually one protein at a time. Limited to no multiplexing
Spatial information	Yes; MS can be operated as a "mass microscope"	Yes, antibody-based optical imaging
Cost	Affordable via MS Facilities or high start-up cost	Low start-up cost; high consumables cost; requirement to validate antibodies

HRMS is not affected by batch-to-batch variability encountered in antibodies (Aebersold et al., 2013; Bradbury & Pluckthun, 2015). It follows that HRMS detection is also not challenged by antibody cross-reactivity with other proteins or a lack of antibody reactivity among homologous proteins between species. With methods being readily transferable from one lab to another and the availability of commercial bench-top HRMS instruments, HRMS has become an indispensable instrumental tool of protein research (Aebersold et al., 2013; Bradbury & Pluckthun, 2015).

2.2 Workflow for HRMS proteomics

Protein measurements using HRMS depend on several interconnected steps (Fig. 2). The three common approaches in proteomics—they are called bottom-up, middle-down, and top-down—have been the focus of recent reviews (Han, Aslanian, & Yates, 2008; Pino, Rose, O'Broin, Shah, & Schilling, 2020). Fig. 2 presents main steps of the most common, bottom-up (shotgun), approach. In this strategy, proteins are digested into typically 8–30 amino acid long peptides using proteolytic enzymes, such as TPCK-modified trypsin. The resulting peptides are optionally barcoded to enable multiplexed quantification among multiple samples and desalted using commercially available kits (e.g., C18 ZipTips, Pierce; C18 spin-columns, Thermo Fisher). Peptides are then sequenced by HRMS executing tandem MS. Protein-specific (called proteotypic) peptides are mapped against proteins for identification and quantification (Zhang et al., 2013). The bottom-up proteomic strategy is well established, sensitive, and supported by bench-top mass spectrometers available on the market and a variety of commercially and freely available data analysis software packages.

During bottom-up proteomics, peptides are typically separated. Separation decreases molecular complexity in the temporal dimension, helping HRMS perform sequencing by carrying out tandem MS measurements. (Ultra)High-performance liquid chromatography (UPLC/HPLC) and capillary electrophoresis (CE) have facilitated the detection of thousands of peptides in *Xenopus* by reducing chemical complexity temporally to tailor to the duty cycle of HRMS. In UPLC, 200 ng–1 μg of peptide (5–10 μg for replicate analysis) is typically separated based on hydrophobicity on a reversed-phase column (e.g., a C18 stationary phase), typically identifying ~1500–3500 proteins with an ~2-h LCMS separation experiment (called a "run") (Baxi, Lombard-Banek, Moody, & Nemes, 2018; Peshkin et al., 2015; Sun et al., 2014). For deeper detection of the proteome, molecular complexity may be reduced via longer separations and/or fractionation off/on-line

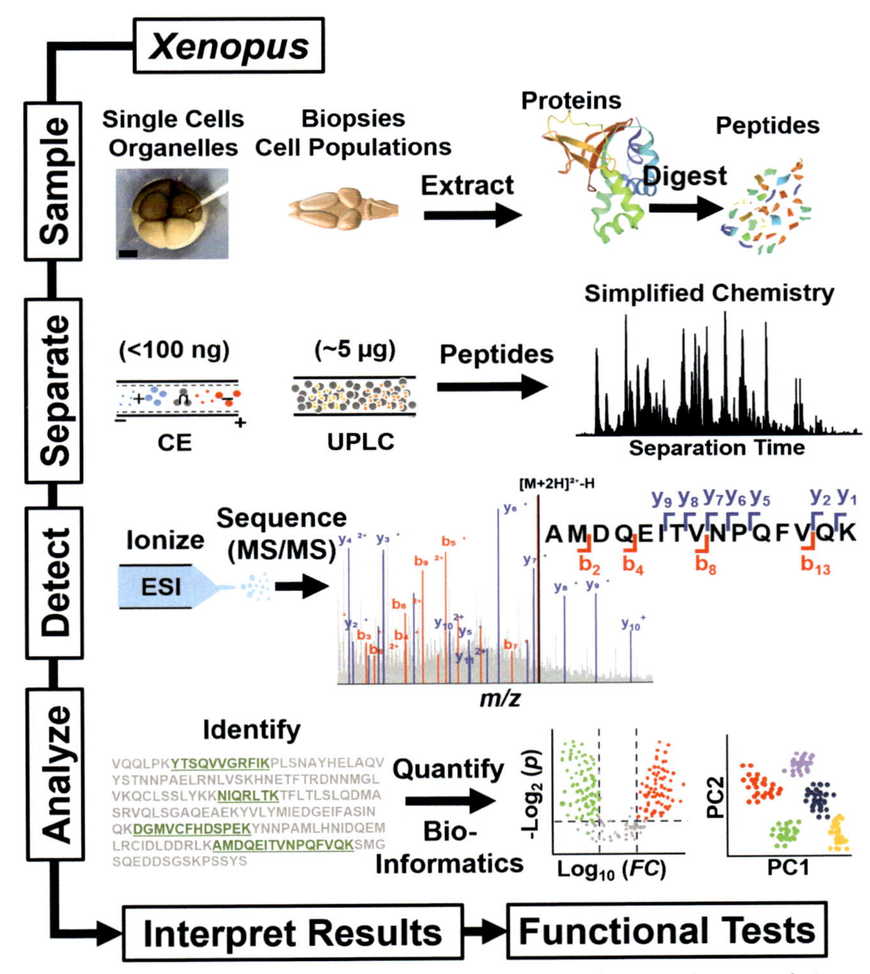

Fig. 2 General steps of HRMS-based proteomics in *X. laevis* embryos and tissues. Proteins are identified and quantified based on the detection of proteotypic peptides. The sequenced peptide demonstrated is from protein Cop9. Discovery and targeted approaches enable high-throughput screening and biomarker discovery. Key: FC, fold change; p, statistical *p*-value (Student's *t*-test shown). *Adapted with permission from Onjiko, R. M., Portero, E. P., Moody, S. A., & Nemes, P. (2017a). In situ microprobe single-cell capillary electrophoresis mass spectrometry: Metabolic reorganization in single differentiating cells in the live vertebrate (Xenopus laevis) embryo.* Analytical Chemistry, 89, *7069–7076.*

(e.g., high pH reversed phase or strong cation exchange chromatography) followed by reversed-phase LC and HRMS analysis of each fraction. Further details about experimental strategies by LC-HRMS are available in the references (Issaq, 2001; Zhang et al., 2013). Alternatively, using CE, peptides may be resolved based on differences in electrophoretic mobility (Fig. 2). With higher separation power (resolution) and sensitivity, we and others have found CE to complement UPLC for protein amounts limited to <100 ng (reviewed in DeLaney, Sauer, Vu, & Li, 2019; Fonslow & Yates, 2009; Lombard-Banek et al., 2017). UPLC and CE have uncovered previously unknown gene translational differences between cells of the embryo (see later), reviewed in Lombard-Banek et al. (2017). Separation approaches using orthogonal principles of separation can be used sequentially to further simplify sample complexity.

Peptides are detected next. The molecules exiting separation by UPLC/CE must be charged, usually by entraining them into a device performing electrospray ionization (ESI). The ionization source is attached to the mass spectrometer, where the ESI-generated ions are analyzed, and detected. Modern-day instruments offer high mass resolving powers for molecular identifications. At present, the average spectral resolution is ~40,000–100,000 full width at half maximum (FWHM) on commercial time-of-flight and 250,000–1,000,000 FWHM on commercial orbitrap ESI mass spectrometers. These instruments are equipped with compartments where peptide ions can be isolated and broken down (called dissociated or fragmented) into structure-specific fragments via tandem MS (aka. MS/MS or MS^2). The resulting fragment ions are analyzed, typically to accurate mass with <5 ppm error.

Tandem HRMS directly sequences the peptide ions and can identify PTMs. Fig. 2 shows the tandem mass spectrum of a peptide signal detected at m/z 1747.8683. Upon higher-energy collisional dissociation (HCD), this peptide produced characteristic fragment (called *b*- and *y*-type) ions (see insets). These mass and sequence data were matched to the peptide sequence: AMDQEITVNPQFVQK (from the enzyme Cop9) with a mass accuracy of 10 ppm and a stringent <1% false discovery rate (FDR). To deepen sequence coverage and help identify PTMs and their location on the peptide backbone, additional types of fragmentation techniques are available. Collision-induced dissociation (CID), electron transfer dissociation (ETD), electron capture dissociation (ECD), or their combination (e.g., EThCD) have demonstrated complementary performance for sequence determination (reviewed in Chu et al., 2015). HRMS-MS^2 marries high accuracy and precision for identifying peptides (recall Table 1).

The experimentally detected peptide sequences are mapped to the proteome of the organism under study using bioinformatic tools (software). The proteome is typically predicted in silico based on the genome and/or experimentally measured mRNA data. The confidence of identifications is enhanced by filtering spectral matches to a stringent <1% FDR, usually against a reverse-sequence decoy database. Common contaminant proteins are also annotated and removed during this step, such as human keratin proteins (e.g., from personnel hair and skin), *E. coli* proteins (e.g., from bacterial cultures processed in the same lab space), and porcine trypsin (used for digesting proteins). A variety of software packages are available to facilitate bioinformatic data analysis. Open-source (free of charge) applications include Skyline (Pino, Searle, et al., 2020), MSFragger (Kong, Leprevost, Avtonomov, Mellacheruvu, & Nesvizhskii, 2017), and MaxQuant (Tyanova, Temu, & Cox, 2016). Commercial products, such as ProteomeDiscoverer (ThermoFisher Scientific), provide a user-friendly environment by integrating open-source bioinformatic algorithms and gene ontology tools into a single platform. These and other bioinformatic resources are reviewed in (Chen, Hou, Tanner, & Cheng, 2020). The resulting list of protein identifications may be imported to knowledge bases to evaluate gene ontology, study pathways/networks, and protein interactions, as well as regulatory mechanisms of neural development and developmental diseases that have been annotated for humans. Popular resources include DAVID (Huang, Sherman, & Lempicki, 2009), PantherDB (http://pantherdb.org/), Xenbase (http://www.xenbase.org), STRING (Jensen et al., 2009), and Ingenuity Pathway Analysis (QIAGEN, Redwood City, CA).

There are several ways to quantify proteins by HRMS. The concentration of a protein is in stoichiometric relationship with the concentration of its proteotypic peptide. As the peptide is quantified based on ion signal abundance by the mass spectrometer, it is possible to quantify using HRMS any protein between samples that yield proteotypic peptide ion signals. To enhance quantitative precision, calculated protein signals may be normalized to the total, mean, or median signal abundance of the proteins. Alternatively, house-keeping genes with comparable expression may be used for normalization. With "label-free quantification" (LFQ), this workflow enables the relative or absolute quantification of proteins (Cox et al., 2014; Lindemann et al., 2017). Further, specialized "mass tags" (TMT, ThermoFisher Scientific; iTRAQ, Sigma-Aldrich) permit multiplexing quantification among samples at once, enhancing sensitivity, reproducibility, while reducing instrumental analysis time (see later). Chemical (isobaric mass tags) based labeling strategies and stable isotope labeling (such as SILAC) have enabled the detection of newly translated proteins during neural development in *Xenopus* (Ong et al., 2002). Applications of mass tag and chemical labeling based HRMS approaches to answer questions in developmental neurobiology are described later in this chapter. As a result, modern HRMS provides accurate and reproducible quantification of hundreds-to-thousands of different proteins, without requiring the use of functional probes.

2.3 Proteomics in *Xenopus* by HRMS

Xenopus brings many advantages for HRMS-based proteomics. The embryos can be generated year-round, and a female frog lays about 200–500 embryos each time, thus providing a convenient way to scale up workflows. Another advantage of the embryo is size and protein content. The early cleavage-stage embryo offers ~130 μg of protein, sufficient for high-sensitivity HRMS proteomic measurements (Sun et al., 2016). External fertilization and fast development allow timing proteomics experiments to developmental milestones (Baxi et al., 2018; Peshkin et al., 2015; Sun et al., 2014). In *Xenopus*, pigmentation and stereotypical cell divisions permit ready identification and manipulation of cells. Reproducible fate maps are available for embryos at the 16- (Moody, 1987a, 1987b) and 32-cell (Dale & Slack, 1987) stage. We have integrated this information with ultrasensitive HRMS to peer into proteomes of neural, epidermal, or hindgut fated single cells and cell clones in 4–128-cell embryos (Lombard-Banek et al., 2021; Lombard-Banek, Moody, & Nemes, 2016). Moreover, undifferentiated cells of the early embryo can be explanted and cultured in the presence of morphogens to induce cell differentiation. Recently, *Xenopus* animal cap explants helped generate a sufficient number of ciliated cells for deep HRMS-based proteomic analysis of RNA-associated proteins required for ciliary beating (Drew et al., 2020).

2.4 Challenges in *Xenopus* proteomics

Xenopus is not without challenges for HRMS. *Xenopus laevis* has an allotetraploid genome, in which over 56% of genes have a long and a short copy (gene duplication), with minor sequence variations for the same gene. The proteoforms resulting from this genetic background are anticipated to expand the proteome compared to mammalian animal models (Fig. 1A) (Session et al., 2016). The most recently sequenced version of the genome (version 9.2) reports 46,582 protein-coding genes in comparison to ~30,000 genes in the mouse (Waterston et al., 2002). The ~30,000 genes in the human genome have been estimated to generate over a million proteoforms (Jensen, 2004). Consequently, a given *X. laevis* sample can be expected to yield over a million proteoforms. With a diploid genome containing 21,891 protein coding genes (genome version 10), *Xenopus tropicalis* offers a reduced molecular space and an attractive alternative to test functional significance through gene editing (e.g., CRISPR/Cas9) (Nakayama et al., 2014). Larger size and ~5-times more material in *X. laevis* embryos make this model beneficial for cell biological and single-cell analyses (Harland & Grainger, 2011). During the bottom-up proteomic workflow, tryptic digestion cleaves each protein into multiple peptides, thus expanding the molecular space for analysis (Figs. 2 and 3A).

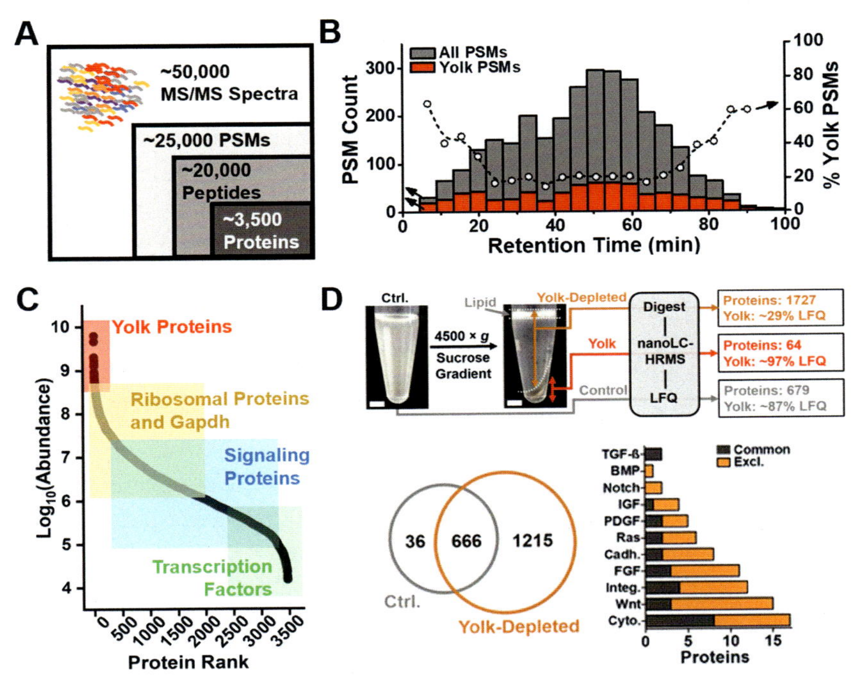

Fig. 3 Challenges and remedies in MS proteomics in *Xenopus*. (A) Vast molecular complexity in proteomic digests tax the sequencing of peptide signals via tandem MS (MS/MS). (B) Chemical separation reduces molecular complexity to enhance peptide identifications. Abundant yolk peptides limit sequencing throughput. (C) A broad dynamic range of concentration challenges the identification of low-abundance proteins. For example, abundant yolk proteins overwhelm signal abundance, hindering detection sensitivity. Shaded boxes represent the types of proteins that are typically present at a given abundance range. (D) Removal of yolk platelets (*top panel*) by sucrose pelleting (*bottom panel*) ca. triples the number of identifiable proteins, permitting the quantification of important signaling pathways. *Adapted with permission from Baxi, A. B., Lombard-Banek, C., Moody, S. A., & Nemes, P. (2018). Proteomic characterization of the neural ectoderm fated cell clones in the Xenopus laevis embryo by high-resolution mass spectrometry.* ACS Chemical Neuroscience, *9, 2064–2073.*

Such molecular complexity inherent to bottom-up proteomics taxes detection sensitivity and the success of peptide sequencing in HRMS, which in turn can hinder protein identification. In a routine bottom-up proteomics run, we sequence ~25,000 peptides over 100–120 min of separation. Fig. 3A presents one such experiment series from our lab, in which ~50,000 precursor peptide ions were analyzed to be able to ascribe ~50% to peptide sequences (<1% FDR) from ~3500 proteins that were measured between multiple runs (<1% FDR). Fig. 3B shows the separation of peptides over 100 min from a single run on a complex *Xenopus* digest.

Further, as in other vertebrate models, the concentration of proteins spans over a vast dynamic range. Fig. 3C ranks proteins by abundance from our measurement of *X. laevis* whole embryos (at gastrulation stages). The HRMS signal intensities reveal a 6–7 log-order dynamic range of concentration in the embryo. Detection of low-abundance proteins, such as transcription factors, is challenged by abundant signals, which saturate the duty cycle of MS/MS measurements. In the cleavage–to–neurula-stage *Xenopus* embryos, the yolk protein vitellogenin (forms Vtga1, Vtga2, Vtgb1, Vtgb2) constitutes up to ~90% of the proteome. These proteins are packaged in yolk platelets, distributed throughout the cell among all the cells in the embryo (Baxi et al., 2018; Peuchen, Sun, & Dovichi, 2016). Abundant yolk proteins interfere with and diminish the detection of lower abundance proteins. As shown in Fig. 3D, pelleting of the yolk platelets via soft centrifugation in sucrose before processing the cell lysate for bottom-up proteomics enhanced detection of the proteome in the *X. laevis* oocyte (Wuehr et al., 2014) and gastrula (Baxi et al., 2018). An ~70% reduction in yolk protein concentration (called "deyolking"), expanded detection from ~700 proteins to ~1900 proteins, an ~3-fold improvement in protein identification. Many of the proteins detected after deyolking are transcription factors, players of signaling pathways, and secondary messengers in the dissected neural ectoderm (Baxi et al., 2018). In single *X. laevis* blastomeres, where deyolking cannot be readily performed due to a limitation in sample size, sensitivity was enhanced by freeing up tandem HRMS duty cycle through programmed exclusion of a portion of tryptic peptide ions resulting from vitellogenin (Lombard-Banek, Moody, Manzin, & Nemes, 2019).

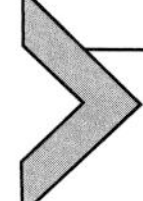

3. HRMS proteomics for developmental neurobiology in *Xenopus*

HRMS-based proteomics has expanded the bioanalytical toolbox of developmental neurobiology. This technology has enabled us and others to identify and quantify hundreds to thousands of proteins with previously unavailable accuracy, precision, and confidence. This metadata can be used to peer into protein/pathway dynamics and to seek out new insights and non-canonical mechanisms driving neural development. The data generated from untargeted experiments in turn has empowered the design of targeted experiments aimed at determining the significance of genes and gene products in specific biological contexts. In what follows, we highlight select studies that illustrate the study of neurodevelopmental questions using *Xenopus* by applying HRMS proteomic approaches in organs, tissues, cells, and single cells (see Table 2).

3.1 Organs

External development and large batch size make *Xenopus* helpful for studying protein expression during neural and sensory organ development. Dynamic changes can be traced by one of many approaches enabling relative and/or absolute protein quantification (see earlier). Designer mass tags (TMTs, iTRAQ, etc.) allow for the relative quantification of proteins among multiple samples (e.g., tissues, biological conditions, replicates). The approach is known to empower comparative analyses with high accuracy and precision (O'Connell, Paulo, O'Brien, & Gygi, 2018). We have recently used TMT-based relative quantification to assess proteome dynamics during five key steps of inner ear morphogenesis in *X. laevis* (Baxi et al., 2020). Sustained enrichment of integrin, collagen, and extracellular matrix proteins marked formation of the inner ear.

3.2 Tissues

Tissues may be readily manipulated and dissected from embryos, larvae, and tadpoles for HRMS proteomics. As shown in Fig. 4, we recently dissected the newly established neural ectoderm at the onset of gastrulation by tracing the lineage of the neural fated cell in the early embryo (Baxi et al., 2018). This survey of the neural ectodermal proteome captured proteins with critical roles during neural development, including molecular signaling via the FGF, Wnt, RhoGTPase, and Notch pathways. A similarly lineage-specific approach also allowed us to characterize the proteome of the developing Spemann organizer (Quach et al., 2019). This signaling center is transiently established at the onset of gastrulation and plays critical roles in patterning the embryo, including the induction of the neural ectoderm. We uncovered that the Spemann organizer tissues were enriched in mitochondrial proteins that participate in oxidative phosphorylation (Baxi et al., 2021). These and

Table 2 Examples of proteomics by HRMS supporting neurodevelopmental studies in *Xenopus*.

Targeted cell/tissue type	Modes of separation and MS	Type of quantification	No. of proteins identified	References
Retina	2D DIGE-RPLC-MS	Label-free	~2000	Wang et al. (2010)
Otic vesicles	High-pH RPLC-Low-pH RPLC-MS	Relative (TMT)	~5000	Baxi, Moody, and Nemes (2020)
Spinal cord tissues	SCX-RPLC-MS	Relative (iTRAQ)	~6000	Lee-Liu, Sun, Dovichi, and Larrain (2018)
Neural ectoderm tissues (beginning of gastrulation)	High-pH RPLC-Low-pH RPLC-MS	Label-free	~2500	Baxi et al. (2018)
Spemann organizer tissue (beginning of gastrulation)	High-pH RPLC-Low-pH RPLC-MS	Relative (TMT)	~3000	Baxi, Quach, Li, and Nemes (2021) and Quach, Baxi, and Nemes (2019)
Axons	RPLC-MS	Metabolic labeling (pSILAC)	~350 labeled proteins	Cagnetta, Frese, Shigeoka, Krijgsveld, and Holt (2018)
Axons	MudPIT-MS	Metabolic labeling (BONCAT)	~4800	Schiapparelli et al. (2019)
Neural, epidermal, and endodermal fated single cells	CE-MS	Label-free	~438	Lombard-Banek, Moody, et al. (2016) and Lombard-Banek, Reddy, Moody, and Nemes (2016)
Neural fated single cells	CE-MS	Label-free	~738	Lombard-Banek et al. (2021)

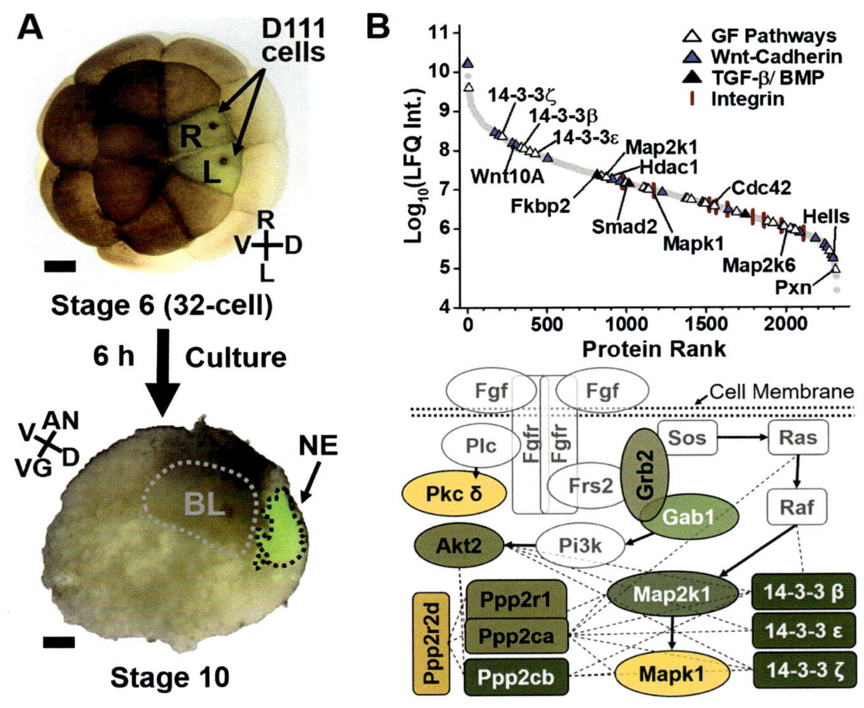

Fig. 4 Lineage tracing and HRMS to profile protein expression during neural induction. (A) Injection of neural fated (D111) cells of the *Xenopus* embryo and subsequent dissection of the neural ectoderm at the beginning of gastrulation (stage 10). (B) Dynamic range of quantified proteins in the tissue (*top panel*). Representative proteins with important roles in neurodevelopmental signaling pathways are highlighted. Kinases and second messengers from the FGF pathway are represented (*bottom panel*). *Adapted with permission from Baxi, A. B., Lombard-Banek, C., Moody, S. A., & Nemes, P. (2018). Proteomic characterization of the neural ectoderm fated cell clones in the Xenopus laevis embryo by high-resolution mass spectrometry.* ACS Chemical Neuroscience, *9, 2064–2073.*

other large-scale proteomic studies from transiently formed tissues generate an important resource of thousands of proteins that can be functionally evaluated for their role in neural induction and development.

Indeed, HRMS and *Xenopus laevis* enabled the investigation of proteome dynamics during regeneration after spinal cord injury (Fig. 5). Multiplexed relative proteomic analysis following spinal cord injury in regenerative (early tadpole) and non-regenerative (post-metamorphosis) stages of spinal cords (Fig. 5A) facilitated the identification of 172 proteins that were significantly upregulated during spinal cord regeneration in the early *Xenopus* tadpole (Lee-Liu et al., 2018). Proteomic studies of organogenesis and regeneration

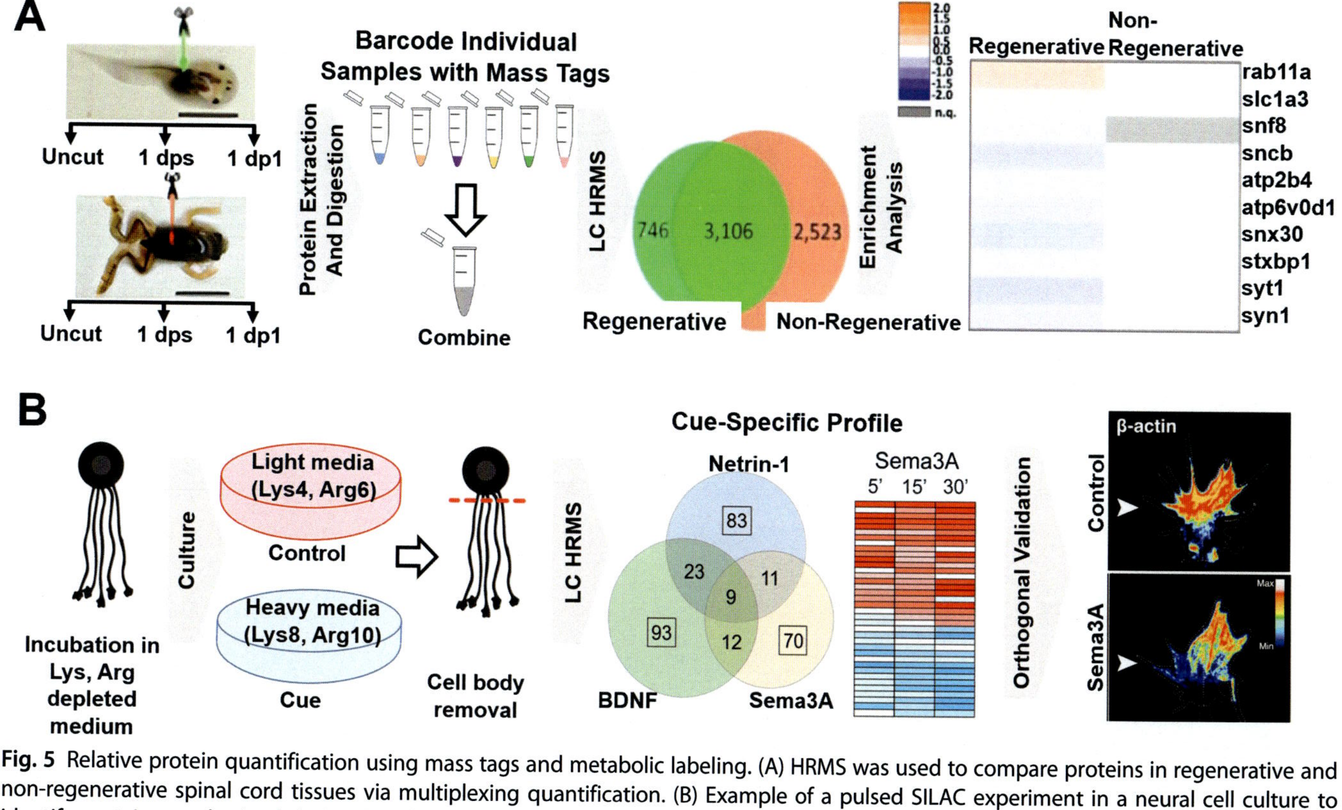

Fig. 5 Relative protein quantification using mass tags and metabolic labeling. (A) HRMS was used to compare proteins in regenerative and non-regenerative spinal cord tissues via multiplexing quantification. (B) Example of a pulsed SILAC experiment in a neural cell culture to identify proteins synthesized during cue-driven axon guidance. The repulsive gradient with Sema3A was found to lead to an enrichment of beta-actin on the opposite side of the axon growth cone. *Adapted with permission from Cagnetta, R., Frese, C. K., Shigeoka, T., Krijgsveld, J., & Holt, C. E. (2018). Rapid cue-specific remodeling of the nascent axonal proteome.* Neuron, 99, *29–46; Lee-Liu, D., Sun, L. L., Dovichi, N. J., & Larrain, J. (2018). Quantitative proteomics after spinal cord injury (SCI) in a regenerative and a nonregenerative stage in the frog Xenopus laevis.* Molecular & Cellular Proteomics, 17, *592–606.*

from tadpole and later stages of *Xenopus* benefit from an overall decreased abundance of yolk proteins that are consumed during development, contrary to yolk-laden tissues from earlier stages of development studied in the next section.

3.3 Cells

Xenopus tissues can be readily explanted to monitor axon growth along with pulsed metabolic labeling to study protein translation. This labeling approach, called pulsed stable isotope labeling by amino acids in cell culture (pSILAC), relies on the integration of heavy isotope labeled amino acids to monitor protein translation. Fig. 5B shows one such study in which retinal ganglion cells were explanted in media containing heavy isotope labeled essential amino acids to quantify protein remodeling during the growth of retinal axons in the presence of axon guiding cues (Cagnetta et al., 2018). This study design uncovered nascent proteins produced in axons in response to chemo-attractive cues, such as Netrin, BDNF, and Sema3A (Fig. 5B). Results from this study agreed with immunofluorescence imaging that demonstrated that proteins synthesized in response to repulsive gradients (Sema3A) are localized at the tip of the axon growth cone, revealing their direct involvement in axon growth.

Alternatively, bio-orthogonal non-canonical amino acid tagging (BONCAT) enables the quantification of protein synthesis by incorporation of non-canonical amino acids such as L-azidohomoalanine (AHA) in newly synthesized proteins. The AHA can be equipped with biotin tags for enrichment and detection by HRMS. AHA was injected into the visual system to label neuronal proteins synthesized in response to different visual cues to study neural plasticity in *Xenopus* tadpoles (Schiapparelli et al., 2019). Ultimately, a group of candidate plasticity proteins were identified that changed significantly between different visual experiences and primarily involved functions such as RNA splicing, protein translation, as well as chromatin remodeling. Teasing apart the nascent proteome through metabolic labeling and untargeted proteomic analysis via HRMS has the potential to yield valuable insights into neural development.

HRMS has also been valuable in confirming and identifying proteins in *Xenopus* (Fig. 6). This technology has sufficient sensitivity and throughput to identify proteins in large numbers of spots from 2D gel electrophoresis. As shown in Fig. 6A, a study of photosensory cells in the retina using 2D gel electrophoresis implicated a 16-kDa protein in the organization of the cell

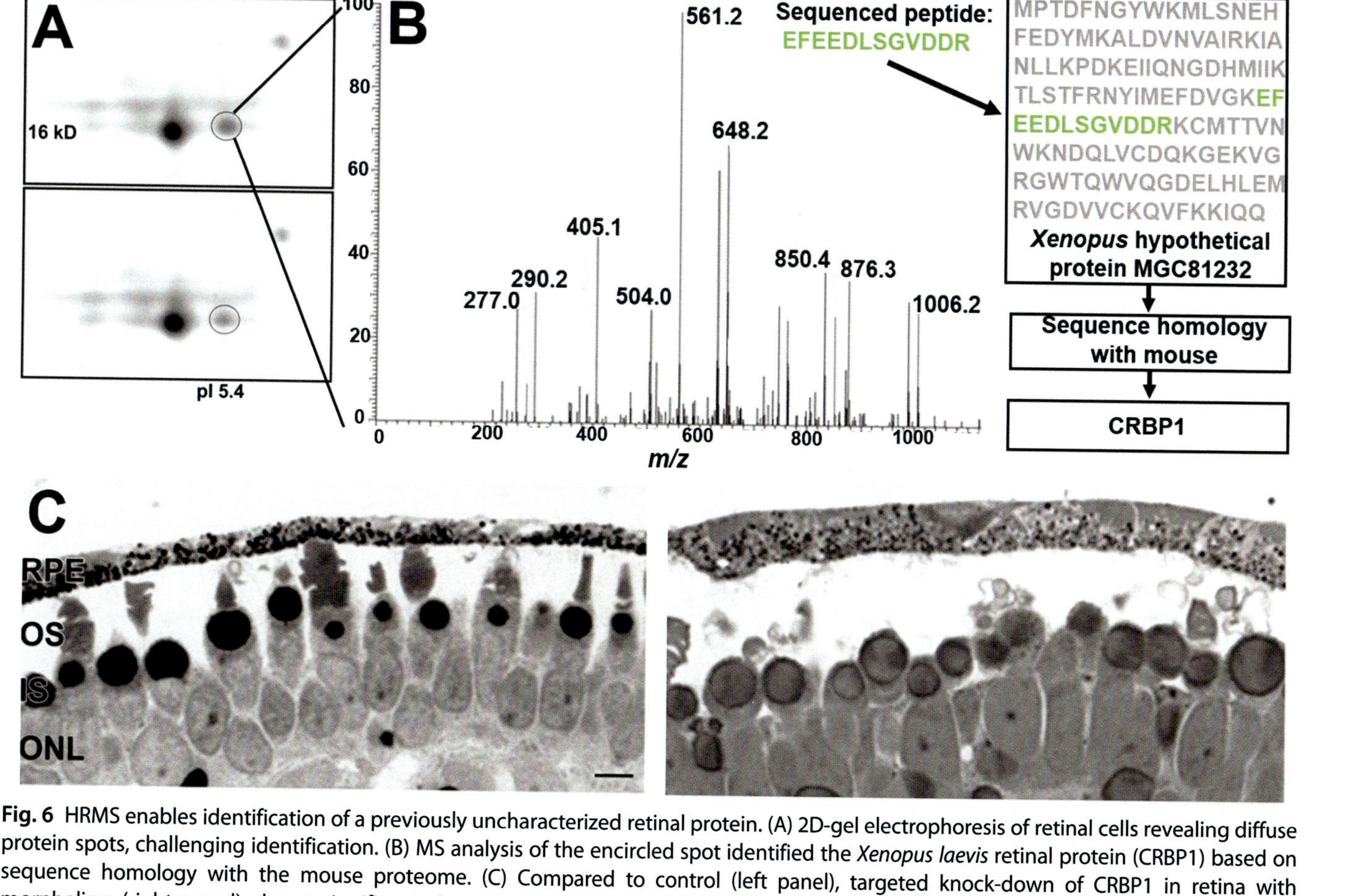

Fig. 6 HRMS enables identification of a previously uncharacterized retinal protein. (A) 2D-gel electrophoresis of retinal cells revealing diffuse protein spots, challenging identification. (B) MS analysis of the encircled spot identified the *Xenopus laevis* retinal protein (CRBP1) based on sequence homology with the mouse proteome. (C) Compared to control (left panel), targeted knock-down of CRBP1 in retina with morpholino (right panel) shows significant disruption of the outer segment. *Adapted with permission from Wang, X. F., Tong, Y. A., Giorgianni, F., Beranova-Giorgianni, S., Penn, J. S., Jablonski, M. M. (2010). Cellular retinol binding protein 1 modulates photoreceptor outer segment folding in the isolated eye.* Developmental Neurobiology, 70, *623–635.*

membrane. Using HRMS (Fig. 6B), it was possible to make a confident identification for the *Xenopus laevis* retinal protein (CRBP1) based on the high-precision sequencing of proteotypic peptides generated in a bottom-up workflow. Validation of protein identity by HRMS, in turn, permitted follow-up biological experiments to spatially interrogate expression using immunohistochemistry. Targeted knockdown of CRBP1, shown in Fig. 6C, revealed the molecular role of organizing the membrane of the cells (Wang et al., 2010). Integration of HRMS with high-resolution imaging and functional experiments can be a viable tool for exploring molecular mechanisms of tissue differentiation in space and time.

3.3.1 Single cells

Xenopus has also been used to explore molecular mechanisms with cellular and subcellular resolution (Fig. 7). Large cell sizes in the early developing embryo facilitate cell handling, sample collection, and detection sensitivity. For example, an average cell in the 16-cell *X. laevis* embryo measures ~250-μm diameter (nonspherical cell shape, see Fig. 7A) and contains ~16-μg of protein. With ~90% protein content dominated by yolk proteins, a single cell is estimated to yield ~1 μg of non-yolk proteins. These protein amounts are sufficient for high-sensitivity mass spectrometers, opening the door to single-cell proteomics in the developing embryo (Lombard-Banek, Moody, et al., 2016; Sun et al., 2016). Further, cells of the early embryo can be identified based on fate maps (see earlier) to target proteomic measurements to specific cell lineages and tissues (Onjiko et al., 2017b). Cell clones may be marked by fluorescent dyes or expressing fluorescent proteins (Fig. 4). The quantitative proteomic studies from identified embryonic cells have revealed previously unknown proteomic cell heterogeneity at a surprisingly early stage of development. This information is useful for follow-up studies to decipher the functional roles of the proteins during early induction of the neural fate.

Results from such experiments can drive new technologies and our current understanding of developmental processes forward. By analyzing the molecular composition of identified cells in the early stage embryo, we discovered small molecules (called metabolites) that alter normal cell fate decisions between neural and epidermal tissues in the embryo (Onjiko, Moody, & Nemes, 2015). Using our home-built HRMS instruments, we have also characterized the proteomic state of early embryonic cells that are fated to form neural, ectodermal, and endodermal tissues (Lombard-Banek et al., 2019; Lombard-Banek, Moody, & Nemes, 2016;

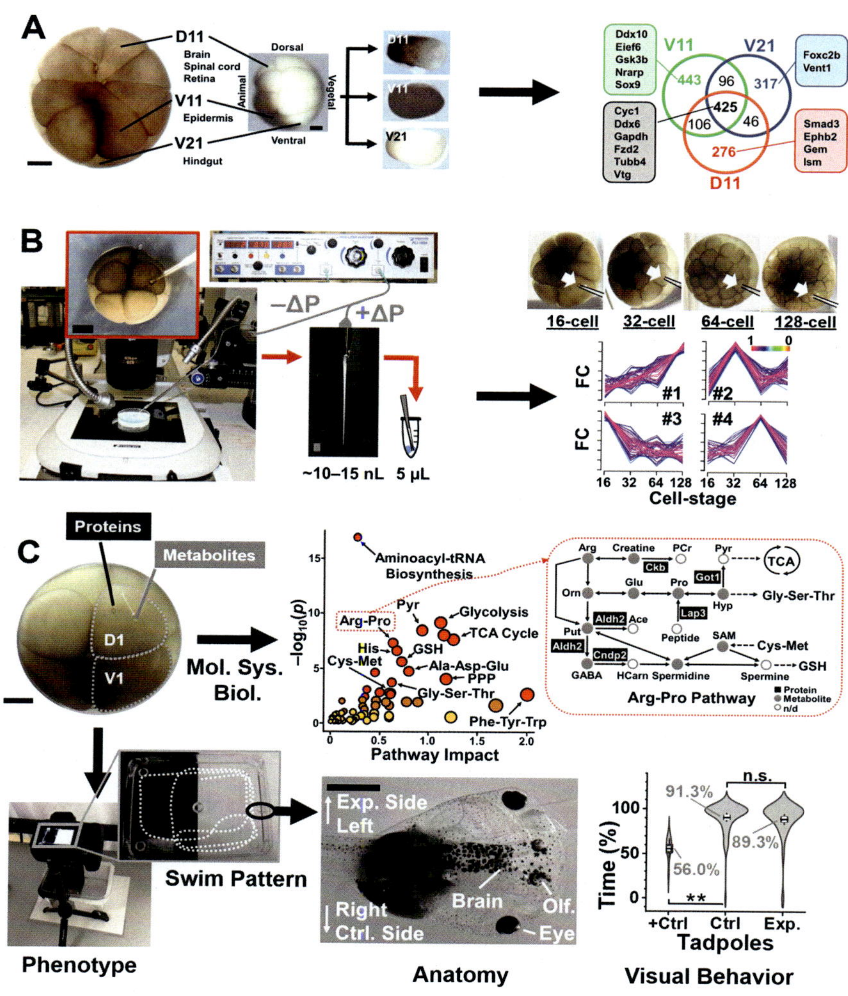

Fig. 7 Single-cell proteo-metabolomics of neural tissue fated cells in live embryos. (A) Whole-cell dissection HRMS (*left panel*) revealing proteomic differences between D11, V11, and V21 blastomeres fated to respectively form neural, epidermal, and endodermal tissues (*right panel*). Scale = 250 μm. (B) In situ microsampling HRMS of the neural-fated cell (*left panel*) revealing proteome reorganization in descendent cell clones (*right panel*). Scale = 500 μm (black), 200 μm (gray). (C) Microsampling and HRMS detecting proteo-metabolomic networks (*top panel*) in the D1 and V1 cells of live embryos, without affecting the anatomy or visual behavior of the future tadpoles (*bottom panel*). Positive control for visual assay (+Ctrl) via double-axotomy of the optic nerves. Scale = 250 μm (embryo), 1 mm (tadpole). *Adapted with permission from Lombard-Banek, C., Li, J., Portero, E. P., Onjiko, R. M., Singer, C. D., Plotnick, D., et al. (2021). In vivo subcellular mass spectrometry enables proteo-metabolomic single-cell systems biology in a*

Lombard-Banek, Moody, et al., 2016). Fig. 7A compares the proteomic profiles of the cells (Lombard-Banek, Moody, et al., 2016). These data revealed that the chordate embryo sets up proteomic cell asymmetry along the dorsal-ventral and animal-vegetal axes at a rather early stage of development, when cell heterogeneity is only detectable along the animal-vegetal axes by deep sequencing of the transcriptome. Further, the single cells appeared to execute proteomic programs that were indicative of their tissue fate. For example, the neural-destined dorsal cell was enriched in geminin, a protein important for regulating transcription of neuronal-fate promoting genes. These results highlight the enrichment of potentially fate-promoting proteins and metabolites in the neural-fated domain of the embryo early on.

Fast molecular events can be studied using subcellular HRMS. We have recently developed a capillary microsampling approached to swiftly (<5 s) collect proteins from cells in the live embryo (Lombard-Banek et al., 2021, 2019; Onjiko, Plotnick, Moody, & Nemes, 2017; Onjiko et al., 2017a, 2017b; Portero & Nemes, 2019). As illustrated in Fig. 7B, a small volume (∼10 nL) was aspirated from identified cells and analyzed by HRMS, resulting in the identification of ∼800 protein groups (representing method sensitivity of ∼700 zmol). Spatially and temporally scalable proteomic analysis also enabled us to monitor protein reorganization during the formation of the neural tissue fated cell clone (Lombard-Banek et al., 2019). Proteins were grouped based on their expression profile over the cell clone, revealing progressive and transient mechanisms of regulation. These quantitative HRMS datasets provide previously unavailable information on the proteomic state of spatially and temporally evolving clones, supplying candidate genes or gene products to determine function.

---

chordate embryo developing to a normally behaving tadpole (X. laevis). Angewandte Chemie, International Edition; *Lombard-Banek, C., Moody, S. A., Manzin, M. C., & Nemes, P. (2019). Microsampling capillary electrophoresis mass spectrometry enables single-cell proteomics in complex tissues: Developing cell clones in live Xenopus laevis and zebrafish embryos.* Analytical Chemistry, 91, *4797–4805; Lombard-Banek, C., Moody, S. A., & Nemes, P. (2016). Single-cell mass spectrometry for discovery proteomics: Quantifying translational cell heterogeneity in the 16-cell frog* (Xenopus) *embryo.* Angewandte Chemie, International Edition, 55, *2454–2458; Onjiko, R. M., Moody, S. A., & Nemes, P. (2015). Single-cell mass spectrometry reveals small molecules that affect cell fates in the 16-cell embryo.* Proceedings of the National Academy of Sciences of the United States of America, 112, *6545–6550; Onjiko, R. M., Portero, E. P., Moody, S. A., & Nemes, P. (2017a). In situ microprobe single-cell capillary electrophoresis mass spectrometry: Metabolic reorganization in single differentiating cells in the live vertebrate* (Xenopus laevis) *embryo.* Analytical Chemistry, 89, *7069–7076.*

Further, subcellular HRMS has also enabled in vivo molecular systems biology of identified cells in live embryos that develop to behaving tadpoles after analysis (Fig. 7C). We have recently tailored capillary microsampling to minimize damage to the cell and its membrane as a miniscule portion of the cell content was aspirated. The approach caused no damage to the neighboring cells or embryonic development. In vivo microsampling with CE-ESI-MS permitted the detection of both proteins and metabolites in single ventral and dorsal cells of the 8-cell embryo (Lombard-Banek et al., 2021). As shown in Fig. 7C, the results revealed proteo-metabolomic networks with differential quantitative activities between the neural (D1) and epidermal (V1) fated cells. The microsampling approach preserved the viability of the sampled cell and the embryo. In our study, 95% of the sampled embryos successfully developed into sentient tadpoles that had indistinguishable anatomy from their wild-type siblings. Further, the tadpoles also presented indistinguishable visual behavior in a background color preference assay that was validated based on double-axotomy of the optic nerves (positive Control, see Fig. 7C). In vivo subcellular proteo-metabolomic CE-HRMS expands the molecular toolbox of molecular systems biology, cell and developmental biology, and neurobiology.

4. Conclusions

Xenopus and HRMS-based proteomics empower developmental neurobiology. This biological model and bioanalytical technology offer important advantages in experimentation and data fidelity. Detection and quantification of large numbers, hundreds-to-thousands, of proteins (and their PTMs) over a broad linear dynamic range of concentration complements molecular information available from transcriptomics to investigate complex molecular dynamics in the developing system. HRMS also enables the detection of small molecules for holistic systems molecular biology, including molecules partaking in energy balance (e.g., adenosine triphosphate), redox control (e.g., glutathione and oxidized glutathione), and signaling (e.g., retinoic acid). Additionally, the technology affords an imaging modality (not discussed here) to document the distribution of molecules in variable spatial resolution, as reviewed in (Lombard-Banek et al., 2017). In one such form of HRMS, MALDI-TOF was used to map neuropeptides in the crab central nervous system (DeLaney et al., 2021) and small molecules in sections of *Xenopus* embryos and tadpoles (Wang et al., 2019). The integration of *Xenopus* with multi-'omics, including HRMS-based proteomics and metabolomics, raises a potential to study molecular mechanisms of development in previously unavailable details at the realms of the organism, organ, tissue, and the cell.

Acknowledgments

This research was partially supported by the National Science Foundation under award IOS-1832968 CAREER (to P.N.) and the National Institutes of Health under awards R35GM124755 (to P.N.) and F31DC018742 (to A.B.B).

References

Aebersold, R., Burlingame, A. L., & Bradshaw, R. A. (2013). Western blots versus selected reaction monitoring assays: Time to turn the tables? *Molecular & Cellular Proteomics, 12*, 2381–2382.

Arth, A. C., Tinker, S. C., Simeone, R. M., Ailes, E. C., Cragan, J. D., & Grosse, S. D. (2017). Inpatient hospitalization costs associated with birth defects among persons of all ages—United States, 2013. *MMWR. Morbidity and Mortality Weekly Report, 66*, 41–46.

Baxi, A. B., Lombard-Banek, C., Moody, S. A., & Nemes, P. (2018). Proteomic characterization of the neural ectoderm fated cell clones in the *Xenopus laevis* embryo by high-resolution mass spectrometry. *ACS Chemical Neuroscience, 9*, 2064–2073.

Baxi, A. B., Moody, S. A., & Nemes, P. (2020). NanoLC-MS based discovery proteomic analysis of the frog inner ear. In *68th ASMS Conference on Mass Spectrometry and Allied Topics Online Meeting*.

Baxi, A. B., Quach, V. M., Li, J., & Nemes, P. (2021). Proteo-metabolomic study of the spemann's organizer in the vertebrate (frog) embryo. In *17th annual conference US HUPO*.

Blum, M., & Ott, T. (2018). *Xenopus*: An undervalued model organism to study and model human genetic disease. *Cells, Tissues, Organs, 205*, 303–313.

Borodinsky, L. N. (2017). *Xenopus laevis* as a model organism for the study of spinal cord formation, development, function and regeneration. *Frontiers in Neural Circuits, 11*, 1–9.

Bradbury, A., & Pluckthun, A. (2015). Standardize antibodies used in research. *Nature, 518*, 27–29.

Cagnetta, R., Frese, C. K., Shigeoka, T., Krijgsveld, J., & Holt, C. E. (2018). Rapid cue-specific remodeling of the nascent axonal proteome. *Neuron, 99*, 29–46.

Chen, C., Hou, J., Tanner, J. J., & Cheng, J. L. (2020). Bioinformatics methods for mass spectrometry-based proteomics data analysis. *International Journal of Molecular Sciences, 21*, 25.

Chu, I. K., Siu, C. K., Lau, J. K. C., Tang, W. K., Mu, X. Y., Lai, C. K., et al. (2015). Proposed nomenclature for peptide ion fragmentation. *International Journal of Mass Spectrometry, 390*, 24–27.

Cox, J., Hein, M. Y., Luber, C. A., Paron, I., Nagaraj, N., & Mann, M. (2014). Accurate proteome-wide label-free quantification delayed normalization and maximal peptide ratio extraction, termed MaxLFQ. *Molecular & Cellular Proteomics, 13*, 2513–2526.

Dale, L., & Slack, J. M. W. (1987). Fate map for the 32-cell stage of *Xenopus laevis*. *Development, 99*, 527–551.

DeLaney, K., Hu, M., Hellenbrand, T., Dickinson, P. S., Nusbaum, M. P., & Li, L. (2021). Mass spectrometry quantification, localization, and discovery of feeding-related neuropeptides in *cancer borealis*. *ACS Chemical Neuroscience, 12*, 782–798.

DeLaney, K., Sauer, C. S., Vu, N. Q., & Li, L. J. (2019). Recent advances and new perspectives in capillary electrophoresis-mass spectrometry for single cell "omics". *Molecules, 24*, 1–21.

Demircan, T., Sibai, M., & Altuntas, E. (2020). Proteome data to explore the axolotl limb regeneration capacity at neotenic and metamorphic stages. *Data in Brief*, *29*, 1–7.

Drew, K., Lee, C., Cox, R. M., Dang, V., Devitt, C. C., McWhite, C. D., et al. (2020). A systematic, label-free method for identifying RNA-associated proteins in vivo provides insights into vertebrate ciliary beating machinery. *Developmental Biology*, *467*, 108–117.

Duan, G. Y., & Walther, D. (2015). The roles of post-translational modifications in the context of protein interaction networks. *PLoS Computational Biology*, *11*, 1–23.

Duncan, A. R., & Khokha, M. K. (2016). *Xenopus* as a model organism for birth defects—Congenital heart disease and heterotaxy. *Seminars in Cell & Developmental Biology*, *51*, 73–79.

Federspiel, J. D., Tandon, P., Wilczewski, C. M., Wasson, L., Herring, L. E., Venkatesh, S. S., et al. (2019). Conservation and divergence of protein pathways in the vertebrate heart. *PLoS Biology*, *17*, 1–31.

Fonslow, B. R., & Yates, J. R. (2009). Capillary electrophoresis applied to proteomic analysis. *Journal of Separation Science*, *32*, 1175–1188.

Geng, X. F., Guo, J. L., Zang, X. Y., Chang, C. F., Shang, H. T., Wei, H., et al. (2019). Proteomic analysis of eleven tissues in the Chinese giant salamander (*Andrias davidianus*). *Scientific Reports*, *9*, 1–10.

Han, X. M., Aslanian, A., & Yates, J. R. (2008). Mass spectrometry for proteomics. *Current Opinion in Chemical Biology*, *12*, 483–490.

Harland, R. M., & Grainger, R. M. (2011). *Xenopus* research: Metamorphosed by genetics and genomics. *Trends in Genetics*, *27*, 507–515.

Hashimoto, Y., Greco, T. M., & Cristea, I. M. (2019). Contribution of mass spectrometry-based proteomics to discoveries in developmental biology. In A. G. Woods, & C. C. Darie (Eds.), *Advancements of mass spectrometry in biomedical research* (2nd ed., pp. 143–154). Ag, Cham: Springer International Publishing.

Huang, D. W., Sherman, B. T., & Lempicki, R. A. (2009). Bioinformatics enrichment tools: Paths toward the comprehensive functional analysis of large gene lists. *Nucleic Acids Research*, *37*, 1–13.

Hwang, W. Y., Marquez, J., & Khokha, M. K. (2019). *Xenopus*: Driving the discovery of novel genes in patient disease and their underlying pathological mechanisms relevant for organogenesis. *Frontiers in Physiology*, *10*, 1–7.

Issaq, H. J. (2001). The role of separation science in proteomics research. *Electrophoresis*, *22*, 3629–3638.

Jensen, O. N. (2004). Modification-specific proteomics: Characterization of post-translational modifications by mass spectrometry. *Current Opinion in Chemical Biology*, *8*, 33–41.

Jensen, L. J., Kuhn, M., Stark, M., Chaffron, S., Creevey, C., Muller, J., et al. (2009). STRING 8—A global view on proteins and their functional interactions in 630 organisms. *Nucleic Acids Research*, *37*, D412–D416.

Kong, A. T., Leprevost, F. V., Avtonomov, D. M., Mellacheruvu, D., & Nesvizhskii, A. I. (2017). MSFragger: Ultrafast and comprehensive peptide identification in mass spectrometry-based proteomics. *Nature Methods*, *14*, 513–520.

Lee-Liu, D., Sun, L. L., Dovichi, N. J., & Larrain, J. (2018). Quantitative proteomics after spinal cord injury (SCI) in a regenerative and a nonregenerative stage in the frog *Xenopus laevis*. *Molecular & Cellular Proteomics*, *17*, 592–606.

Lindemann, C., Thomanek, N., Hundt, F., Lerari, T., Meyer, H. E., Wolters, D., et al. (2017). Strategies in relative and absolute quantitative mass spectrometry based proteomics. *Biological Chemistry*, *398*, 687–699.

Lombard-Banek, C., Li, J., Portero, E. P., Onjiko, R. M., Singer, C. D., Plotnick, D., et al. (2021). In vivo subcellular mass spectrometry enables proteo-metabolomic single-cell

systems biology in a chordate embryo developing to a normally behaving tadpole (*X. laevis*). *Angewandte Chemie, International Edition*. https://doi.org/10.1002/anie.202100923.
Lombard-Banek, C., Moody, S. A., Manzin, M. C., & Nemes, P. (2019). Microsampling capillary electrophoresis mass spectrometry enables single-cell proteomics in complex tissues: Developing cell clones in live *Xenopus laevis* and zebrafish embryos. *Analytical Chemistry*, *91*, 4797–4805.
Lombard-Banek, C., Moody, S. A., & Nemes, P. (2016). Single-cell mass spectrometry for discovery proteomics: Quantifying translational cell heterogeneity in the 16-cell frog (*Xenopus*) embryo. *Angewandte Chemie, International Edition*, *55*, 2454–2458.
Lombard-Banek, C., Portero, E. P., Onjiko, R. M., & Nemes, P. (2017). New-generation mass spectrometry expands the toolbox of cell and developmental biology. *Genesis*, *55*, 1–34.
Lombard-Banek, C., Reddy, S., Moody, S. A., & Nemes, P. (2016). Label-free quantification of proteins in single embryonic cells with neural fate in the cleavage-stage frog (*Xenopus laevis*) embryo using capillary electrophoresis electrospray ionization high-resolution mass spectrometry (CE-ESI-HRMS). *Molecular & Cellular Proteomics*, *15*, 2756–2768.
Moody, S. A. (1987a). Fates of the blastomeres of the 16-cell stage *Xenopus* embryo. *Developmental Biology*, *119*, 560–578.
Moody, S. A. (1987b). Fates of the blastomeres of the 32-cell-stage *Xenopus* embryo. *Developmental Biology*, *122*, 300–319.
Nakayama, T., Blitz, I. L., Fish, M. B., Odeleye, A. O., Manohar, S., Cho, K. W. Y., et al. (2014). Cas9-based genome editing in *Xenopus tropicalis*. In J. A. Doudna, & E. J. Sontheimer (Eds.), *Use of Crispr/Cas9, Zfns, and Talens in generating site-specific genome alterations* (pp. 355–375). San Diego: Elsevier Academic Press Inc.
O'Connell, J. D., Paulo, J. A., O'Brien, J. J., & Gygi, S. P. (2018). Proteome-wide evaluation of two common protein quantification methods. *Journal of Proteome Research*, *17*, 1934–1942.
Ong, S. E., Blagoev, B., Kratchmarova, I., Kristensen, D. B., Steen, H., Pandey, A., et al. (2002). Stable isotope labeling by amino acids in cell culture, SILAC, as a simple and accurate approach to expression proteomics. *Molecular & Cellular Proteomics*, *1*, 376–386.
Onjiko, R. M., Moody, S. A., & Nemes, P. (2015). Single-cell mass spectrometry reveals small molecules that affect cell fates in the 16-cell embryo. *Proceedings of the National Academy of Sciences of the United States of America*, *112*, 6545–6550.
Onjiko, R. M., Plotnick, D. O., Moody, S. A., & Nemes, P. (2017). Metabolic comparison of dorsal versus ventral cells directly in the live 8-cell frog embryo by microprobe single-cell CE-ESI-MS. *Analytical Methods*, *9*, 4964–4970.
Onjiko, R. M., Portero, E. P., Moody, S. A., & Nemes, P. (2017a). In situ microprobe single-cell capillary electrophoresis mass spectrometry: Metabolic reorganization in single differentiating cells in the live vertebrate (*Xenopus laevis*) embryo. *Analytical Chemistry*, *89*, 7069–7076.
Onjiko, R. M., Portero, E. P., Moody, S. A., & Nemes, P. (2017b). Microprobe capillary electrophoresis mass spectrometry for single-cell metabolomics in live frog (*Xenopus laevis*) embryos. *Journal of Visualized Experiments*, 1–9, e56956. https://doi.org/10.3791/56956.
Peshkin, L., Wuehr, M., Pearl, E., Haas, W., Freeman, R. M., Gerhart, J. C., et al. (2015). On the relationship of protein and mRNA dynamics in vertebrate embryonic development. *Developmental Cell*, *35*, 383–394.

Peuchen, E. H., Sun, L. L., & Dovichi, N. J. (2016). Optimization and comparison of bottom-up proteomic sample preparation for early-stage *Xenopus laevis* embryos. *Analytical and Bioanalytical Chemistry, 408*, 4743–4749.

Pino, L. K., Rose, J., O'Broin, A., Shah, S., & Schilling, B. (2020). Emerging mass spectrometry-based proteomics methodologies for novel biomedical applications. *Biochemical Society Transactions, 48*, 1953–1966.

Pino, L. K., Searle, B. C., Bollinger, J. G., Nunn, B., MacLean, B., & MacCoss, M. J. (2020). The skyline ecosystem: Informatics for quantitative mass spectrometry proteomics. *Mass Spectrometry Reviews, 39*, 229–244.

Portero, E. P., & Nemes, P. (2019). Dual cationic-anionic profiling of metabolites in a single identified cell in a live *Xenopus laevis* embryo by microprobe CE-ESI-MS. *Analyst, 144*, 892–899.

Pratt, K. G., & Khakhalin, A. S. (2013). Modeling human neurodevelopmental disorders in the *Xenopus* tadpole: From mechanisms to therapeutic targets. *Disease Models & Mechanisms, 6*, 1057–1065.

Presler, M., Van Itallie, E., Klein, A. M., Kunz, R., Coughlin, M. L., Peshkin, L., et al. (2017). Proteomics of phosphorylation and protein dynamics during fertilization and meiotic exit in the *Xenopus* egg. *Proceedings of the National Academy of Sciences of the United States of America, 114*, E10838–E10847.

Quach, V. M., Baxi, A. B., & Nemes, P. (2019). Proteomic characterization of the spemann organizer in Xenopus laevis (frog) embryos. In *67th ASMS Conference on Mass Spectrometry and Allied Topics*.

Sater, A. K., & Moody, S. A. (2017). Using *Xenopus* to understand human diseases and developmental disorders. *Genesis, 55*, 1–14.

Schiapparelli, L. M., Shah, S. H., Ma, Y. H., McClatchy, D. B., Sharma, P., Li, J. L., et al. (2019). The retinal ganglion cell transportome identifies proteins transported to axons and presynaptic compartments in the visual system in vivo. *Cell Reports, 28*, 1935–1947.

Session, A. M., Uno, Y., Kwon, T., Hapman, J. A. C., Toyoda, A., Takahashi, S., et al. (2016). Genome evolution in the allotetraploid frog *Xenopus laevis*. *Nature, 538*, 336–343.

Smits, A. H., Lindeboom, R. G. H., Perino, M., van Heeringen, S. J., Veenstra, G. J. C., & Vermeulen, M. (2014). Global absolute quantification reveals tight regulation of protein expression in single *Xenopus* eggs. *Nucleic Acids Research, 42*, 9880–9891.

Sun, L. L., Bertke, M. M., Champion, M. M., Zhu, G. J., Huber, P. W., & Dovichi, N. J. (2014). Quantitative proteomics of *Xenopus laevis* embryos: Expression kinetics of nearly 4000 proteins during early development. *Scientific Reports, 4*, 1–9.

Sun, L. L., Dubiak, K. M., Peuchen, E. H., Zhang, Z. B., Zhu, G. J., Huber, P. W., et al. (2016). Single cell proteomics using frog (*Xenopus laevis*) blastomeres isolated from early stage embryos, which form a geometric progression in protein content. *Analytical Chemistry, 88*, 6653–6657.

Tyanova, S., Temu, T., & Cox, J. (2016). The maxquant computational platform for mass spectrometry-based shotgun proteomics. *Nature Protocols, 11*, 2301–2319.

Virag, D., Dalmadi-Kiss, B., Vekey, K., Drahos, L., Klebovich, I., Antal, I., et al. (2020). Current trends in the analysis of post-translational modifications. *Chromatographia, 83*, 1–10.

Walther, T. C., & Mann, M. (2010). Mass spectrometry-based proteomics in cell biology. *The Journal of Cell Biology, 190*, 491–500.

Wang, M., Dubiak, K., Zhang, Z. B., Huber, P. W., Chen, D. D. Y., & Dovichi, N. J. (2019). MALDI-imaging of early stage *Xenopus laevis* embryos. *Talanta, 204*, 138–144.

Wang, X. F., Tong, Y. A., Giorgianni, F., Beranova-Giorgianni, S., Penn, J. S., & Jablonski, M. M. (2010). Cellular retinol binding protein 1 modulates photoreceptor outer segment folding in the isolated eye. *Developmental Neurobiology, 70*, 623–635.

Waterston, R. H., Lindblad-Toh, K., Birney, E., Rogers, J., Abril, J. F., Agarwal, P., et al. (2002). Initial sequencing and comparative analysis of the mouse genome. *Nature*, *420*, 520–562.
Wuehr, M., Freeman, R. M., Presler, M., Horb, M. E., Peshkin, L., Gygi, S. P., et al. (2014). Deep proteomics of the *Xenopus laevis* egg using an mRNA-derived reference database. *Current Biology*, *24*, 1467–1475.
Xenopus Community. (2020). *Xenopus community white paper*. https://www.xenbase.org/: Xenbase.
Zhang, Y. Y., Fonslow, B. R., Shan, B., Baek, M. C., & Yates, J. R. (2013). Protein analysis by shotgun/bottom-up proteomics. *Chemical Reviews*, *113*, 2343–2394.

Amphibian models for regeneration and disease

CHAPTER EIGHT

Salamanders: The molecular basis of tissue regeneration and its relevance to human disease

Claudia Marcela Arenas Gómez and Karen Echeverri*
Marine Biological Laboratory, Eugene Bell Center for Regenerative Biology and Tissue Engineering, University of Chicago, Woods Hole, MA, United States
*Corresponding author: e-mail address: kecheverri@mbl.edu

Contents

Abstract

Salamanders are recognized for their ability to regenerate a broad range of tissues. They have also have been used for hundreds of years for classical developmental biology studies because of their large accessible embryos. The range of tissues these animals can regenerate is fascinating, from full limbs to parts of the brain or heart, a potential that is missing in humans. Many promising research efforts are working to decipher the molecular blueprints shared across the organisms that naturally have the capacity to regenerate different tissues and organs. Salamanders are an excellent example of a vertebrate that can functionally regenerate a wide range of tissue types. In this review, we outline some of the significant insights that have been made that are aiding in understanding the cellular and molecular mechanisms of tissue regeneration in salamanders and discuss why salamanders are a worthy model in which to study regenerative biology and how this may benefit research fields like regenerative medicine to develop therapies for humans in the future.

Current Topics in Developmental Biology, Volume 145
ISSN 0070-2153
https://doi.org/10.1016/bs.ctdb.2020.11.009

1. Introduction

Tissue regeneration is widely distributed across both plant and animal phylogenies (Birnbaum & Sánchez Alvarado, 2008; Brockes & Kumar, 2008). The main goal of tissue regeneration is to restore the morphological and functional features of tissue after an injury. In contrast the majority of the mammals are experts at repairing wounds which results in scar tissue but have low regenerative capacity with the exception of tissues such as the liver, bone marrow, gut epithelium, that can recover after a moderate injury (Brockes & Kumar, 2008; Londono, Sun, Tuan, & Lozito, 2018). Among vertebrates, salamanders are one of the organisms that have the outstanding ability to regenerate different tissues and organs such as the limb, heart, spinal cord, and lens (Birnbaum & Sánchez Alvarado, 2008; Brockes & Kumar, 2008; Dinsmore & American Society of Zoologists, 2008; Tanaka, 2016; Tanaka & Ferretti, 2009). Different species of salamanders have been used to give some molecular and cellular insights into the mechanisms that promote a regenerative response (Joven, Elewa, & Simon, 2019) (Fig. 1).

The most widely used salamander species in the developmental and regenerative biology fields are *Ambystoma mexicanum* (Mexican Axolotl) and *Notophthalmus viridescens* (Eastern Newt) (Joven et al., 2019) (Fig. 2A). However not all salamander species have the same regenerative ability, important insights into these divergent features have come from studying other species such as *Pleurodeles waltl* (Elewa et al., 2017), *Cynops pyrrhogaster* (Nakamura et al., 2014; Tsutsumi, Inoue, Yamada, & Agata, 2015), *Bolitglossa ramose* (Arenas Gómez, Gomez Molina, Zapata, & Delgado, 2017), *Hynobius chinensis* (Che, Sun, Wang, & Xu, 2014) and *Andrias davidianus* (Geng et al., 2015) (Fig. 1). Considering the differences in how highly related salamanders execute a regenerative program, it is clear that is will be important to understand more broadly the extent of diversity in the pro-regenerative programs found in salamanders, especially considering there are around 739 different species widely distributed around the world (AmphibiaWeb, 2019).

To date researchers have made important discoveries regarding some cellular and molecular mechanisms that promote regenerative responses in salamanders. We now know that there is a nerve dependency to regeneration, remodeling of the extracellular matrix plays a crucial; possibly instructive role in directing cells and that the timing of the immune cells arriving to the injury site is important (Calve, Odelberg, & Simon, 2010; Campbell & Crews, 2008; Godwin, Pinto, & Rosenthal, 2013; Kumar, Godwin, Gates, Garza-Garcia, & Brockes, 2007; Tsai, Baselga-Garriga, & Melton, 2019).

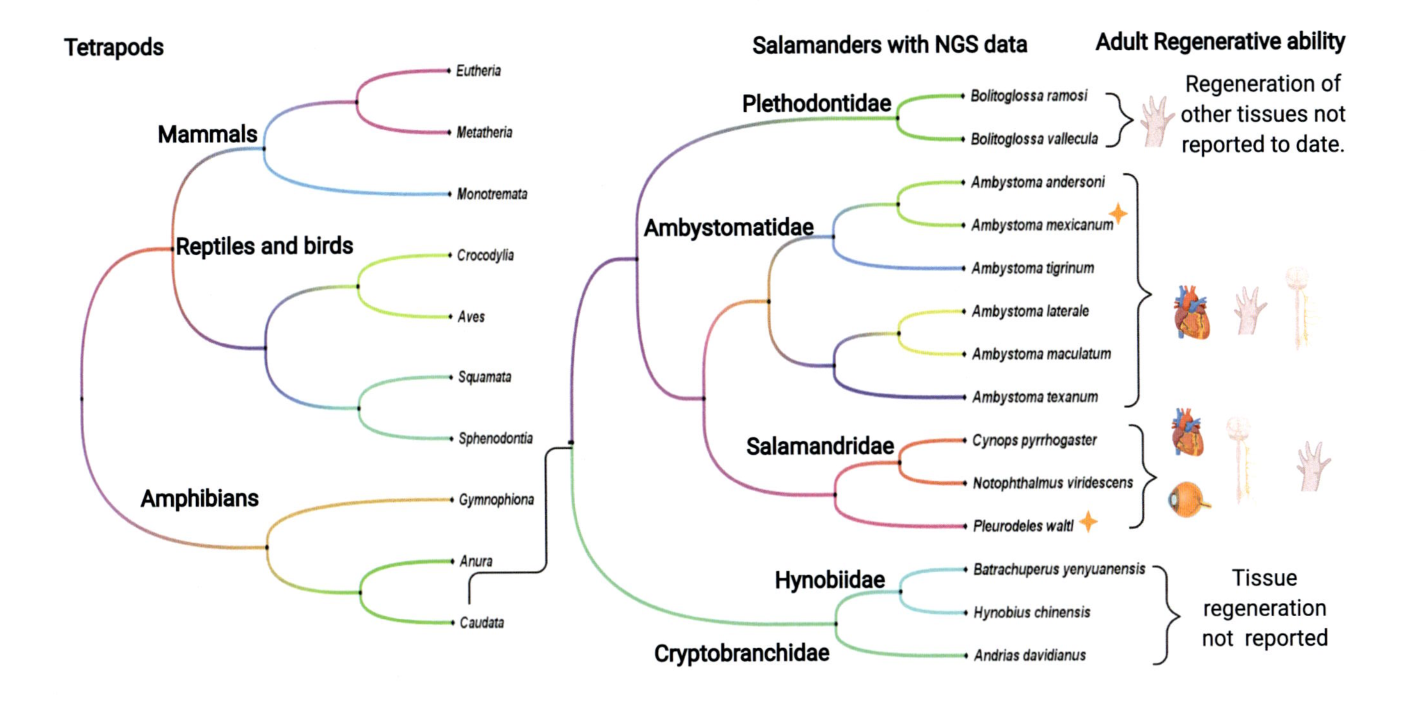

Fig. 1 Salamanders species with Next-Generation Sequencing Data (NGS) and their regenerative capacities as adults. Salamanders are to date the only known tetrapod with the capacity to fully regenerate limbs. From the 10 salamanders families only species from 5 families have transcriptional profiling data from regenerating tissue available. *A. mexicanum* and *P. waltl* (orange star) are the only salamanders with sequenced genomes to date.

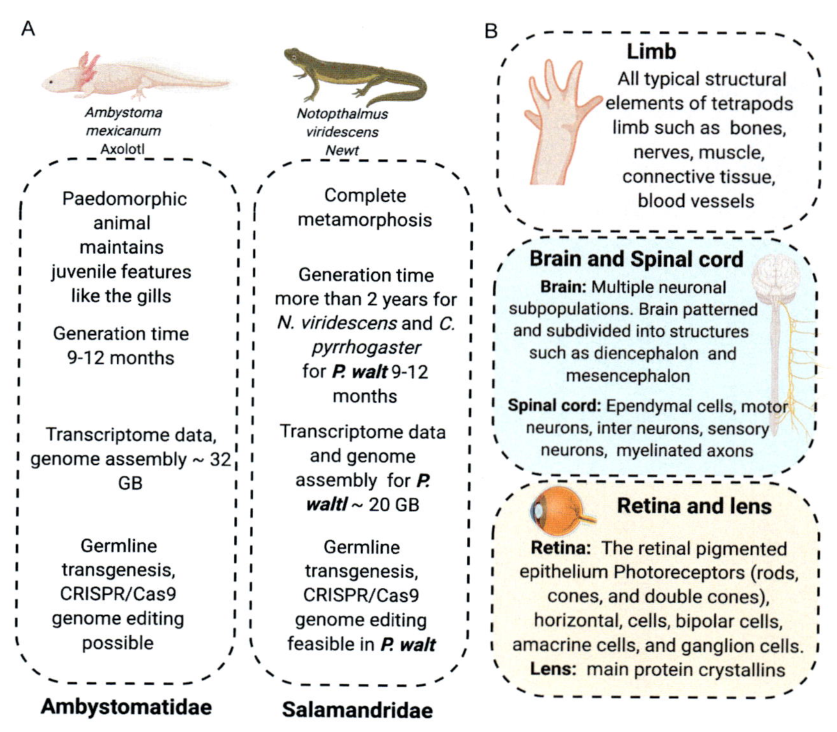

Fig. 2 Main salamanders model used in field of regenerative biology. (A) *A. mexicanum*, *N. viridescens* and *P. waltl* are the main salamanders that have been established in the field to understand tissue regeneration. *P. waltl* has a genome assembled and the generation time is shorter than *N. viridescens*. (B) Main morphological features in the limb, spinal cord, brain, lens and retina that make salamanders comparable with human tissues.

This knowledge may provide useful information to the field of regenerative medicine and tissue engineering in order to design novel therapies for diseases or to reduce scarring after wounding in humans (Brockes & Gates, 2014; Brockes & Kumar, 2005). Additionally these organisms may help us to understand complex diseases such as cancer, because the molecular and cellular landscape expressed during tissue regeneration is comparable with the environment during an oncogenic process (Brockes, 1998; Fior, 2014; Oviedo & Beane, 2009; Vieira, Wells, & McCusker, 2020). Finally, salamanders also could be useful organisms to understand the aging process (Vieira et al., 2020; Yun, 2015, 2018). It is known that most salamanders keep the regenerative potential throughout life; however, it is clear that younger animals regenerate faster and make fewer mistakes, suggesting that even in salamanders the aging process affects regenerative ability; however, old salamanders still regenerate much better than humans (Monaghan et al., 2014; Yun, 2015).

Here, we will review the knowledge that we have learnt from salamanders about tissue regeneration in a range of tissues and organs and how this response differs in humans. We will give an overview of the main molecular and cellular mechanisms known to be involved and how modern genetic tools have been aiding to elucidate this amazing biological process in order to translate it in the future to the field of regenerative medicine.

The urodele amphibians, commonly referred to as salamanders, have been studied for centuries. The 18th century was considered a golden era to study tissue regeneration in different research organisms, among them, Réaumur's research on insect appendage regeneration, Tremblay's work on Hydra regeneration, Bonnet's work on worm regeneration and Lazzaro Spallanzani first documented salamander limb regeneration in 1768 (Dinsmore & American Society of Zoologists, 2008; Spallanzani, 1768). At that time the technology was not yet developed to enable researchers to decipher the molecular and cellular mechanisms of regeneration in salamanders. In the past 30 years, research on tissue regeneration in salamanders has started to reemerge thanks to the development of new tools that allow us to label and track cells in vivo and to explore molecular interactions in the genome; this is enabling researchers to decipher different cellular and molecular process that are key to promoting functional regeneration. In the next chapter, we will discuss some of the recent research on regeneration in salamanders such as limb, and neural tissue regeneration, these are tissues which have high homology to their mammalian counterparts (Fig. 2B).

2. Limb regeneration

Salamanders are to date the only known tetrapod able to regenerate limbs after an injury throughout their lives. Regenerative ability in salamanders appears to be an ancient trait, fossil records suggest that it has been conserved for approximately 300 million years. Using fossil records Fröbisch et al. identified that limb regeneration occured in *Micromelerpeton*, a distant relative of modern amphibians (Fröbisch, Bickelmann, & Witzmann, 2014). The salamander species, *A. mexicanum* and *N. viridescens*, have been the main salamanders used to study the molecular and cellular mechanism that promote limb regeneration (Brockes & Kumar, 2005, 2008; Haas & Whited, 2017; Joven et al., 2019; Simon & Tanaka, 2013; Tanaka, 2016; Tsonis & Fox, 2009). The main stages of limb regeneration after an amputation can be summarized as wound healing, blastema formation, blastema patterning and finally cell differentiation to replace all the lost cell types reviewed in (Stocum, 1979, 1991) (Fig. 3A).

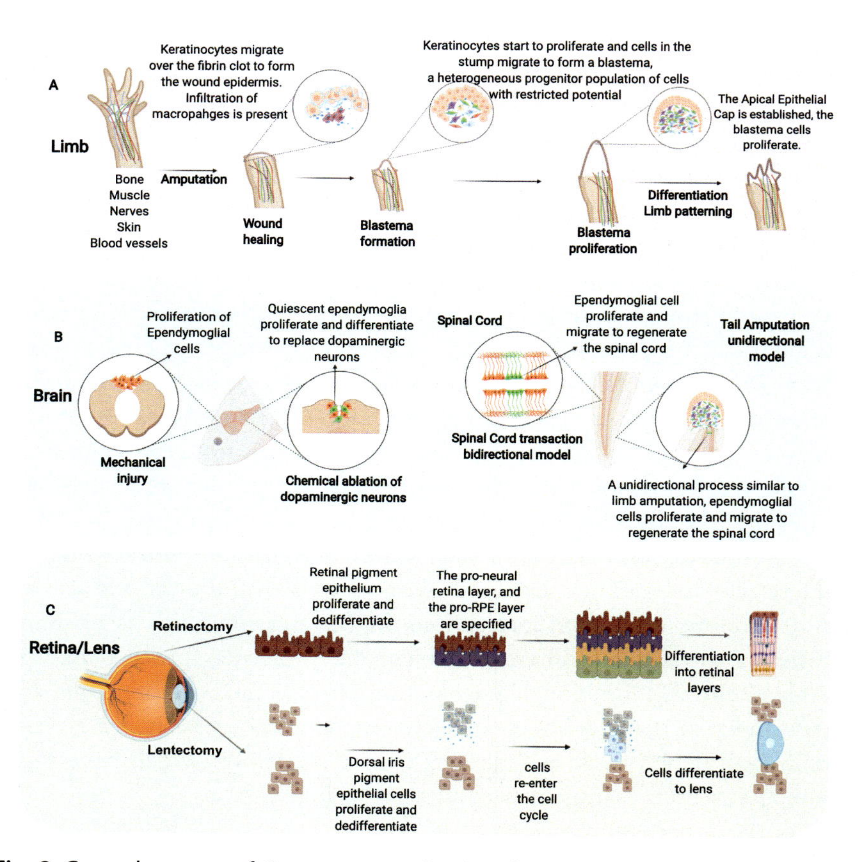

Fig. 3 General process of tissue regeneration in salamanders. (A) Limb regeneration main stages, wound healing, blastema formation, blastema patterning and finally cell differentiation to replace all the lost cell types. (B) Brain and spinal cord regeneration is largely dependent on activation of ependymoglial cells. (C) Retina and lens regeneration is led by the transdifferentiation of the retinal pigmented epithelium (RPE) after a retinectomy and the dorsal iris pigment epithelial cells after a lentectomy.

After amputation, the wound healing phase is characterized by the formation of a fibrin clot and a fast migration of non-proliferative keratinocytes (Arenas Gómez, Sabin, & Echeverri, 2020; Chalkley, 1954; Murawala, Tanaka, & Currie, 2012; Thornton, 1954). Successful scar-free wound healing is necessary for regeneration to proceed. During the wound healing phase, cell migrate to close the wound and from the distal epithelium commonly referred to as the wound epithelium. Throughout the migration process the cells differentially regulate gene expression and potentially begin to express and/or secrete proteins that activate surrounding cells to form a blastema (Campbell & Crews, 2008). Older studies have shown that covering a wound with uninjured skins inhibits the regeneration process, this suggests that the changes in gene expression in response to injury are crucial for the scar free regenerative process to proceed (Tassava & Garling, 1979; Tassava & Loyd, 1977). The wound epithelium formed during regeneration is considered a specialized structure as when the keratinocytes start to migrate they start to produce their own ECM composed mainly of laminin, collagen type IV, collagen type XII, MMP3, and MMP9 (Campbell & Crews, 2008). During re-epithelization of the wound site other crucial processes are advancing in the stump tissue such as histolysis of remnant injured tissues like dermis, muscle, bond, and remodeling of ECM. Equally important, an immune response characterized by the increase of anti-inflammatory cytokines (e.g., IL-4, IL-10, IL-13), proinflammatory cytokines (e.g., TNF-α, IL-17) and macrophages chemotactic molecules (e.g., CCL4, CCL3, CXCL12) is activated by 1 day post-amputation (dpa) until 15 dpa (Godwin et al., 2013; Tsai et al., 2019).

Once the wound epithelium is formed the basal keratinocytes start to proliferate; they form an epithelium layer that is called the apical epithelial cap (AEC), which is thought to play a similar role to that of the apical ectodermal ridge during limb development (Chalkley, 1954; Iten & Bryant, 1973; Thornton, 1954). The AEC is also referred to in the literature as the regenerative epithelium (Campbell & Crews, 2008; Campbell et al., 2011; Christensen & Tassava, 2000; Satoh, Graham, Bryant, & Gardiner, 2008).

This structure, AEC or regenerative epithelium, is an important source of molecules that are thought to be essential in recruiting cells to the injury site to form a blastema; which is defined as a mound of proliferating cells that will eventually differentiate and replace the lost limb (Campbell et al., 2011; Satoh, Bryant, & Gardiner, 2012; Satoh, Graham, et al., 2008). The blastema is a pool of heterogeneous progenitor cells with mainly restricted potential that are going to be the source of the cells to re-establish the tissues that form the limb (Gardiner, Muneoka, & Bryant, 1986; Kragl et al., 2013; Muneoka, Fox, & Bryant, 1986; Nye, Cameron, Chernoff, & Stocum, 2003; Satoh, Bryant, & Gardiner, 2008). One of the big questions in the field is, what are the main cellular and molecular mechanism that contribute to the generation of the progenitor cells that form the blastema? One of the key components is the remodeling of the ECM, where spatiotemporal changes of Tenascin (TN), Fibronectin (FN), and Hyaluronic acid (HA) are essential during the early days to promote regeneration.

In newts, a transitional matrix composed of TN and FN favors promotion of cell proliferation and an ECM rich in HA, TN, FN infiltrates the damaged basal membrane of the skeletal muscle to trigger dedifferentiation of muscle cells (Calve et al., 2010). Muscle dedifferentiation is a process which occurs in response to injury, causing a multi-nucleated muscle fiber to fragment and give rise to mono-nucleated cells that enter into the blastema (Hay, 1959, 1966). Muscle dedifferentiation is an important source of new cells to regenerate muscle during limb regeneration in newts. Interestingly the closely related axolotls use their resident +*pax7* stem cells (satellite cells) to regenerate muscle in the context of limb regeneration. This difference was elegantly shown using the power of transgenic animals via a *Cre-loxP* genetic fate mapping of skeletal muscle during limb regeneration in *A. mexicanum* and *N. viridescens* (Sandoval-Guzman et al., 2014). It remains to be seen how widespread these two mechanisms of regenerating muscle are among other species of salamanders. Axolotls in this scenario appear more similar to mouse and humans which both can repair small pieces of muscle damage by activating their resident muscle stem cells, the Pax7 positive satellite cell (Lepper, Partridge, & Fan, 2011; Sambasivan et al., 2011). However, limb regeneration requires more than just muscle cells, so where do the other cells come from? The pool of progenitors cells in the blastema during limb regeneration express some but not all of the factors used to make induced pluripotent stem cells, *c-myc, oct4, sox2* and *klf4*, commonly referred to as the Yamanaka factors (Takahashi & Yamanaka, 2006). Newts express three of the four Yamanaka factors used to induced pluripotent stem cells (iPSCs) (klf4, sox2, c-myc) (Maki et al., 2009) while axolotls limb blastema cells only express two of them (klf4 and c-myc), nevertheless further investigation is required to understand the dynamic of this factors in salamanders (Knapp et al., 2013). The cross-species conservation of the genetic factors necessary for forming pluripotent stem cells suggests that genetic factors necessary for regeneration are conserved but potentially not activated in mammals in response to injury. However, the exact role of these factors in regeneration has yet to be deciphered, the upregulation of these factors in newt might suggest they play a role in muscle dedifferentiation or that the cells in the newt blastema are pluripotent whilst the cells in the axolotl blastema, which only express klf4 and c-myc may be more restricted in their potential.

The blastema is a structure unique to many animals that regenerate, so where do all the cells come from to form a blastema? Early skin grafting work suggests that the dermal fibroblasts from the skin are the main cell contributor to the blastema cells (Muneoka et al., 1986). This work was further supported using grafting between transgenic GFP axolotls and non-transgenic animals, which also showed that many cells in the blastema come from the dermal fibroblasts in the skin (Kragl et al., 2009). This transgenic approach, of grafting embryonic tissue between transgenic and non-transgenic animals, also allowed the contribution of other cells types to the blastema to be mapped, suggesting that most cells remain lineage restricted during regeneration but that plasticity was observed in fibroblasts (Kragl et al., 2009). More recent work, using a *Cre-loxP-Prrx1* reporter, that labels the fibroblasts of the connective tissue, combined with single-cell transcriptome (scRNA-seq) demonstrated the cell heterogeneity of the connective tissue that contributes to the limb blastema. *Prrx1* is a homeobox gene expressed in the connective tissue precursors during limb development and regeneration. Interestingly, at an early stage point, the gene expression profile present in the dedifferentiated connective tissue cells differed for the limb development bud, whereas at 11 dpa the blastema cells have a similar genetic state as that of a limb bud (Gerber et al., 2018). This supports the idea that the early stages of forming a blastema requires a distinct molecular circuitry that is different from how limb development is executed.

Once the blastema is established these cells must then proliferate, work from several groups has shown that the interplay between the AEC and the nerve is crucial to stimulate the proliferation of the blastema cells. Initially the AEC had been proposed as the main source of mitogenic factors to induce proliferation in the blastema; however, other research has shown that there are factors released from the nerve that are essential for proliferation (Stocum, 2017). Among these factors is the expression of the anterior gradient protein (nAG) first described in *N. viridesence*. nAG is expressed in the first days (5–8 dpa) in the Schwann cells and at 10 dpa the expression switches to the AEC. The expression of nAG depends on the nerve, when the regenerative limb is denervated the expression of nAG is abolished (Grassme et al., 2016; Kumar et al., 2007). However, during limb development, nAG is expressed in the glands of the apical ectodermal ridge and is not nerve dependent (Kumar & Delgado, 2011). This is one interesting example of one of the different mechanisms between limb development and limb regeneration. When nAG starts to be expressed in the AEC the blastema cells express the cell surface protein PROD1 which is expressed in a proximal to distal gradient (da Silva, Gates, & Brockes, 2002). nAG was identified as the nerve dependent ligand for PROD1 (Kumar et al., 2007). To date PROD1 is thought to be a taxon-specific protein found in salamanders and its expression is a key to enable the interaction with nAG to activate blastema cell proliferation (Garza-Garcia, Driscoll, & Brockes, 2010). In newts PROD1 interacts with the EGFR by a GPI-anchor, which is a residue in the α-helical region of the protein crucial for the interaction and action of PROD1 in newts. Curiously, axolotls also have a PROD1 gene but it's lacks a GPI anchor domain, however, it also interacts with the EGFR (Blassberg, Garza-Garcia, Janmohamed, Gates, & Brockes, 2011). It is known that PROD1 is expressed in different salamander families, however, how conserved this cell-signaling axis is in other species of salamanders is an open question (Garza-Garcia et al., 2010; Geng et al., 2015).

Several other factors involved in blastema cell proliferation in *A. mexicanum* have been reported such as FGF8, BMP (Satoh, Makanae, Nishimoto, & Mitogawa, 2016), FGF2 (Mullen, Bryant, Torok, Blumberg, & Gardiner, 1996), Neuregulin 1 (Farkas, Freitas, Bryant, Whited, & Monaghan, 2016). Neuregulin 1 (NRG1) is a nerve derived growth factor which can rescue proliferation in a denervated limb but intriguingly is also expressed by other cells types at different stages of the regenerative process. NRG1 and its receptors (ERBB2, ERBB3) are expressed by the basal keratinocytes of the AEC and by the mesenchymal cells of the blastema; however, the expression of NRG1 in the blastema cells alone does not suffice to trigger a mitotic response. The input of motor and sensory nerves are crucial to increase expression of NRG1 to induce the proliferation of the blastema cells (Farkas et al., 2016). This gives insight to the complexity of the crosstalk between the AEC, nerve, and blastema cells to trigger the correct amount of cell division to enable a new limb to be regenerated.

During the blastema formation, another process is taking place the establishment of the axial patterns of the regenerative limb, which is necessary to restore the 3D limb structure (Brockes & Kumar, 2005; Bryant et al., 2017; Simon & Tanaka, 2013). One of the key molecules that helps to reestablish the positional hierarchy of the cells in a proximal to distal gradient is retinoic acid (RA) (Crawford & Stocum, 1988; Maden, 1982). Retinoic acid has been shown to change the molecular positional identity of cells in a blastema; a hand blastema exposed to specific level of retinoic reprograms and regenerates all upper arm elements and then the hand (Crawford & Stocum, 1988, Maden, 1982). The gene Prod1 was originally identified as a cell surface gene that is RA responsive (da Silva et al., 2002). Prod1 has also been identified to be expressed in a proximodistal pattern and when overexpressed in distal cells in a blastema changes their identity to proximal (Echeverri & Tanaka, 2005). Many Hox genes play important roles in patterning the blastema, the homeobox containing genes *meis1* and *meis2* have been identified as target genes of RA proximalizing activity during limb regeneration (Mercader, Tanaka, & Torres, 2005). *Meis1* and *meis2* have been shown in in vitro assays to regulate the axolotl PROD1 promoter (Shaikh, Gates, & Brockes, 2011). Furthermore, other homeobox genes such as *hoxa9*, *hoxa13*, *hoxa11* are also expressed in a proximal to distal gradient in the limb blastema (Roensch, Tazaki, Chara, & Tanaka, 2013). It is well-established now that many genes that are used during limb development are re-used during limb regeneration, like genes that belong to the signaling pathways of WNT, BMP and Shh (Bryant, Endo, & Gardiner, 2002; Imokawa & Yoshizato, 1997; Knapp et al., 2013; Monaghan et al., 2012; Nacu, Gromberg, Oliveira, Drechsel, & Tanaka, 2016; Nacu & Tanaka, 2011; Satoh et al., 2016). Retinoic acid has been shown to play similar key roles in mammalian limb bud development, excess RA in limb development leads to defects in expression of shh and meis genes and subsequent mis-patterning of the limb, again illustrating the high degree of conservation of pathways between non-regenerative and regenerative animals (Dudley, Ros, & Tabin, 2002; Giguere, Ong, Evans, & Tabin, 1989; Logan, Simon, & Tabin, 1998). However, it is also becoming clear that the circuitry that activates and regulates these genes during regeneration is a unique combination and is not simply a recapitulation of limb development. Recent single cell approaches have given new insights into the molecular program and signatures of cells in the blastema, including intriguing data suggesting that connective tissue cells in the limb revert to a homogenous progenitor state in response to limb amputation (Gerber, Gerber, et al., 2018). Interestingly, a related paper using similar techniques suggests that the blastema contains a fibroblast-like progenitor cell (Leigh et al., 2018). Importantly both of these studies now show that at the molecular level the blastema essentially becomes an embryonic like limb bud, suggesting that regeneration re-uses the embryonic framework for building a limb. Most of the genes that have been identified to play crucial roles in blastema formation are present in humans; however, we do not form a blastema in response to injury. This lack of blastema appears to be one of the key missing aspects of response to injury in salamanders versus mammals and a key to regenerative therapies may lie in identifying how we direct cells toward blastema formation rather than to scar formation.

3. Regeneration of neural tissue

3.1 Spinal cord and brain regeneration

An area of immense interest for regenerative therapies is nervous system tissue. Millions of people around the world live with neurodegenerative disease and thousands more are diagnosed each year, with very limited therapies available. In contrast; salamanders can functionally regenerate complex networks of neural tissue, ranging from the spinal cord and brain to tissues in the eye (Freitas, Yandulskaya, & Monaghan, 2019; Diaz Quiroz & Echeverri, 2013; Joven et al., 2019; Tazaki, Tanaka, & Fei, 2017) (Fig. 3B–C). Nervous tissue regeneration is not limited to salamanders, there are many examples of vertebrates than can regenerate the nervous systems like lamprey, zebrafish and Xenopus (Davis, Troxel, Kohler, Grossmann, & McClellan, 1993; Edwards-Faret et al., 2017; Freed, de Medinaceli, & Wyatt, 1985; Ghosh & Hui, 2018; Jacyniak, McDonald, & Vickaryous, 2017). Xenopus is an interesting example as their extensive nervous system regeneration is limited to their pre-metamorphic state, loss of regenerative ability has to some extent been co-related to the development of a more complex immune systems and to lack of activation of neural progenitor cells among the amphibians (Gibbs, Chittur, & Szaro, 2011; Lee-Liu, Mendez-Olivos, Munoz, & Larrain, 2017).

Salamanders, like *A. mexicanum* and *N. viridescens*, are competent to regenerate the spinal cord through-out life (Butler & Ward, 1965, 1967; Clarke, Alexander, & Holder, 1988; Diaz Quiroz & Echeverri, 2013; Piatt, 1955; Tazaki et al., 2017). Different injury models have been used to understand the cellular and molecular profiles that drive this process. The tail amputation model which is unidirectional, is an injury model most comparable to limb amputation where the formation of a blastema is a key point in the process (Iten & Bryant, 1976). In contrast the spinal cord transection or ablation model is bidirectional, where cells from both sides of the injury side contribute to the regeneration of the ependymal tube and new neurons, but no clear blastema is formed; however, there is extensive cell death of different non-neural cell types around the injury site (Sabin, Santos-Ferreira, Essig, Rudasill, & Echeverri, 2015). This type of injury model is closer to spinal cord injuries that occur in humans; yet the outcomes are polar opposites.

Following a spinal cord injury (SCI) in salamander, the cells that line the central canal often referred to as glial cells, ependymal cells or ependymoglial cells respond to the injury signal by migrating and proliferating (Albors et al., 2015; Egar & Singer, 1972; O' Hara, Ega, & Eag, 1992; O'Hara & Chernoff, 1994; Sabin et al., 2015). These cells have a characteristic oval shape, send long processes out to the plial surface and express both glial acidic fibrillar protein (GFAP), the classic glial cell marker and Sox2, the traditional neural stem cell marker (Fei et al., 2014; McHedlishvili, Epperlein, Telzerow, & Tanaka, 2007; McHedlishvili et al., 2012; O'Hara, Egar, & Chernoff, 1992). These cells act as neural stem cells after injury; they migrate and proliferate to first repair the lesion and then differentiate to replace lost glial cells and neurons. This provides the environment to allow the axon growth to ultimately regain sensory and motor function comparable to a pre-injury state. One major difference between salamander and mammals after SCI is that mammals form a glial scar, which is the main barrier to axonal regrowth (Bradbury & Burnside, 2019; Bradbury & McMahon, 2006; Fitch & Silver, 2008). Understanding the early signals that direct the cells that line the central canal toward a regenerative response versus formation of a glial scar may be key to unlocking regenerative potential, as previous work in mammals has shown that the severed axons have the potential to regrow if given the right environment (Adams & Gallo, 2018; Bradbury & Burnside, 2019).

A significant difference between humans and pro-regenerative vertebrates is the response of the glial cells to injury. Humans activate their glial cells to form a glial scar which prevents more injury from occurring but also reprograms the glial cells to reactive astrocytes that express many proteins that are inhibitory to axonal regrowth including vimentin, GFAP, chondroitin sulfate proteoglycans (CSPGs) reviewed in Adams and Gallo (2018), Dyck and Karimi-Abdolrezaee (2015), Fitch and Silver (2008), Silver (2016), Silver and Miller (2004), Tran, Warren, and Silver (2018). In contrast axolotl glial cells ramp up their cell division such that division happen faster, this process is dependent on the planar cell polarity pathway (Albors et al., 2015). Axolotls glial cells are activated to divide and migrate in a zone of 500 μm adjacent to the injury site in both tail amputation and spinal cord ablation models (McHedlishvili et al., 2007; Sabin et al., 2015). Axolotl glial cells express GFAP, a traditional marker of glial cells that is transcriptionally upregulated after injury in humans and has become a hallmark of reactive gliosis; however, axolotl downregulate GFAP in response to injury. Recent work from Sabin et al. has identified a key highly conserved transcriptional complex that differs in its make up between axolotls and humans and is involved in the regulation of GFAP. The AP-1 transcription factor made up of the heterodimer c-Fos and c-Jun is activated after injury in human or mouse glial cells and binds to the promoter of GFAP activating its transcription (Gao et al., 2013). Axolotls also from an AP-1 transcription factor after injury but they form the complex via heterodimerization of c-Fos and JunB, which in contrast leads to the downregulation of GFAP. If c-Jun is activated in glial cells then genes involved in reactive gliosis are activated like CSPGs, vimentin and collagen and axon regeneration is inhibited (Sabin, Jiang, Gearhart, Stewart, & Echeverri, 2019). The formation of non-canonical AP-1 transcription factor may be a key step that prevents glial scar formation in axolotls and promotes a regenerative response. How conserved this molecular circuitry is for regulating the glial cell response to injury is known but it will be interesting in the future to examine it more broadly across several salamander species. As we move toward translating knowledge from salamanders to humans this molecular complex gives us a starting point of highly conserved genes to start to modulate in vitro in mouse or human cells. Today it is unknown if salamanders spinal cords have the same complexity of glial cells and astrocytes that humans have. It will be important to address this question by taking advantage of the advances in molecular technologies like single cell sequencing to identify cell signatures for all cells within an adult salamander spinal cord.

Thus far much of the more recent work in salamander spinal cord regeneration has focused on the role of the glial cells. A major question in the field is how is function restored. Earlier work in field has attempted to track the axonal regeneration using retrograde tracing. This works suggests that retrograde neurons are restored to their almost original number after complete transection (Clarke et al., 1988). How regrowing axons find their targets and how the functional circuits are restored is completely unknown. It is essential in the future to re-examine these aspects of regeneration with current technologies to understand how plasticity is there in making new connections and how is information relayed to the brain to reconnect circuits.

3.2 Brain regeneration

Salamanders can also functionally repair lesions to the brain. As in SCI, different injury models have been established to understand brain regeneration in salamanders. This includes, specific brain region extirpation (e.g., unilateral forebrain extirpation) and selective ablation of neuronal subtypes (Joven et al., 2019). When a brain extirpation is performed the GFAP positive ependymal glial cells are the cells that proliferate and differentiate to recover the neuronal diversity and form the new inter-neuronal connections that restore function but interestingly are not a faithful replication of the original (Amamoto et al., 2016; Maden, Manwell, & Ormerod, 2013; Urata, Yamashita, Inoue, & Agata, 2018). Axolotls can regenerate the full diversity of neurons that were present before injury, but although they regain function, they do not regenerate the same circuitry. Work in adult axolotl pallium has shown that while after mechanical injury new born neurons organise the same architecture of the brain they appear to fail to regenerate the long distance axonal tracts and exact circuitry that was present before injury (Amamoto et al., 2016). However, it is possible that the establishment of these long tract circuitry occurs over a much longer time period and longer observation points are necessary.

Work in the adult newt *P. waltl* using a model of excision of a quarter of the mesencephalon has given some interesting molecular insights into regeneration over the period of 1.6 years. This long-term observation of the brain regeneration process suggests there is an immediate response which is imperfect, followed by a longer regenerative response that leads to a better regenerative outcome. This data suggests that the rostral caudal region exhibits a self-organizing regenerative ability dependent upon the Pax7+ ependymoglia cells, while the isthmic region may represent a very early neurogenic niche (Urata et al., 2018).

Similarly, during the ablation of specific regions of the brain, the reactivation of quiescent resident GFAP+ ependymoglial cells are crucial for both axolotl and newt brain regeneration (Berg et al., 2010; Maden et al., 2013). This process is under the regulation of at least one neurotransmitter; dopamine. In newts dopamine appears to be essential to keep the ependymal glial cells in a quiescent state, ablation of dopamine neurons activates the ependymal glial cells to proliferate and undergo neurogenesis (Berg, Kirkham, Wang, Frisén, & Simon, 2011). Additionally, a Parkinson-like model has been developed in salamanders, where specific ablation of dopaminergic neurons was achieved using 6-hydroxydopamine. Thirty days post-ablation the neurons were regenerated, suggesting that the main cellular process involved was the reactivation of ependymoglial cells from the ventricular region that are GFAP and Sox2+ cells that mature into dopaminergic neurons to restore the affected area (Parish, Beljajeva, Arenas, & Simon, 2007). Salamanders are a promising model to understand the cellular and molecular mechanisms involved in the regeneration of the nervous system. This work illustrates the high conservation of cell identities and signaling molecules between salamanders and humans and strengthens the possibilities for developing novel therapeutic interventions for neurodegenerative diseases based on this research (Hedlund et al., 2016).

3.3 Lens and retina regeneration

In salamanders, the capacity to regenerate optic structures varies dramatically among different species. Newts such as *N. viridescens*, *P. walt,* and

C. pyrrogaster retain the capacity to regenerate the lens and retina throughout their lives. However, salamanders like *A. mexicanum* lose this capacity 2 weeks post-hatching (Suetsugu-Maki et al., 2012) and different species of the family Plethodontidae are unable to regenerate lens as adults (Henry & Hamilton, 2018; Henry & Tsonis, 2010; Stone, 1967). This has led to newts being one of the main models that have been used in the field to understand this regenerative process. However, to be able to understand where in the evolution this trait was lost or potentially gained in certain salamanders then a comparative analysis across many salamander species is necessary to decipher the molecular circuitry that defines lens and retina regeneration.

The main cellular process used for lens and retina regeneration is transdifferentiation, where differentiated somatic cells dedifferentiate to convert into another cell type with a different developmental trajectory (Tsonis & Del Rio-Tsonis, 2004) (Fig. 3C). After a lentectomy, the dorsal iris pigment epithelial cells (PECs) start to proliferate and dedifferentiate, at the 4 day post-lentectomy (dpl) the cells re-enter the cell cycle and by 8 dpl begin to form a vesicle, at 15 dpl lens fiber start to be formed and lens regeneration is considered complete by 25–30 dpl (Eguchi, 1963; Tsonis & Del Rio-Tsonis, 2004; Tsonis, Madhavan, Tancous, & Del Rio-Tsonis, 2004). One of the initial signals after injury is the expression of thrombin from the dorsal iris, the ventral iris also has PECs but the lens never regenerates from this side, suggesting that the absence of thrombin expression in the ventral iris is part of the reason for lack of regeneration from this tissue (Imokawa, Simon, & Brockes, 2004). Interestingly, in vitro isolation of PECs from different organisms from dorsal and ventral iris including mammals, are capable of being induced to transdifferentiate in specific conditions to structures that resemble lentoids (Tsonis, 2006; Tsonis, Jang, Del Rio-Tsonis, & Eguchi, 2001). This suggests that elucidating the circuitry to induce the correct molecular environment that promotes lens regeneration is a key to understand this process.

Another important signal is FGF, when FGFR signaling is inhibited, lens regeneration is abolished. Like thrombin; FGF is also not expressed in the ventral iris, suggesting FGF signaling is crucial for the lens regeneration from the dorsal iris (Del Rio-Tsonis, Trombley, McMahon, & Tsonis, 1998; Hayashi, Mizuno, Ueda, Okamoto, & Kondoh, 2004; Rio-Tsonis, Jung, Chiu, & Tsonis, 1997). Using RNA-seq and comparing the genes expression profile between the dorsal and ventral iris after 4–8 dpl, Sousounis et al. identified a group of genes that are upregulated specifically in the dorsal iris. Among these genes are those related to cell cycle, cytoskeleton, transcriptional apparatus, and the immune system (Sousounis et al., 2013). The axolotl loses the ability to regenerate the lens as it matures, the timing correlates with the development of a more mature immune system and other hallmarks of aging. Transcriptional profiling studies on different ages of axolotl lens tissue reported that components of the immune system such as genes associated with macrophages, basophils and B-cells among others were upregulated mainly in the old larvae, which are unable to regenerate lens; however, there is no functional data to provide evidence that this loss is due to increased immune cell activity (Sousounis, Athippozhy, Voss, & Tsonis, 2014). In newts a model has been proposed where the leucocytes are attracted to the fibrin clot which is composed of thrombin and transmembrane protein tissue factor after a lentectomy and there the leucocytes could activate the expression of the FGF2 which induces the re-entry to the cell cycle of the PECs (Godwin, Liem, & Brockes, 2010). This suggests that the immune system has a positive role during lens regeneration in newts. It would be interesting to know if there is a difference in immune cell composition between the axolotls and newts in the context of lens regeneration. However, there may be other reasons to explain the lack of regeneration in axolotls, for example, 2 weeks after hatching axolotl larvae PECs start to express several regulators of cellular differentiation such *notch* and *bmp*, which may be implicated in the loss of the capacity to transdifferentiate of PECs. Additionally factors related to aging have been found to be upregulated in older axolotl larvae such as genes that restrict DNA synthesis and cell proliferation (Sousounis et al., 2014). In contrast aging does not effect lens regeneration in newts. In a long-term project using *C. pyrrhogaster* the lens was removed 18 times from the same animals for 16 years and by the time of the last tissue collection, the animals were at least 30 years old. The author compared the gene expression profile and the structural properties of the lens of young animals and old animals and no changes were observed (Eguchi et al., 2011). This suggests that age is not a limitation for lens regeneration in this species of newts but more studies are needed to understand the influence of aging-related genes in the context of lens regeneration in other salamanders. This is an emerging field of research in the study of tissue regeneration in salamanders and will discussed later in this chapter.

Salamanders can also regenerate the retina after extirpation of retina is performed. As in lens regeneration, transdifferentiation of the retinal pigment epithelium (RPE) is the source of cells to regrow the new retina.

The RPE are quiescent cells and when an injury happens these cells are activated to transdifferentiate, start dividing again; many of the signaling molecules used during development like the Fgfs and Bmps play an important role in retina regeneration (Chiba, 2014). A second cell source has also been reported to contribute to the retina regeneration, these are the retinal progenitor cells present in the ciliary marginal zone (Chiba, 2014; Chiba et al., 2006; Grigoryan & Markitantova, 2016).

The RPE in humans and newt show close similarities in their intrinsic properties; however, in humans after a retinal injury the RPE has a different response whereby they lose their epithelial characteristics, migrate, and proliferate by transforming into mesenchymal cells (epithelial-mesenchymal transition, EMT) such as myofibroblasts, these cells are the ones that are going to contribute to retinal disorders. Interestingly when the human RPE undergoes the EMT these cells express *c-myc*, *klf4*, *pax6* and *mitf* which make them behave as "stem cells," however, they do not express *sox2*, which is a crucial transcription factor for neural specification in the retina (Salero et al., 2012). Islam et al. (2014) reported that the RPE in newts doesn't contain retinal stem/progenitor cells. Instead after a retinal injury RPE cells re-enter the cell cycle (5–10 days post-injury) and are reprogrammed via upregulation of reprogramming factors such as *c-myc*, *klf4*, *pax6*, *sox2*, and *mitf* to transform into multipotent cells termed retinal pigment epithelium stem cells. These cells have the potential to differentiate into two cell populations, the pro-neural retina layer, and the pro-RPE layer to rebuild the new retina (Islam et al., 2014). The exact cell reprogramming mechanisms used by salamander retinal cells is one of the big questions that remains in the field; interestingly they use only some of the factors that are needed to generate human iPSC. However, they appear to have all of the classical reprogramming genes like Klf4, Oct4, Myc and Sox2 and interestingly axolotl Pou2 can replace human Oct4 to reprogram somatic nuclei to pluripotent stem cells (Tapia et al., 2012). This suggests that there is a high degree of conservation among reprogramming circuitry but that there are different combinations of routes to the same endpoint. Nevertheless the biggest blackbox is still how to redirect cells toward the correct differentiation program to regenerate the correct tissue.

4. Cancer and regeneration: Similar pathways different outputs

Tumors have been described as wounds that never heal (Flier, Underhill, & Dvorak, 1986). However, wound healing and tumorigenesis share crucial mechanisms such as proliferation, migration, angiogenesis, and ECM remodeling (Flier et al., 1986). The formation of spontaneous tumors in salamander is very low (Brunst & Roque, 1969; Harshbarger, Chang, DeLanney, Rose, & Green, 1999; Khudoley & Ellselv, 1979; Shioda, Uchida, & Nakayamą, 2011). Some studies have even suggested that amphibians are resistant to developing malignant neoplasms; however, tumors have been found in various salamanders (Okamoto, 1987, 1997; Rose & Rose, 1952; Rose et al., 1949; Tsonis & Eguchi, 1981, 1982). The similarities between a malignant tumor and a regenerating blastema has been recognized and pondered on by scientists for years. Both are mounds of proliferating cells, but one controls its environment in a very different manner to eventually regenerate specific tissue types while the other often invades the body destroying the tissues with devastating outcomes to human health (Brockes, 1998; Oviedo & Beane, 2009; Pearson & Alvarado, 2009; Sarig & Tzahor, 2017; Vieira et al., 2020).

The permissive microenvironment available for tissue regeneration in salamanders is very similar to the tumor environment, for example, the ECM composition, like fibronectin and tenascin which in a tumor are an important component for cell migration (Gopal et al., 2017; Sun, Londono, Hudnall, Tuan, & Lozito, 2018). Proteins of the ECM are not the only molecular commonalities, a database published recently summarizes the commonly expressed genes during tissue regeneration and cancer in mammals (Zhao, Rotgans, Wang, & Cummins, 2016). This represents an important source to develop studies that help further the understanding of the dynamics of these two-biological processes.

Lens regeneration represents a good model to test different conditions because regeneration only happens in the dorsal iris, the ventral iris doesn't have this capacity (Okamoto, 1987, 1997). One study using different carcinogenic chemicals (e.g., nickel subsulfide) proved that when these chemicals are introduced in lentectomized newt eyes the dorsal iris proceeds normally with the len regeneration. However, in the ventral iris the formation of multiples lens was observed and the production of a melanoma-like ocular tumor were present only in the ventral iris. Surprisingly, when a high concentration of chemical was used the dorsal iris regeneration was stopped but didn't show any formation of a tumor. Similarly, using the limb regeneration model, the blastema was exposed to chemical carcinogens, defects in regeneration and abnormalities in the limb patterning was observed but no uncontrolled growth similar to a tumor formation was seen (Tsonis & Eguchi, 1981, 1982). Other research has shown that if a renal tumor is grafted to the skin of a salamander forelimb, it starts to grown and generates a malignant mass, however, if a limb amputation is performed bissecting the tumor, the regeneration proceeds normally and the tumor heals, suggesting a reversion in the tumorigenic mechanism (Rose et al., 1949). Those observations raise different questions like, why are the dedifferentiated cells in a regeneration field more resistant to forming tumors? Or, is there specific molecular circuitry in the regeneration microenvironment that controls the gene network necesscary to induce regeneration rather than uncontrolled proliferation? One theory that has been proposed to understand this behavior in organisms that have the natural capacity to regenerate and avoid tumor formation is the expression pattern of tumor suppressor genes such as p53, retinoblastoma, Pten and Hippo; reviewed in Pearson and Alvarado (2009) and Pomerantz and Blau (2013). In non-regenerative organisms like mammals, the mutation in these genes is a known trigger of the tumorigenesis process (Wang, Wu, Rajasekaran, & Shin, 2018). However, these tumor suppressor genes in organisms like salamanders could be regulating the proliferation, dedifferentiation, and genomic stability of the cells implicated in the regeneration (Pomerantz & Blau, 2013). These genes are evolutionary conserved, found in different vertebrates and invertebrates organisms (Pearson & Alvarado, 2009). Nonetheless, in vertebrates like mammals the family of genes of tumor suppressors has diversified, for example, the family of p53 which include also p63 and p73, and in humans the tumor suppressor Arf; however, Arf is not found to date in any highly regenerative organisms like amphibians (Pomerantz & Blau, 2013). Understanding how the expansion of these tumor suppressor genes has occurred from salamanders to mammals, and how highly regenerative species use these gene families in a regeneration context may explain why the regeneration blastema is refractive to tumorigenesis at the molecular level. Elucidating this process could help to understand the genetic balance necessary to trigger a regenerative and not a tumorigenesis process in mammals.

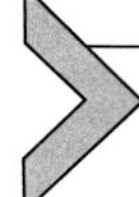

5. Aging, cellular senescence and immune system: Their influence on tissue regeneration

Although humans have some limited capacity for regeneration, this capacity declines with age. As we age humans often accumulate more underlying health problems including high blood pressure, inflammation, high blood sugar levels, which all contribute negatively to our limited regenerative ability. A common question in the regeneration field is, do salamanders regenerate throughout life? The best example that they do, is the above discussion of lens regeneration where newts have been shown to perfectly regenerate the lens 18 times over a 30-year life span. The data on other types of regeneration in salamanders is much more limited. Most salamanders do regenerate throughout life, however, some studies on limb regeneration suggest that older animals make more patterning mistakes in the regeneration of limbs; the data on other forms of regeneration in old animals is lacking. More recently the role of cellular senescence, a process linked to aging has been shown to play an interesting novel role in the limb regeneration process in salamanders.

Cellular senescence is a stress response that stops proliferation, this process is triggered by different factors such as DNA damage, telomere shortening, oxidative stress (Yun, 2018), and is linked to age-related pathologies (e.g., fibrotic diseases). However, it has a role during physiological conditions such as tissue remodeling during embryogenesis and activation of the immune system (Muñoz-Espín & Serrano, 2014). Senescent cells express a senescence-associated secretory phenotype that allows them to secrete a variety of molecules such as growth factors, cytokines, chemokines and matrix remodeling proteins reviewed in (Kuilman & Peeper, 2009), that attract phagocytes to an injury site and have a role in the clearance of dead cells (Muñoz-Espín & Serrano, 2014). Cellular senescence has an important role during aging. In mammals senescent cells accumulate in adult organs as the organism age, for example, in skin, lung, liver, spleen, and kidney (McHugh & Gil, 2018). Senescence is induced in those adult tissues by different factors such as telomere shortening, metabolic dysfunction, and loss of proteostasis (López-Otín, Blasco, Partridge, Serrano, & Kroemer, 2013; McHugh & Gil, 2018). Also, senescence is related with the loss of mammals regenerative capacities, for example, during muscle regeneration in mouse their quiescent resident stem cells turn into senescent cells, which is driven-through epigenetic changes such as the repression of $p16^{INK4a}$ by the positive regulation of Bmi1, which is a component of the Polycomb Repressive Complex 1 (PRC1, Etienne, Liu, Skinner, Conboy, & Conboy, 2020). As mice age, although they have resident muscle stem cells their capacity for regeneration decreases as cellular senescence increases.

In the context of tissue regeneration, different model organisms have reported cellular senescence as a crucial step during regeneration such as fin regeneration in zebrafish (Da Silva-Álvarez et al., 2020) and limb regeneration in salamanders (Yun, Davaapil, & Brockes, 2015). Yun et al. (2015) reported a comparative assay to identify if the response to a stress input can trigger salamander cells to stop proliferating and identified that induced senescent cells display a molecular profile that is similar to that identified in mammal senescent cells (e.g., γH2AX foci, high levels of ROS), showing that the senescent state between mammals and salamanders is comparable. They interestingly reported that during limb regeneration at 7 dpa a high number of senescent cells where observed in the regenerating tissue, and that macrophages are essential to clear those cells from the regenerative microenvironment. They observed no increase in senescent cells in repeated amputations or older animals (3 years), suggesting that the axolotl has a unique mechanism for sensing and clearing senescent cells. In the future, it will be important to know if much older animals, 10 or 20 years old can clear senescent cells as efficiently or does accumulation and slower clearing of them correlate to mistakes in regeneration.

Besides tissue regeneration, cellular senescence is a conserved mechanism during organogenesis to allow tissue growth, remodeling, and patterning in different vertebrates (Muñoz-Espín et al., 2013; Storer et al., 2013) including amphibians. In axolotls, cellular senescence has been observed in development in different tissues including in the kidney (Davaapil, Brockes, & Yun, 2017; Villiard et al., 2017), olfactory epithelium of nerve fascicles and lateral organs (Villiard et al., 2017). Interestingly, in axolotls during limb development senescent cells are not present; they have only been found during limb regeneration, which is another example of how limb regeneration and limb development in salamanders could have molecular and cellular independent pathways (Yun et al., 2015).

The studies that have been done in cellular senescence in salamanders have shown how the transient expression of cellular senescence during tissue regeneration works as a positive input. What induces and controls the timing of this transient state and efficient clearing of these cells at the molecular level is unknown. Additionally, it is unknown if cellular senescence is essential for regeneration of other tissues or organs in salamanders.

5.1 The immune system and tissue regeneration

One of the currently discussed considerations to have successful regenerative therapies is the immunomodulation of the microenvironment of the affected tissue or organ (Pino, Westover, Johnston, Buf, & Humes, 2018), to have a balance between the host defense and the healing of the injury (Godwin, Pinto, & Rosenthal, 2017). A strict correlation between the complexity of the immune system and the capacity to trigger a regenerative response has been explored in different model organisms such as mammals

(Porrello et al., 2011; Seifert et al., 2012), amphibians (Godwin, Debuque, Salimova, & Rosenthal, 2017; Godwin et al., 2013; King, Neff, & Mescher, 2012; Mescher, Neff, & King, 2013; Tsai et al., 2019; Tsai, Baselga-Garriga, & Melton, 2020) and fish (Lai et al., 2017; Tsarouchas et al., 2018). In mammals, it is well known that skin scar-free wound healing is possible in embryonic stages and the decline in this capacity is linked with the development of a more mature immune system as we age. Similarly, scar-free repair in mice is lost during the first days of birth, which correlates with the development of the immune system (Porrello et al., 2011). Also in Xenopus, young larval animals have high regenerative ability but when they undergo metamorphosis, during which time adaptative immune response develops, they largely lose their regenerative ability (King et al., 2012; Mescher et al., 2013).

In salamanders, the role of the immune system has mainly been explored in the context of limb and heart regeneration (Godwin et al., 2013; Tsai et al., 2019, 2020). One of the main cell populations that has been reported as crucial to promote tissue regeneration is macrophages, which secrete anti and pro-inflammatory molecules to immunomodulate the microenvironment of the injury site (Godwin, Kuraitis, & Rosenthal, 2014), which includes the novel role of clearing senescent cells during limb regeneration (Yun et al., 2015). In salamanders, drugs like Clodronate liposomes have been used to deplete the macrophage population during limb regeneration (Godwin, Debuque, et al., 2017). This approach has revealed a role for macrophages in blastema formation but they are not necessary for wound closure. These studies suggest that depletion of macrophages leads to a dysregulation of the expression of genes such as collagens, MMPs, and collagen remodeling enzymes, which are crucial for the remodeling of the extracellular matrix and their disruption leads to the formation of a fibrotic tissue that is not favorable for tissue regeneration (Godwin et al., 2013). This is an example of how the crosstalk between the immune system and the ECM is an important modulator of tissue regeneration, this may be a feature of regeneration that is shared across the Metazoans that have outstanding regenerative capacities, reviewed in Arenas Gómez et al. (2020).

Some very interesting recent research has reported how early blastema cells upregulate Interleukin 8 (IL-8) expression, promoting the recruitment of monocytes and granulocytes (e.g., neutrophils) during wound healing by the interaction with the receptor CXCR-1/2. The knockdown of IL-8 or the inhibition of CXCR-1/2 leads to defects in the blastema formation (Tsai et al., 2019). Furthermore, the wound epidermis, the apical epidermal cap, and the blastema cells express cytokines like *midkine* that module the ECM, tissue histolysis and inflammatory microenvironment during early stages of limb regeneration (Tsai et al., 2020), which shows that the cells of the immune system are not the only ones that can secret cytokines to immunomodulate a regenerative microenvironment.

Many studies in limb regeneration to date have been performed in relatively young animals. It will be important in the future to also look at the interaction of the immune cells, the ECM and the role of cellular senescence in limb regeneration in aged animals, 3–5 years or older. Initial work from Yun et al. (Yun et al., 2015) suggests that senescent cells are cleared from the injury site by macrophages and may be necessary for regeneration. In the future, it will be important to understand the exact mechanism salamanders use to clear senescent cells, if this only evoked in a regeneration scenario or do old salamanders constantly use this system as they age to reduce senescent cells. This may suggest that salamanders have evolved an efficient way to remove senescent cells and compare to what is known to occur in aging and inflammatory disease progression in humans.

6. Modern tools for studying salamanders

For many years, the progression of tools in the salamander field was limited by the lack of a sequenced genome. One of the challenges with salamanders is that many have big genomes; in the case of *A. mexicanum* it has a size of ~32 GB, 10 times bigger than humans (Keinath et al., 2015). In the beginning, useful databases manually curated from contigs assembled from Expressed Sequences Tags (EST) collected mainly from *A. mexicanum* (Habermann et al., 2004) and *A. tigrinum* (Putta et al., 2004), were the main data available to perform molecular analysis. This helped with pushing forth the molecular analysis of regeneration and the next break through was techniques to transiently label cells in vivo and track their fate during regeneration (Echeverri & Tanaka, 2003). This was swiftly followed by the development of the first transgenic axolotls (Sobkow, Epperlein, Herklotz, Straube, & Tanaka, 2006) and newts (Casco-Robles et al., 2011; Ueda, Kondoh, & Mizuno, 2005), which opened up the possibility of using cell type specific promoters to image cells during regeneration (Casco-Robles et al., 2011; Hayashi et al., 2013; Joven & Simon, 2018; Khattak et al., 2014, 2013; Khattak & Tanaka, 2015; Sandoval-Guzman et al., 2014; Ueda et al., 2005).

In recent years major advances in computational abilities to handle big data sets has finally allowed the assembly of the huge axolotl genome, finally published in 2018 (Nowoshilow et al., 2018). This has been followed by publication of another genome project in axolotl (Smith et al., 2019) and the sequencing of the Pleurodeles genome (Elewa et al., 2017) and represents a new era in the research of salamanders.

The availability of salamander genomes has also aided in the use of Next Generation Sequencing (NGS) platforms such as RNA-Seq (Dwaraka, Smith, Woodcock, & Voss, 2019) or Single Cell Seq (Gerber, Gerber, et al., 2018; Leigh et al., 2018), which are powerful tools to facilitate the analysis of gene expression profiles during tissue regeneration at different time points, different experimental conditions and compare gene expression profile among different species of salamanders or other organisms.

This approach facilitates the identification of conserved molecular signatures during the time course of tissue regeneration. Transcriptomics, have also enabled the development of de novo reference transcriptome of salamanders that are endangered species such as *Hynobius chinensis*, *Andrias davidianus*, and *Batrachuperus yenyuanensis*, which are important to understand the evolution of gene families or specific traits in salamanders for the adaptation (Che et al., 2014; Huang, Ren, Xiong, Gao, & Sun, 2017; Li et al., 2015; Xiong, Lv, Huang, & Liu, 2019). The study of many different salamander species at the molecular levels enables the identification of conserved or divergent molecular circuitry for different types of regeneration (Arenas Gómez et al., 2017; Dwaraka & Voss, 2019). Ultimately sequenced genomes allow us to probe more deeply not just the pathways that are need to be activated or repressed to promote a regenerative response in salamanders but importantly allow us to dissect the possible conservation of coding sequences of genes between salamanders and humans and enables the comparison of enhancers and other regulatory regions that may be crucial to understand how different cells response to an injury signal.

The other area that having a sequenced genome is essential for is genome editing. The CRISPR/Cas system has changed the way knock-outs are carried out due to the efficiency of the system. Recent work has shown that this systems works very effectively in axolotl embryos for both gene knock outs and knockins (Fei et al., 2016, 2018, 2017, 2014; Flowers, Sanor, & Crews, 2017; Flowers, Timberlake, McLean, Monaghan, & Crews, 2014; Sanor, Flowers, & Crews, 2020).

In Newts, *C. pyrrhogaster* was one of the first in which transgenesis was established, however, it is not an ideal system as it takes a long time to reach sexual maturation (more than 3 years) (Hayashi et al., 2013; Ueda et al., 2005). Other species look more promising such as *P. waltl*, which have greater potential to regenerate lens throughout their lives, a reference genome is available and they have a shorter period to arrive at sexual maturation (9–12 months) (Joven et al., 2019). Both knockin and mutations are possible in Pleurodeles using CRISPR/cas based editing (Cai, Peng, Ren, & Wang, 2019; Elewa et al., 2017; Molla & Yang, 2019).

At the moment, *A. mexicanum* and *P. waltl* are the salamanders of choice in which to perform genome editing due to the range of established techniques and availability of genomic resources. In order to expand the toolkit available for studying development and regeneration in these animals more advances in knock-in strategies using CRISPR/Cas technology are needed, especially inducible systems which allow the function of a gene to be altered in a tissue and time specific manner. It is well known that salamanders have been a hard model to establish as a genetically tractable system, however, the recent advances in genome sequencing and genome editing make them a more accessible research organism to address molecular and cellular questions about development and regeneration.

7. Potential for translation from salamanders to humans

The question often asked about salamander regeneration studies is; how relevant is this to humans and can this knowledge ever be used to promote regeneration in humans?

Currently, the field of tissue engineering and regenerative medicine (TERM) have been creating different strategies to facilitate the replacement of damaged organs (e.g., artificial organs). These strategies include new cell-based therapies for diseases with limited treatments available (e.g., neurodegenerative diseases), the creation of 3D scaffolds mixed with cells that promote the local repair of damaged tissue (e.g., skin), building organs using bioreactors, and in the case of appendage amputations (e.g., limbs) the replacement with biomechanical structures; reviewed in (Baddour, Sousounis, & Tsonis, 2012; Binan, Ajji, De Crescenzo, & Jolicoeur, 2014; Bumbaširević et al., 2020; Han et al., 2020; Vig et al., 2017; Wang, 2019) (reviewed by Bumbaširević et al., 2020; Han et al., 2020; Kim, Lee, & Kim, 2013; Ravichandran, Liu, & Teoh, 2018; Vig et al., 2017; Wang, 2019). However, these strategies have limitations, especially with integrating the ex vivo structure to the in vivo system, some of those limitations include the re-vascularization, innervations and the reassembly of the mechanical properties of the tissue or organ (Dimmeler, Ding, Rando, & Trounson, 2014). However, insights into how to connect the vasculature and nervous systems between old and newly formed tissue may be gained by carefully studying how animals like salamanders integrate the old with the new.

Model organisms in tissue regeneration such as salamanders could help us to understand the molecular and cellular pathways needed for a successful regenerative response. It is clear that the modulation of the microenvironment where the regeneration takes place is a key element for successful tissue regeneration, including, the modulation of the extracellular matrix, the regulation of the immune system, the molecular factors required for innervation, and the cellular mechanisms activated in order to repopulate the tissue.

In this chapter, the regeneration of some of the tissues and organs in salamanders was included, these recent advances show the potential to translate this knowledge to develop new therapies in humans. For example; in the case of the eye, the conservation of the cell types and genetic pathways illustrate the potential for this knowledge to be used to design novel translational therapies (Barbosa-Sabanero et al., 2012; Carido et al., 2014; Chiba, 2014; Del Rio-Tsonis et al., 1998; Hayashi et al., 2004, 2013; Haynes, Gutierrez, Aycinena, Tsonis, & Del Rio-Tsonis, 2007; Islam et al., 2014; Tsonis & Del Rio-Tsonis, 2004; Zhu et al., 2013; Zhu, Schreiter, & Tanaka, 2016). Similarly, the research on spinal cord regeneration in salamanders has identified critical molecular pathways that are conserved in mammals but which axolotls specifically activate to promote regeneration instead of glial scar formation (Diaz Quiroz, Tsai, Coyle, Sehm, & Echeverri, 2014; Sabin et al., 2019, 2015), this knowledge of the axolotl spinal cord may help to inform potential cell based therapeutics in the future (Albors et al., 2015; Diaz Quiroz, Li, Aparicio, & Echeverri, 2016; Meinhardt et al., 2014). Similarly, the research done in brain regeneration using newts identified the role of dopamine in the modulation of the fate of the ependymoglial cells, a strategy that has been tested in mammals (Berg et al., 2010, 2011; Hedlund et al., 2016).

One of the biggest challenges in regenerative medicine may be the restoration of a full appendage like a leg, research on salamander limb regeneration is invaluable to this field. Salamanders effortlessly regenerate and integrate a new limb with the existing infrastructures and seamlessly reconnect the old and new vasculature and neural networks to regenerate a perfectly functioning limb. Research over several decades has given significant insights into how the salamanders do this, we now understand better where the cells come from to form a new limb and how much plasticity there is in given cell populations (Kragl et al., 2009; Muneoka & Bryant, 1984; Muneoka et al., 1986; Stocum, 1980, 1983, 1998, 2017). In the case of limbs, the skeletal system is very close to humans, however, in amphibians one of the differences is a group of bones calls the *basale commune*, which is an amalgamation in the base of digits I-II. Also, the polarity during digit development is different, in salamanders it is in a preaxial order (order of digit formation is IV-(V)-III-II-I) and in the other tetrapods it is postaxial (II-I-III-IV(-V)) (Fröbisch & Shubin, 2011). Despite this difference in development; the cellular and morphological composition is highly conserved cross-species, highlighting the possibilities for cross-species translational potential.

Researchers are making progress in understanding how cells know how much tissue to regenerate and how that new tissue is patterned to regenerate the correct structures, interestingly salamanders share many features with humans, like the use of Pax7 satellite cells to regenerate limb muscle and role of shh and Hox genes in patterning the limb, suggesting potential avenues for translating knowledge cross-species (Carlson, Komine, Bryant, & Gardiner, 2001; da Silva et al., 2002; Gardiner, Blumberg, Komine, & Bryant, 1995; Gerber et al., 2018; Imokawa & Yoshizato, 1997; Knapp et al., 2013; Mercader et al., 2005; Nacu et al., 2013, 2016; Roensch et al., 2013; Sandoval-Guzman et al., 2014; Sugiura, Wang, Barsacchi, Simon, & Tanaka, 2016).

8. Conclusion and perspectives

Salamanders represent a group of crucial vertebrates in which to understand the diversity of cellular and molecular mechanisms that have evolved to regenerate different tissue types. Understanding within one group of animals the multiple different pathways that can be deployed in different manners may help us understand why humans are very limited in their regenerative abilities. There are still many unknowns in the salamander regeneration field which are essential to decipher, for example, how do the injured cells know exactly how much tissue to regenerate? Despite many years of research, we still do not have a molecular answer to this question. Control of tissue size will be an essential element to understand both in terms of regenerating whole structures like limbs but also with regard to cell replacement in specific organs like the heart or brain. Also understanding how to control the connection of newly differentiated tissue to older tissue via vasculature and neuronal circuitry to restore functional tissue is essential for translational medicine to be possible. Salamander regeneration is a field with a long history; however, in terms of understanding how to translate the molecular knowledge of regenerating a limb to humans is still in its infancy.

References

Adams, K. L., & Gallo, V. (2018). The diversity and disparity of the glial scar. *Nature Neuroscience*, *21*(1), 9–15.

Albors, A. R., Tazaki, A., Rost, F., Nowoshilow, S., Chara, O., & Tanaka, E. M. (2015). Planar cell polarity-mediated induction of neural stem cell expansion during axolotl spinal cord regeneration. *eLife*, *4*(NOVEMBER2015), 1–29.

Amamoto, R., Huerta, V. G. L., Takahashi, E., Dai, G., Grant, A. K., Fu, Z., et al. (2016). Adult axolotls can regenerate original neuronal diversity in response to brain injury. *eLife*, *5*, e13998. https://doi.org/10.7554/eLife.13998.

AmphibiaWeb. (2019). *AmphibiaWeb:Information on amphibian biology and conservation.*

Arenas Gómez, C. M., Gomez Molina, A., Zapata, J. D., & Delgado, J. P. (2017). Limb regeneration in a direct-developing terrestrial salamander, Bolitoglossa ramosi (Caudata: Plethodontidae). *Regeneration*, *4*(4), 227–235.

Arenas Gómez, C. M., Sabin, K. Z., & Echeverri, K. (2020). Wound healing across the animal kingdom: Crosstalk between the immune system and the extracellular matrix. *Developmental Dynamics: An Official Publication of the American Association of the Anatomists*, *249*(7), 834–846.

Baddour, J. A., Sousounis, K., & Tsonis, P. A. (2012). Organ repair and regeneration: An overview. *Birth Defects Research. Part C, Embryo Today*, *96*(1), 1–29.

Barbosa-Sabanero, K., Hoffmann, A., Judge, C., Lightcap, N., Tsonis, P. A., & Del Rio-Tsonis, K. (2012). Lens and retina regeneration: New perspectives from model organisms. *The Biochemical Journal*, *447*(3), 321–334.

Berg, D. A., Kirkham, M., Beljajeva, A., Knapp, D., Habermann, B., Ryge, J., et al. (2010). Efficient regeneration by activation of neurogenesis in homeostatically quiescent regions of the adult vertebrate brain. *Development*, *137*(24), 4127–4134.

Berg, D. A., Kirkham, M., Wang, H., Frisén, J., & Simon, A. (2011). Dopamine controls neurogenesis in the adult salamander midbrain in homeostasis and during regeneration of dopamine neurons. *Cell Stem Cell*, *8*(4), 426–433.

Binan, L., Ajji, A., De Crescenzo, G., & Jolicoeur, M. (2014). Approaches for neural tissue regeneration. *Stem Cell Reviews and Reports*, *10*(1), 44–59.

Birnbaum, K. D., & Sánchez Alvarado, A. (2008). Slicing across kingdoms: Regeneration in plants and animals. *Cell*, *132*(4), 697–710.

Blassberg, R. A., Garza-Garcia, A., Janmohamed, A., Gates, P. B., & Brockes, J. P. (2011). Functional convergence of signalling by GPI-anchored and anchorless forms of a salamander protein implicated in limb regeneration. *Journal of Cell Science*, *124*(Pt. 1), 47–56.

Bradbury, E. J., & Burnside, E. R. (2019). Moving beyond the glial scar for spinal cord repair. *Nature Communications*, *10*(1), 3879.

Bradbury, E. J., & McMahon, S. B. (2006). Spinal cord repair strategies: Why do they work? *Nature Reviews. Neuroscience*, 7(8), 644–653.

Brockes, J. P. (1998). Regeneration and cancer. *Biochimica et Biophysica Acta, Reviews on Cancer*, *1377*(1), 1–11.

Brockes, J. P., & Gates, P. B. (2014). Mechanisms underlying vertebrate limb regeneration: Lessons from the salamander. *Biochemical Society Transactions*, *42*(3), 625–630.

Brockes, J. P., & Kumar, A. (2005). Appendage regeneration in adult vertebrates and implications for regenerative medicine. *Science (New York, N.Y.)*, *310*(5756), 1919–1923.

Brockes, J. P., & Kumar, A. (2008). Comparative aspects of animal regeneration. *Annual Review of Cell and Developmental Biology*, *24*, 525–549.
Brunst, V. V., & Roque, A. L. (1969). A spontaneous teratoma in an axolotl (Siredon mexicanum). *Cancer Research*, *29*(1), 223–229.
Bryant, S. V., Endo, T., & Gardiner, D. M. (2002). Vertebrate limb regeneration and the origin of limb stem cells. *International Journal of Developmental Biology*, *46*(7), 887–896.
Bryant, D. M., Johnson, K., Ditommaso, T., Regev, A., Haas, B. J., & Whited, J. L. (2017). A tissue-mapped axolotl De novo transcriptome enables identification of limb regeneration factors. *Cell Reports*, *18*, 762–776.
Bumbaširević, M., Lesic, A., Palibrk, T., Milovanovic, D., Zoka, M., Kravić-Stevović, T., et al. (2020). The current state of bionic limbs from the surgeon's viewpoint. *EFORT Open Reviews*, *5*(2), 65–72.
Butler, E. G., & Ward, M. B. (1965). Reconstitution of the spinal cord following ablation in urodele larvae. *Journal of Experimental Zoology*, *160*(1), 47–65.
Butler, E. G., & Ward, M. B. (1967). Reconstitution of the spinal cord after ablation in adult Triturus. *Developmental Biology*, *15*(5), 464–486.
Cai, H., Peng, Z., Ren, R., & Wang, H. (2019). Efficient gene disruption via base editing induced stop in newt Pleurodeles waltl. *Genes*, *10*(11), 837.
Calve, S., Odelberg, S. J., & Simon, H.-G. (2010). A transitional extracellular matrix instructs cell behavior during muscle regeneration. *Developmental Biology*, *344*(1), 259–271.
Campbell, L., & Crews, C. (2008). Wound epidermis formation and function in urodele amphibian limb regeneration. *Cellular and Molecular Life Sciences*, *65*, 73–79.
Campbell, L. J., Suarez-Castillo, E. C., Ortiz-Zuazaga, H., Knapp, D., Tanaka, E. M., & Crews, C. M. (2011). Gene expression profile of the regeneration epithelium during axolotl limb regeneration. *Developmental Dynamics*, *240*(7), 1826–1840.
Carido, M., Zhu, Y., Postel, K., Benkner, B., Cimalla, P., Karl, M. O., et al. (2014). Characterization of a mouse model with complete RPE loss and its use for RPE cell transplantation. *Investigative Ophthalmology & Visual Science*, *55*(8), 5431–5444.
Carlson, M., Komine, Y., Bryant, S., & Gardiner, D. M. (2001). Expression of Hoxb 13 and Hoxc 10 in developing and regenerating axolotl limbs and tails. *Developmental Biology*, *229*, 396–406.
Casco-Robles, M. M., Yamada, S., Miura, T., Nakamura, K., Haynes, T., Maki, N., et al. (2011). Expressing exogenous genes in newts by transgenesis. *Nature Protocols*, *6*(5), 600–608.
Chalkley, D. T. (1954). A quantitative histological analysis of forelimb regeneration in triturus viridescens. *Journal of Morphology*, *94*(1), 21–70.
Che, R., Sun, Y., Wang, R., & Xu, T. (2014). Transcriptomic analysis of endangered Chinese salamander: Identification of immune, sex and reproduction-related genes and genetic markers. *PLoS One*, *9*(1), e87940.
Chiba, C. (2014). The retinal pigment epithelium: An important player of retinal disorders and regeneration. *Experimental Eye Research*, *123*, 107–114.
Chiba, C., Hoshino, A., Nakamura, K., Susaki, K., Yamano, Y., Kaneko, Y., et al. (2006). Visual cycle protein RPE65 persists in new retinal cells during retinal regeneration of adult newt. *The Journal of Comparative Neurology*, *495*(4), 391–407.
Christensen, R. N., & Tassava, R. A. (2000). Apical epithelial cap morphology and fibronectin gene expression in regenerating axolotl limbs. *Developmental Dynamics*, *217*(2), 216–224.
Clarke, J. D., Alexander, R., & Holder, N. (1988). Regeneration of descending axons in the spinal cord of the axolotl. *Neuroscience Letters*, *89*(1), 1–6.
Crawford, K., & Stocum, D. L. (1988). Retinoic acid coordinately proximalizes regenerate pattern and blastema differential affinity in axolotl limbs. *Development (Cambridge, England)*, *102*(4), 687–698.

da Silva, S. M., Gates, P. B., & Brockes, J. P. (2002). The newt Ortholog of CD59 is implicated in Proximodistal identity during amphibian limb regeneration. *Developmental Cell, 3*(4), 547–555.

Da Silva-Álvarez, S., Guerra-Varela, J., Sobrido-Cameán, D., Quelle, A., Barreiro-Iglesias, A., Sánchez, L., et al. (2020). Cell senescence contributes to tissue regeneration in zebrafish. *Aging Cell, 19*(1), 1–5.

Davaapil, H., Brockes, J. P., & Yun, M. H. (2017). Conserved and novel functions of programmed cellular senescence during vertebrate development. *Development (Cambridge), 144*(1), 106–114.

Davis, G. R., Troxel, M. T., Kohler, V. J., Grossmann, E. R., & McClellan, A. D. (1993). Time course of locomotor recovery and functional regeneration in spinal transected lampreys: Kinematics and electromyography. *Experimental Brain Research, 97*, 83–95.

Del Rio-Tsonis, K., Trombley, M. T., McMahon, G., & Tsonis, P. A. (1998). Regulation of lens regeneration by fibroblast growth factor receptor 1. *Developmental Dynamics: An Official Publication of the American Association of the Anatomists, 213*(1), 140–146.

Diaz Quiroz, J. F., & Echeverri, K. (2013). Spinal cord regeneration: Where fish, frogs and salamanders lead the way, can we follow? *The Biochemical Journal, 451*(3), 353–364.

Diaz Quiroz, J., Li, Y., Aparicio, C., & Echeverri, K. (2016). Development of a 3D matrix for modeling mammalian spinal cord injury in vitro. *Neural Regeneration Research, 11*(11), 1810.

Diaz Quiroz, J. F., Tsai, E., Coyle, M., Sehm, T., & Echeverri, K. (2014). Precise control of mi R-125b levels is required to create a regeneration-permissive environment after spinal cord injury: A cross-species comparison between salamander and rat. *Disease Models & Mechanisms*, 7(6), 601–611.

Dimmeler, S., Ding, S., Rando, T. A., & Trounson, A. (2014). Translational strategies and challenges in regenerative medicine. *Nature Medicine, 20*(8), 814–821.

Dinsmore, C. E., & American Society of Zoologists. (2008). *A history of regeneration research: Milestones in the evolution of a science*. Cambridge University Press.

Dudley, A. T., Ros, M. A., & Tabin, C. J. (2002). A re-examination of proximodistal patterning during vertebrate limb development. *Nature, 418*, 539–544.

Dwaraka, V. B., Smith, J. J., Woodcock, M. R., & Voss, S. R. (2019). Comparative transcriptomics of limb regeneration: Identification of conserved expression changes among three species of Ambystoma. *Genomics, 111*(6), 1216–1225.

Dwaraka, V. B., & Voss, S. R. (2019). Towards comparative analyses of salamander limb regeneration. *Journal of Experimental Zoology Part B: Molecular and Developmental Evolution*. https://doi.org/10.1002/jez.b.22902.

Dyck, S. M., & Karimi-Abdolrezaee, S. (2015). Chondroitin sulfate proteoglycans: Key modulators in the developing and pathologic central nervous system. *Experimental Neurology, 269*, 169–187.

Echeverri, K., & Tanaka, E. M. (2003). Electroporation as a tool to study in vivo spinal cord regeneration. *Developmental Dynamics, 226*(2), 418–425.

Echeverri, K., & Tanaka, E. (2005). Proximodistal patterning during limb regeneration. *Developmental Biology, 279*, 391–401.

Edwards-Faret, G., Muñoz, R., Méndez-Olivos, E. E., Lee-Liu, D., Tapia, V. S., & Larraín, J. (2017). Spinal cord regeneration in Xenopus laevis. *Nature Protocols, 12*(2), 372–389.

Egar, M., & Singer, M. (1972). The role of ependyma in spinal cord regeneration in the urodele, Triturus. *Experimental Neurology, 37*(2), 422–430.

Eguchi, G. (1963). Electon microscopic studies on lens regeneration I mechanism of depigmentation of the iris. *Embryologia, 8*(1), 45–62.

Eguchi, G., Eguchi, Y., Nakamura, K., Yadav, M. C., Millan, J. L., & Tsonis, P. A. (2011). Regenerative capacity in newts is not altered by repeated regeneration and ageing. *Nature Communications, 2*, 384.

Elewa, A., Wang, H., Talavera-López, C., Joven, A., Brito, G., Kumar, A., et al. (2017). Reading and editing the Pleurodeles waltl genome reveals novel features of tetrapod regeneration. *Nature Communications*, *8*(1), 1–9.

Etienne, J., Liu, C., Skinner, C. M., Conboy, M. J., & Conboy, I. M. (2020). Skeletal muscle as an experimental model of choice to study tissue aging and rejuvenation. *Skeletal Muscle*, *10*(1), 1–16.

Farkas, J. E., Freitas, P. D., Bryant, D. M., Whited, J. L., & Monaghan, J. R. (2016). Neuregulin-1 signaling is essential for nerve-dependent axolotl limb regeneration. *Development (Cambridge)*, *143*(15), 2724–2731.

Fei, J.-F., Knapp, D., Schuez, M., Murawala, P., Zou, Y., Pal Singh, S., et al. (2016). Tissue- and time-directed electroporation of CAS9 protein–gRNA complexes in vivo yields efficient multigene knockout for studying gene function in regeneration. *npj Regenerative Medicine*, *1*(1), 1–9.

Fei, J. F., Lou, W. P., Knapp, D., Murawala, P., Gerber, T., Taniguchi, Y., et al. (2018). Application and optimization of CRISPR-Cas 9-mediated genome engineering in axolotl (Ambystoma mexicanum). *Nature Protocols*, *13*(12), 2908–2943.

Fei, J.-F., Schuez, M., Knapp, D., Taniguchi, Y., Drechsel, D. N., & Tanaka, E. M. (2017). Efficient gene knockin in axolotl and its use to test the role of satellite cells in limb regeneration. *Proceedings of the National Academy of Sciences of the United States of America*, *114*(47), 12501–12506.

Fei, J. F., Schuez, M., Tazaki, A., Taniguchi, Y., Roensch, K., & Tanaka, E. M. (2014). CRISPR-mediated genomic deletion of sox 2 in the axolotl shows a requirement in spinal cord neural stem cell amplification during tail regeneration. *Stem Cell Reports*, *3*(3), 444–459.

Fior, J. (2014). Salamander regeneration as a model for developing novel regenerative and anticancer therapies. *Journal of Cancer*, *5*, 715–719.

Fitch, M. T., & Silver, J. (2008). CNS injury, glial scars, and inflammation: Inhibitory extracellular matrices and regeneration failure. *Experimental Neurology*, *209*(2), 294–301.

Flier, J. S., Underhill, L. H., & Dvorak, H. F. (1986). Tumors: Wounds that do not heal. *New England Journal of Medicine*, *315*(26), 1650–1659.

Flowers, G. P., Sanor, L. D., & Crews, C. M. (2017). Lineage tracing of genome-edited alleles reveals high fidelity axolotl limb regeneration. *eLife*, *6*, 1–15.

Flowers, G. P., Timberlake, A. T., McLean, K. C., Monaghan, J. R., & Crews, C. M. (2014). Highly efficient targeted mutagenesis in axolotl using Cas9 RNA-guided nuclease. *Development*, *141*(10), 2165–2171.

Freed, W. J., de Medinaceli, L., & Wyatt, R. J. (1985). Promoting functional plasticity in the damaged nervous system. *Science*, *227*(4694), 1544–1552.

Freitas, P. D., Yandulskaya, A. S., & Monaghan, J. R. (2019). Spinal cord regeneration in amphibians: A historical perspective. *Developmental Neurobiology*, *79*(5), 437–452.

Fröbisch, N. B., Bickelmann, C., & Witzmann, F. (2014). Early evolution of limb regeneration in tetrapods: Evidence from a 300-million-year-old amphibian. *Proceedings of the Royal Society B: Biological Sciences*, *281*(1794), 20141550. https://doi.org/10.1098/rspb.2014.1550.

Fröbisch, N. B., & Shubin, N. H. (2011). Salamander limb development: Integrating genes, morphology, and fossils. *Developmental Dynamics*, *240*(5), 1087–1099.

Gao, K., Wang, C. R., Jiang, F., Wong, A. Y. K., Su, N., Jiang, J. H., et al. (2013). Traumatic scratch injury in astrocytes triggers calcium influx to activate the JNK/c-Jun/AP-1 pathway and switch on GFAP expression. *Glia*, *61*(12), 2063–2077.

Gardiner, D. M., Blumberg, B., Komine, Y., & Bryant, S. V. (1995). Regulation of HoxA expression in developing and regenerating axolotl limbs. *Development*, *121*(6), 1731–1741.

Gardiner, D. M., Muneoka, K., & Bryant, S. V. (1986). The migration of dermal cells during blastema formation in axolotls. *Developmental Biology*, *118*(2), 488–493.

Garza-Garcia, A. A., Driscoll, P. C., & Brockes, J. P. (2010). Evidence for the local evolution of mechanisms underlying limb regeneration in salamanders. *Integrative and Comparative Biology*, *50*(4), 528–535.

Geng, J., Gates, P. B., Kumar, A., Guenther, S., Garza-Garcia, A., Kuenne, C., et al. (2015). Identification of the orphan gene prod 1 in basal and other salamander families. *EvoDevo*, *6*, 9.

Geng, X., Wei, H., Shang, H., Zhou, M., Chen, B., Zhang, F., et al. (2015). Proteomic analysis of the skin of Chinese giant salamander (Andrias davidianus). *Journal of Proteomics*, *119*, 196–208.

Gerber, T., Gerber, T., Murawala, P., Knapp, D., Masselink, W., Schuez, M., et al. (2018). Single-cell analysis uncovers convergence of cell identities during axolotl limb regeneration. *Science*, *0681*(September), 1–19.

Gerber, T., Murawala, P., Knapp, D., Masselink, W., Schuez, M., Hermann, S., et al. (2018). Single-cell analysis uncovers convergence of cell identities during axolotl limb regeneration. *Science (New York, N.Y.)*, *362*(6413).

Ghosh, S., & Hui, S. P. (2018). Axonal regeneration in zebrafish spinal cord. *Regeneration*, *5*(1), 43.

Gibbs, K., Chittur, S., & Szaro, B. (2011). Metamorphosis and the regenerative capacity of spinal cord axons in Xenopus laevis. *The European Journal of Neuroscience*, *33*(1), 9–25.

Giguere, V., Ong, E. S., Evans, R. M., & Tabin, C. J. (1989). Spatial and temporal expression of the retinoic acid receptor in the regenerating amphibian limb. *Nature*, *337*(6207), 566–569 (published erratum appears in Nature 1989 Sep 7;341(6237):80).

Godwin, J. W., Debuque, R., Salimova, E., & Rosenthal, N. A. (2017). Heart regeneration in the salamander relies on macrophage-mediated control of fibroblast activation and the extracellular landscape. *NPJ Regen Med*, *2*, 22. https://doi.org/10.1038/s41536-017-0027-y.

Godwin, J., Kuraitis, D., & Rosenthal, N. (2014). Extracellular matrix considerations for scar-free repair and regeneration: Insights from regenerative diversity among vertebrates. *The International Journal of Biochemistry & Cell Biology*, *56*, 47–55.

Godwin, J. W., Liem, K. F., & Brockes, J. P. (2010). Tissue factor expression in newt iris coincides with thrombin activation and lens regeneration. *Mechanisms of Development*, *127*(7–8), 321–328.

Godwin, J., Pinto, A. R., & Rosenthal, N.a. (2013). Macrophages are required for adult salamander limb regeneration. *Proceedings of the National Academy of Sciences of the United States of America*, *110*(23), 9415–9420.

Godwin, J. W., Pinto, A. R., & Rosenthal, N. A. (2017). Chasing the recipe for a pro-regenerative immune system. *Seminars in Cell & Developmental Biology*, *61*, 71–79.

Gopal, S., Veracini, L., Grall, D., Butori, C., Schaub, S., Audebert, S., et al. (2017). Fibronectin-guided migration of carcinoma collectives. *Nature Communications*, *8*(1), 14105.

Grassme, K. S., Garza-Garcia, A., Delgado, J. P., Godwin, J. W., Kumar, A., Gates, P. B., et al. (2016). Mechanism of action of secreted newt anterior gradient protein. *PLoS One*, *11*(4), e0154176.

Grigoryan, E., & Markitantova, Y. (2016). Cellular and molecular preconditions for retinal pigment epithelium (RPE) natural reprogramming during retinal regeneration in Urodela. *Biomedicine*, *4*(4), 28.

Haas, B. J., & Whited, J. L. (2017). Advances in decoding axolotl limb regeneration. *Trends in Genetics*, *33*(8), 553–565.

Habermann, B., Bebin, A., Herklotz, S., Volkmer, M., Eckelt, K., Pehlke, K., et al. (2004). An Ambystoma mexicanum EST sequencing project: Analysis of 17,352 expressed sequence tags from embryonic and regenerating blastema cDNA libraries. *Genome Biology*, *5*(9), R67.

Han, F., Wang, J., Ding, L., Hu, Y., Li, W., Yuan, Z., et al. (2020). Tissue engineering and regenerative medicine: Achievements, future, and sustainability in Asia. *Frontiers in Bioengineering and Biotechnology*, *8*, 83.

Harshbarger, J. C., Chang, S. C., DeLanney, L. E., Rose, F. L., & Green, D. E. (1999). Cutaneous mastocytomas in the neotenic caudate amphibians Ambystoma mexicanum (axolotl) and Ambystoma tigrinum (tiger salamander). *Journal of Cancer Research and Clinical Oncology*, *125*(3–4), 187–192.

Hay, E. D. (1959). Electron microscopic observations of muscle dedifferentiation in regenerating Ambystoma limbs. *Developmental Biology*, *1*, 555–585.

Hay, E. (1966). *Regeneration*. New York: Holt, Rinehart, Wilson.

Hayashi, T., Mizuno, N., Ueda, Y., Okamoto, M., & Kondoh, H. (2004). FGF2 triggers iris-derived lens regeneration in newt eye. *Mechanisms of Development*, *121*(6), 519–526.

Hayashi, T., Yokotani, N., Tane, S., Matsumoto, A., Myouga, A., Okamoto, M., et al. (2013). Molecular genetic system for regenerative studies using newts. *Development, Growth & Differentiation*, *55*(2), 229–236.

Haynes, T., Gutierrez, C., Aycinena, J. C., Tsonis, P. A., & Del Rio-Tsonis, K. (2007). BMP signaling mediates stem/progenitor cell-induced retina regeneration. *Proceedings of the National Academy of Sciences of the United States of America*, *104*(51), 20380.

Hedlund, E., Belnoue, L., Theofilopoulos, S., Salto, C., Bye, C., Parish, C., et al. (2016). Dopamine receptor antagonists enhance proliferation and neurogenesis of midbrain Lmx1a-expressing progenitors. *Scientific Reports*, *6*(1), 26448.

Henry, J. J., & Hamilton, P. W. (2018). Diverse evolutionary origins and mechanisms of lens regeneration. *Molecular Biology and Evolution*, *35*(7), 1563–1575.

Henry, J. J., & Tsonis, P. A. (2010). Molecular and cellular aspects of amphibian lens regeneration. *Progress in Retinal and Eye Research*, *29*(6), 543–555.

Huang, Y., Ren, H. T., Xiong, J. L., Gao, X. C., & Sun, X. H. (2017). Identification and characterization of known and novel microRNAs in three tissues of Chinese giant salamander base on deep sequencing approach. *Genomics*, *109*(3–4), 258–264.

Imokawa, Y., Simon, A., & Brockes, J. P. (2004). A critical role for thrombin in vertebrate lens regeneration. *Philosophical Transactions of the Royal Society of London. Series B, Biological Sciences*, *359*(1445), 765–776.

Imokawa, Y., & Yoshizato, K. (1997). Expression of sonic hedgehog gene in regenerating newt limb blastemas recapitulates that in developing limbbuds. *Proceedings of the National Academy of Sciences*, *94*(17), 9159–9164.

Islam, M. R., Nakamura, K., Casco-Robles, M. M., Kunahong, A., Inami, W., Toyama, F., et al. (2014). The newt reprograms mature RPE cells into a unique multipotent state for retinal regeneration. *Scientific Reports*, *4*, 1–8.

Iten, L. E., & Bryant, S. V. (1973). Forelimb regeneration from different levels of amputation in the newt, Notophthalmus viridescens: Length, rate, and stages. *Wilhelm Roux' Archiv für Entwicklungsmechanik der Organismen*, *173*(4), 263–282.

Iten, L. E., & Bryant, S. V. (1976). Regeneration from different levels along the tail of the newt, Notophthalmus viridescens. *The Journal of Experimental Zoology*, *196*(3), 293–306.

Jacyniak, K., McDonald, R. P., & Vickaryous, M. K. (2017). Tail regeneration and other phenomena of wound healing and tissue restoration in lizards. *Journal of Experimental Biology*, *220*(16), 2858–2869.

Joven, A., Elewa, A., & Simon, A. (2019). Model systems for regeneration: Salamanders. *Development (Cambridge)*, *146*(14), (dev167700-dev167700).

Joven, A., & Simon, A. (2018). Homeostatic and regenerative neurogenesis in salamanders. *Progress in Neurobiology*, *170*(December 2017), 81–98.
Keinath, M. C., Timoshevskiy, V. A., Timoshevskaya, N. Y., Tsonis, P. A., Voss, S. R., & Smith, J. J. (2015). Initial characterization of the large genome of the salamander Ambystoma mexicanum using shotgun and laser capture chromosome sequencing. *Scientific Reports*, *5*, 16413.
Khattak, S., Murawala, P., Andreas, H., Kappert, V., Schuez, M., Sandoval-Guzman, T., et al. (2014). Optimized axolotl (Ambystoma mexicanum) husbandry, breeding, metamorphosis, transgenesis and tamoxifen-mediated recombination. *Nature Protocols*, *9*(3), 529–540.
Khattak, S., Schuez, M., Richter, T., Knapp, D., Haigo, S. L., Sandoval-Guzman, T., et al. (2013). Germline transgenic methods for tracking cells and testing gene function during regeneration in the axolotl. *Stem Cell Reports*, *1*(1), 90–103.
Khattak, S., & Tanaka, E. M. (2015). Transgenesis in axolotl (Ambystoma mexicanum). *Methods in Molecular Biology*, *1290*, 269–277. https://doi.org/10.1007/978-1-4939-2495-0_21. PMID: 25740493.
Khudoley, V. V., & Ellselv, V. V. (1979). Multiple melanomas in the axolotl Ambystoma mexlcanum 1. *Journal of the National Cancer Institute*, *63*(I), 101–103.
Kim, S. U., Lee, H. J., & Kim, Y. B. (2013). Neural stem cell-based treatment for neurodegenerative diseases. *Neuropathology*, *33*(5), 491–504. https://doi.org/10.1111/neup.12020.
King, M. W., Neff, A. W., & Mescher, A. L. (2012). The developing Xenopus limb as a model for studies on the balance between inflammation and regeneration. *Anatomical record (Hoboken, N.J.: 2007)*, *295*(10), 1552–1561.
Knapp, D., Schulz, H., Rascon, C. A., Volkmer, M., Scholz, J., Nacu, E., et al. (2013). Comparative transcriptional profiling of the axolotl limb identifies a tripartite regeneration-specific gene program. *PLoS One*, *8*(5), e61352.
Kragl, M., Knapp, D., Nacu, E., Khattak, S., Maden, M., Epperlein, H. H., et al. (2009). Cells keep a memory of their tissue origin during axolotl limb regeneration. *Nature*, *460*(7251), 60–65.
Kragl, M., Roensch, K., Nusslein, I., Tazaki, A., Taniguchi, Y., Tarui, H., et al. (2013). Muscle and connective tissue progenitor populations show distinct Twist1 and Twist3 expression profiles during axolotl limb regeneration. *Developmental Biology*, *373*(1), 196–204.
Kuilman, T., & Peeper, D. S. (2009). Senescence-messaging secretome: SMS-ing cellular stress. *Nature Reviews Cancer*, *9*(2), 81–94.
Kumar, A., & Delgado, J. P. (2011). The aneurogenic limb identifies developmental cell interactions underlying vertebrate limb regeneration. *Proceedings of the National Academy of Sciences*, *108*(33), 13588–13593.
Kumar, A., Godwin, J. W., Gates, P. B., Garza-Garcia, A. A., & Brockes, J. P. (2007). Molecular basis for the nerve dependence of limb regeneration in an adult vertebrate. *Science (New York, N.Y.)*, *318*(5851), 772–777.
Lai, S. L., Marin-Juez, R., Moura, P. L., Kuenne, C., Lai, J. K. H., Tsedeke, A. T., et al. (2017). Reciprocal analyses in zebrafish and medaka reveal that harnessing the immune response promotes cardiac regeneration. *eLife*, *6*, e25605. https://doi.org/10.7554/eLife.25605.
Lee-Liu, D., Mendez-Olivos, E. E., Munoz, R., & Larrain, J. (2017). The African clawed frog Xenopus laevis: A model organism to study regeneration of the central nervous system. *Neuroscience Letters*, *652*, 82–93.
Leigh, N. D., Dunlap, G. S., Johnson, K., Mariano, R., Oshiro, R., Wong, A. Y., et al. (2018). Transcriptomic landscape of the blastema niche in regenerating adult axolotl limbs at single-cell resolution. *Nature Communications*, *9*(1), 5153. https://doi.org/10.1038/s41467-018-07604-0.

Lepper, C., Partridge, T. A., & Fan, C.-M. (2011). An absolute requirement for Pax7-positive satellite cells in acute injury-induced skeletal muscle regeneration. *Development (Cambridge, England)*, *138*(17), 3639–3646.

Li, F., Wang, L., Lan, Q., Yang, H., Li, Y., Liu, X., et al. (2015). RNA-Seq analysis and gene discovery of Andrias davidianus using Illumina short read sequencing. *PLoS One*, *10*(4), 1–16.

Logan, M., Simon, H. G., & Tabin, C. (1998). Differential regulation of T-box and homeobox transcription factors suggests roles in controlling chick limb-type identity. *Development*, *125*(15), 2825–2835.

Londono, R., Sun, A. X., Tuan, R. S., & Lozito, T. P. (2018). Tissue repair and Epimorphic regeneration: An overview. *Current Pathobiology Reports*, *6*, 61–69.

López-Otín, C., Blasco, M. A., Partridge, L., Serrano, M., & Kroemer, G. (2013). The hallmarks of aging. *Cell*, *153*, 1194–1217.

Maden, M. (1982). Vitamin a and pattern formation in the regenerating limb. *Nature*, *295*(5851), 672–675.

Maden, M., Manwell, L. A., & Ormerod, B. K. (2013). Proliferation zones in the axolotl brain and regeneration of the telencephalon. *Neural Development*, *8*(1), 1.

Maki, N., Suetsugu-Maki, R., Tarui, H., Agata, K., Del Rio-Tsonis, K., & Tsonis, P. A. (2009). Expression of stem cell pluripotency factors during regeneration in newts. *Developmental Dynamics*, *238*(6), 1613–1616.

McHedlishvili, L., Epperlein, H. H., Telzerow, A., & Tanaka, E. M. (2007). A clonal analysis of neural progenitors during axolotl spinal cord regeneration reveals evidence for both spatially restricted and multipotent progenitors. *Development*, *134*, 2083–2093.

McHedlishvili, L., Mazurov, V., Grassme, K. S., Goehler, K., Robl, B., Tazaki, A., et al. (2012). Reconstitution of the central and peripheral nervous system during salamander tail regeneration. *Proceedings of the National Academy of Sciences of the United States of America*, *109*(34), E2258–E2266.

McHugh, D., & Gil, J. (2018). Senescence and aging: Causes, consequences, and therapeutic avenues. *Journal of Cell Biology*, *217*(1), 65–77.

Meinhardt, A., Eberle, D., Tazaki, A., Ranga, A., Niesche, M., Wilsch-Brauninger, M., et al. (2014). 3D reconstitution of the patterned neural tube from embryonic stem cells. *Stem Cell Reports*, *3*(6), 987–999.

Mercader, N., Tanaka, E. M., & Torres, M. (2005). Proximodistal identity during vertebrate limb regeneration is regulated by Meis homeodomain proteins. *Development (Cambridge, England)*, *132*(18), 4131–4142.

Mescher, A. L., Neff, A. W., & King, M. W. (2013). Changes in the inflammatory response to injury and its resolution during the loss of regenerative capacity in developing Xenopus limbs. *PLoS One*, *8*(11), e80477.

Molla, K. A., & Yang, Y. (2019). CRISPR/Cas-Mediated Base editing: Technical considerations and practical applications. *Trends in Biotechnology*, *37*(10), 1121–1142.

Monaghan, J. R., Athippozhy, A., Seifert, A. W., Putta, S., Stromberg, A. J., Maden, M., et al. (2012). Gene expression patterns specific to the regenerating limb of the Mexican axolotl. *Biology open*, *1*(10), 937–948.

Monaghan, J. R., Stier, A. C., Michonneau, F., Smith, M. D., Pasch, B., Maden, M., et al. (2014). Experimentally induced metamorphosis in axolotls reduces regenerative rate and fidelity. *Regeneration*, *1*(1), 2–14.

Mullen, L. M., Bryant, S. V., Torok, M. A., Blumberg, B., & Gardiner, D. M. (1996). Nerve dependency of regeneration: The role of distal-less and FGF signaling in amphibian limb regeneration. *Development*, *122*(11), 3487–3497.

Muneoka, K., & Bryant, S. V. (1984). Cellular contribution to supernumerary limbs resulting from the interaction between developing and regenerating tissues in the axolotl. *Developmental Biology*, *105*(1), 179–187.

Muneoka, K., Fox, W. F., & Bryant, S. V. (1986). Cellular contribution from dermis and cartilage to the regenerating limb blastema in axolotls. *Developmental Biology*, *116*(1), 256–260.

Muñoz-Espín, D., Cañamero, M., Maraver, A., Gómez-López, G., Contreras, J., Murillo-Cuesta, S., et al. (2013). Programmed cell senescence during mammalian embryonic development. *Cell*, *155*(5), 1104.

Muñoz-Espín, D., & Serrano, M. (2014). Cellular senescence: From physiology to pathology. *Nature Reviews Molecular Cell Biology*, *15*(7), 482–496.

Murawala, P., Tanaka, E. M., & Currie, J. D. (2012). Regeneration: The ultimate example of wound healing. *Seminars in Cell & Developmental Biology*, *23*(9), 954–962.

Nacu, E., Glausch, M., Le, H. Q., Damanik, F. F., Schuez, M., Knapp, D., et al. (2013). Connective tissue cells, but not muscle cells, are involved in establishing the proximo-distal outcome of limb regeneration in the axolotl. *Development*, *140*(3), 513–518.

Nacu, E., Gromberg, E., Oliveira, C. R., Drechsel, D., & Tanaka, E. M. (2016). FGF8 and SHH substitute for anterior-posterior tissue interactions to induce limb regeneration. *Nature*, *533*(7603), 407–410.

Nacu, E., & Tanaka, E. M. (2011). Limb regeneration: A new development? *Annual Review of Cell and Developmental Biology*, *27*, 409–440.

Nakamura, K., Islam, M. R., Takayanagi, M., Yasumuro, H., Inami, W., Kunahong, A., et al. (2014). A transcriptome for the study of early processes of retinal regeneration in the adult newt, Cynops pyrrhogaster. *PLoS One*, *9*(10), e109831.

Nowoshilow, S., Schloissnig, S., Fei, J.-F., Dahl, A., Pang, A. W. C., Pippel, M., et al. (2018). The axolotl genome and the evolution of key tissue formation regulators. *Nature*, *554*(7690), 50–55.

Nye, H. L., Cameron, J. A., Chernoff, E. A., & Stocum, D. L. (2003). Regeneration of the urodele limb: A review. *Developmental Dynamics*, *226*(2), 280–294.

O' Hara, C., Ega, M. W., & Eag, C. (1992). Reorganization of the ependyma during axolotl spinal cord regeneration: Changes in interfilament and fibronectin expression. *Developmental Dynamics*, *193*, 103–115.

O'Hara, C. M., & Chernoff, E. A. (1994). Growth factor modulation of injury-reactive ependymal cell proliferation and migration. *Tissue & Cell*, *26*(4), 599–611.

O'Hara, C., Egar, M., & Chernoff, E. (1992). Reorganization of the ependyma during axolotl spinal cord regeneration: Changes in intermediate filament and fibronectin expression. *Developmental Dynamics*, *193*(2), 103–115.

Okamoto, M. (1987). Induction of ocular tumor by nickel subsulfide in the japanese common newt, cynops pyrrh ogaster. *Cancer Research*, *47*(19), 5092–5140.

Okamoto, M. (1997). Simultaneous demonstration of lens regeneration from dorsal iris and tumour production from ventral iris in the same newt eye after carcinogen administration. *Differentiation*, *61*(5), 285–292.

Oviedo, N. J., & Beane, W. S. (2009). Regeneration: The origin of cancer or a possible cure? *Seminars in Cell & Developmental Biology*, *20*, 557–564.

Parish, C. L., Beljajeva, A., Arenas, E., & Simon, A. (2007). Midbrain dopaminergic neurogenesis and behavioural recovery in a salamander lesion-induced regeneration model. *Development*, *134*(15), 2881–2887.

Pearson, B. J., & Alvarado, A. S. (2009). Regeneration, stem cells, and the evolution of tumor suppression. *Spring*, *LXXIII*, 565–572.

Piatt, J. (1955). Regeneration of the spinal cord in the salamander. *Journal of Experimental Zoology*, *129*(1), 177–207.

Pino, C. J., Westover, A. J., Johnston, K. A., Buf, D. A., & Humes, H. D. (2018). Regenerative medicine and immunomodulatory therapy: Insights from the kidney, heart, brain, and lung. *Kidney International Reports*, *3*(December 2017), 771–783.

Pomerantz, J. H., & Blau, H. M. (2013). Tumor suppressors: Enhancers or suppressors of regeneration? *Development (Cambridge)*, *140*(12), 2502–2512.

Porrello, E. R., Mahmoud, A. I., Simpson, E., Hill, J. A., James, A., Olson, E. N., et al. (2011). Transient regenerative potential of the neonatal mouse heart. *Science*, *331*(6020), 1078–1080.

Putta, S., Smith, J., Walker, J., Rondet, M., Weisrock, D., Monaghan, J., et al. (2004). From biomedicine to natural history research: EST resources for ambystomatid salamanders. *BMC Genomics*, *5*(1), 54.

Ravichandran, A., Liu, Y., & Teoh, S.-H. (2018). Review: Bioreactor design towards generation of relevant engineered tissues: Focus on clinical translation. *Journal of Tissue Engineering and Regenerative Medicine*, *12*(1), e7–e22. https://pubmed.ncbi.nlm.nih.gov/28374578/.

Rio-Tsonis, K. D., Jung, J. C., Chiu, I.-M., & Tsonis, P. A. (1997). Conservation of fibroblast growth factor function in lens regeneration. *Proceedings of the National Academy of Sciences*, *94*(25), 13701–13706.

Roensch, K., Tazaki, A., Chara, O., & Tanaka, E. M. (2013). Progressive specification rather than intercalation of segments during limb regeneration. *Science*, *342*(6164), 1375–1379.

Rose, M., & Rose, F. (1952). Tumor agent transformations in amphibia. *Cancer Research*, *12*(1), 1–12.

Rose, S. M., Wallingford, H. M., Moore, C. R., Price, D., Bard, P., & Mountcastle, V. B. (1949). Transformation of renal tumors of frogs to Normal tissues in regenerating limbs of salamanders. *Science*, *109*(2835), 435–444.

Sabin, K., Jiang, P., Gearhart, M., Stewart, R., & Echeverri, K. (2019). AP-1cFos/Jun B/mi R-200a regulate the pro-regenerative glial cell response during axolotl spinal cord regeneration. *Communications Biology*, *2*(1), 91.

Sabin, K., Santos-Ferreira, T., Essig, J., Rudasill, S., & Echeverri, K. (2015). Dynamic membrane depolarization is an early regulator of ependymoglial cell response to spinal cord injury in axolotl. *Developmental Biology*, *408*(1), 14–25.

Salero, E., Blenkinsop, T. A., Corneo, B., Harris, A., Rabin, D., Stern, J. H., et al. (2012). Adult human RPE can be activated into a multipotent stem cell that produces mesenchymal derivatives. *Cell Stem Cell*, *10*(1), 88–95.

Sambasivan, R., Yao, R., Kissenpfennig, A., Van Wittenberghe, L., Paldi, A., Gayraud-Morel, B., et al. (2011). Pax7-expressing satellite cells are indispensable for adult skeletal muscle regeneration. *Development (Cambridge, England)*, *138*(17), 3647–3656.

Sandoval-Guzman, T., Wang, H., Khattak, S., Schuez, M., Roensch, K., Nacu, E., et al. (2014). Fundamental differences in dedifferentiation and stem cell recruitment during skeletal muscle regeneration in two salamander species. *Cell Stem Cell*, *14*(2), 174–187.

Sanor, L. D., Flowers, G. P., & Crews, C. M. (2020). Multiplex CRISPR/Cas screen in regenerating haploid limbs of chimeric axolotls. *eLife*, *9*, 1–18.

Sarig, R., & Tzahor, E. (2017). The cancer paradigms of mammalian regeneration: Can mammals regenerate as amphibians? *Carcinogenesis*, *38*(4), 359–366.

Satoh, A., Bryant, S. V., & Gardiner, D. M. (2008). Regulation of dermal fibroblast dedifferentiation and redifferentiation during wound healing and limb regeneration in the axolotl. *Development, Growth & Differentiation*, *50*(9), 743–754.

Satoh, A., Bryant, S. V., & Gardiner, D. M. (2012). Nerve signaling regulates basal keratinocyte proliferation in the blastema apical epithelial cap in the axolotl (Ambystoma mexicanum). *Developmental Biology*, *366*(2), 374–381.

Satoh, A., Graham, G. M., Bryant, S. V., & Gardiner, D. M. (2008). Neurotrophic regulation of epidermal dedifferentiation during wound healing and limb regeneration in the axolotl (Ambystoma mexicanum). *Developmental Biology*, *319*(2), 321–335.

Satoh, A., Makanae, A., Nishimoto, Y., & Mitogawa, K. (2016). FGF and BMP derived from dorsal root ganglia regulate blastema induction in limb regeneration in Ambystoma mexicanum. *Developmental Biology*, *417*(1), 114–125.

Seifert, A. W., Kiama, S. G., Seifert, M. G., Goheen, J. R., Palmer, T. M., & Maden, M. (2012). Skin shedding and tissue regeneration in African spiny mice (Acomys). *Nature*, *489*(7417), 561–565.

Shaikh, N., Gates, P. B., & Brockes, J. P. (2011). The Meis homeoprotein regulates the axolotl prod 1 promoter during limb regeneration. *Gene*, *484*(1–2), 69.

Shioda, C., Uchida, K., & Nakayama, H. (2011). Pathological features of olfactory neuroblastoma in an axolotl (Ambystoma mexicanum). *The Journal of Veterinary Medical Science*, *73*(8), 1109–1111.

Silver, J. (2016). The glial scar is more than just astrocytes. *Experimental Neurology*, *286*, 147–149.

Silver, J., & Miller, J. H. (2004). Regeneration beyond the glial scar. *Nature Reviews. Neuroscience*, *5*(2), 146–156.

Simon, A., & Tanaka, E. M. (2013). Limb regeneration. *Wiley Interdisciplinary Reviews: Developmental Biology*, *2*(2), 291–300.

Smith, J. J., Timoshevskaya, N., Timoshevskiy, V. A., Keinath, M. C., Hardy, D., & Voss, S. R. (2019). A chromosome-scale assembly of the axolotl genome. *Genome Research*, *29*(2), 317–324.

Sobkow, L., Epperlein, H. H., Herklotz, S., Straube, W. L., & Tanaka, E. M. (2006). A germline GFP transgenic axolotl and its use to track cell fate: Dual origin of the fin mesenchyme during development and the fate of blood cells during regeneration. *Developmental Biology*, *290*(2), 386–397.

Sousounis, K., Athippozhy, A. T., Voss, S. R., & Tsonis, P.a. (2014). Plasticity for axolotl lens regeneration is associated with age-related changes in gene expression. *Regeneration*, *1*(3), 47–57.

Sousounis, K., Looso, M., Maki, N., Ivester, C. J., Braun, T., & Tsonis, P. A. (2013). Transcriptome analysis of newt lens regeneration reveals distinct gradients in gene expression patterns. *PLoS One*, *8*(4), e61445.

Spallanzani, L. (1768). Prodromo sa un Opera da Imprimersi sopra le Riproduzioni animali. *Modena*, 7.

Stocum, D. L. (1979). Stages of forelimb regeneration in Ambystoma maculatum. *The Journal of Experimental Zoology*, *209*(3), 395–416.

Stocum, D. L. (1980). Intercalary regeneration of symmetrical thighs in the axolotl, Ambystoma mexicanum. *Developmental Biology*, *79*(2), 276–295.

Stocum, D. L. (1983). Amphibian limb regeneration: Distal transformation. *Progress in Clinical and Biological Research*, *110*(Pt A), 467–476.

Stocum, D. L. (1991). Limb regeneration: A call to arms (and legs). *Cell*, *67*(1), 5–8.

Stocum, D. L. (1998). Regenerative biology and engineering: Strategies for tissue restoration [see comments]. *Wound Repair and Regeneration*, *6*(4), 276–290.

Stocum, D. L. (2017). Mechanisms of urodele limb regeneration. *Regeneration*, *4*(October), 159–200.

Stone, L. S. (1967). An investigation recording all salamanders which can and cannot regenerate a lens from the dorsal iris. *Journal of Experimental Zoology*, *164*(1), 87–103.

Storer, M., Mas, A., Robert-Moreno, A., Pecoraro, M., Ortells, M. C., Di Giacomo, V., et al. (2013). Senescence is a developmental mechanism that contributes to embryonic growth and patterning. *Cell*, *155*(5), 1119.

Suetsugu-Maki, R., Maki, N., Nakamura, K., Sumanas, S., Zhu, J., Del Rio-Tsonis, K., et al. (2012). Lens regeneration in axolotl: New evidence of developmental plasticity. *BMC Biology*, *10*(1), 103.

Sugiura, T., Wang, H., Barsacchi, R., Simon, A., & Tanaka, E. M. (2016). MARCKS-like protein is an initiating molecule in axolotl appendage regeneration. *Nature*, *531*(7593), 237–240.

Sun, A. X., Londono, R., Hudnall, M. L., Tuan, R. S., & Lozito, T. P. (2018). Differences in neural stem cell identity and differentiation capacity drive divergent regenerative

outcomes in lizards and salamanders. *Proceedings of the National Academy of Sciences of the United States of America*, *115*(35), E8256–E8265.
Takahashi, K., & Yamanaka, S. (2006). Induction of pluripotent stem cells from mouse embryonic and adult fibroblast cultures by defined factors. *Cell*, *126*(4), 663–676.
Tanaka, E. M. (2016). The molecular and cellular choreography of appendage regeneration. *Cell*, *165*(7), 1598–1608.
Tanaka, E. M., & Ferretti, P. (2009). Considering the evolution of regeneration in the central nervous system. *Nature Reviews. Neuroscience*, *10*(10), 713–723.
Tapia, N., Reinhardt, P., Duemmler, A., Wu, G., Arauzo-Bravo, M. J., Esch, D., et al. (2012). Reprogramming to pluripotency is an ancient trait of vertebrate Oct4 and Pou2 proteins. *Nature Communications*, *3*, 1279.
Tassava, R. A., & Garling, D. J. (1979). Regenerative responses in larval axolotl limbs with skin grafts over the amputation surface. *The Journal of Experimental Zoology*, *208*(1), 97–110.
Tassava, R. A., & Loyd, R. M. (1977). Injury requirement for initiation of regeneration of newt limbs which have whole skin grafts. *Nature*, *268*(5615), 49–50.
Tazaki, A., Tanaka, E. M., & Fei, J. F. (2017). Salamander spinal cord regeneration: The ultimate positive control in vertebrate spinal cord regeneration. *Developmental Biology*, *432*(1), 63–71.
Thornton, C. S. (1954). The relation of epidermal innervation to limb regeneration in Amblystoma larvae. *Journal of Experimental Zoology*, *127*(3), 577–601.
Tran, A. P., Warren, P. M., & Silver, J. (2018). The biology of regeneration failure and success after spinal cord injury. *Physiological Reviews*, *98*(2), 881–917.
Tsai, S. L., Baselga-Garriga, C., & Melton, D. A. (2019). Blastemal progenitors modulate immune signaling during early limb regeneration. *Development*, *146*(1) (dev169128-dev169128).
Tsai, S. L., Baselga-Garriga, C., & Melton, D. A. (2020). Midkine is a dual regulator of wound epidermis development and inflammation during the initiation of limb regeneration. *eLife*, *9*, 1–29.
Tsarouchas, T. M., Wehner, D., Cavone, L., Munir, T., Keatinge, M., Lambertus, M., et al. (2018). Dynamic control of proinflammatory cytokines Il-1β and Tnf-α by macrophages in zebrafish spinal cord regeneration. *Nature Communications*, *9*(1), 4670.
Tsonis, P.a. (2006). How to build and rebuild a lens. *Journal of Anatomy*, *209*(4), 433–437.
Tsonis, P. A., & Del Rio-Tsonis, K. (2004). Lens and retina regeneration: Transdifferentiation, stem cells and clinical applications. *Experimental Eye Research*, *78*(2), 161–172.
Tsonis, P., & Eguchi, G. (1981). Carcinogens on regeneration. *Differentiation*, *20*(1–3), 52–60.
Tsonis, P., & Eguchi, G. (1982). Abnormal limb regeneration without tumor production in adult newts directed by carcinogens, 20-Methylcholanthrene and benzo (α) pyrene: Amphibia/limb regeneration/carcinogen/teratogenesis. *Development, Growth & Differentiation*, *24*(2), 183–190.
Tsonis, P. A., & Fox, T. P. (2009). Regeneration according to Spallanzani. *Developmental Dynamics*, *238*(9), 2357–2363.
Tsonis, P., Jang, W., Del Rio-Tsonis, K., & Eguchi, G. (2001). A unique aged human retinal pigmented epithelial cell line useful for studying lens differentiation in vitro. *International Journal of Developmental Biology*, *45*(5–6), 753–758.
Tsonis, P. A., Madhavan, M., Tancous, E. E., & Del Rio-Tsonis, K. (2004). A newt's eye view of lens regeneration. *The International Journal of Developmental Biology*, *48*(8–9), 975–980.
Tsutsumi, R., Inoue, T., Yamada, S., & Agata, K. (2015). Reintegration of the regenerated and the remaining tissues during joint regeneration in the newt\n Cynops pyrrhogaster. *Regeneration*, *2*(1), 26–36.

Ueda, Y., Kondoh, H., & Mizuno, N. (2005). Generation of transgenic newt Cynops pyrrhogaster for regeneration study. *Genesis (New York, N.Y.: 2000)*, *41*(2), 87–98.
Urata, Y., Yamashita, W., Inoue, T., & Agata, K. (2018). Spatio-temporal neural stem cell behavior leads to both perfect and imperfect structural brain regeneration in adult newts. *Biology Open*, 7(6). https://doi.org/10.1242/bio.033142. bio033142.
Vieira, W. A., Wells, K. M., & McCusker, C. D. (2020). Advancements to the axolotl model for regeneration and aging. *Gerontology*, *66*(3), 212–222.
Vig, K., Chaudhari, A., Tripathi, S., Dixit, S., Sahu, R., Pillai, S., et al. (2017). Advances in skin regeneration using tissue engineering. *International Journal of Molecular Sciences*, *18*(4).
Villiard, É., Denis, J. F., Hashemi, F. S., Igelmann, S., Ferbeyre, G., & Roy, S. (2017). Senescence gives insights into the morphogenetic evolution of anamniotes. *Biology Open*, *6*(6), 891–896.
Wang, X. (2019). Bioartificial organ manufacturing technologies. *Cell Transplantation*, *28*(1), 5–17.
Wang, L.-H., Wu, C.-F., Rajasekaran, N., & Shin, Y. K. (2018). Loss of tumor suppressor gene function in human cancer: An overview. *Cellular Physiology and Biochemistry*, *51*(6), 2647–2693.
Xiong, J., Lv, Y., Huang, Y., & Liu, Q. (2019). The first transcriptome assembly of Yenyuan stream salamander (Batrachuperus yenyuanensis) provides novel insights into its molecular evolution. *International Journal of Molecular Sciences*, *20*(7).
Yun, M. H. (2015). Changes in regenerative capacity through lifespan. *International Journal of Molecular Sciences*, *16*(10), 25392–25432.
Yun, M. H. (2018). Cellular senescence in tissue repair: Every cloud has a silver lining. *International Journal of Developmental Biology*, *62*(6–8), 591–604.
Yun, M. H., Davaapil, H., & Brockes, J. P. (2015). Recurrent turnover of senescent cells during regeneration of a complex structure. *eLife*, *4*, 1–16.
Zhao, M., Rotgans, B., Wang, T., & Cummins, S. F. (2016). REGene: A literature-based knowledgebase of animal regeneration that bridge tissue regeneration and cancer. *Scientific Reports*, *6*(February), 1–11.
Zhu, Y., Carido, M., Meinhardt, A., Kurth, T., Karl, M. O., Ader, M., et al. (2013). Three-dimensional neuroepithelial culture from human embryonic stem cells and its use for quantitative conversion to retinal pigment epithelium. *PLoS One*, *8*(1), e54552.
Zhu, Y., Schreiter, S., & Tanaka, E. M. (2016). Accelerated three-dimensional Neuroepithelium formation from human embryonic stem cells and its use for quantitative differentiation to human retinal pigment epithelium. *Methods in Molecular Biology*, *1307*, 345–355.

CHAPTER NINE

Xenopus as a platform for discovery of genes relevant to human disease

Valentyna Kostiuk and Mustafa K. Khokha*
Pediatric Genomics Discovery Program, Department of Pediatrics and Genetics, Yale University School of Medicine, New Haven, CT, United States
*Corresponding author: e-mail address: mustafa.khokha@yale.edu

Contents

Abstract

Congenital birth defects result from an abnormal development of an embryo and have detrimental effects on children's health. Specifically, congenital heart malformations are a leading cause of death among pediatric patients and often require surgical interventions within the first year of life. Increased efforts to navigate the human genome provide an opportunity to discover multiple *candidate* genes in patients suffering from birth defects. These efforts, however, fail to provide an explanation regarding the

Current Topics in Developmental Biology, Volume 145
ISSN 0070-2153
https://doi.org/10.1016/bs.ctdb.2021.03.005

mechanisms of disease pathogenesis and emphasize the need for an efficient platform to screen *candidate* genes. *Xenopus* is a rapid, cost effective, high-throughput vertebrate organism to model the mechanisms behind human disease. This review provides numerous examples describing the successful use of *Xenopus* to investigate the contribution of patient mutations to complex phenotypes including congenital heart disease and heterotaxy. Moreover, we describe a variety of unique methods that allow us to rapidly recapitulate patients' phenotypes in frogs: gene knockout and knockdown strategies, the use of fate maps for targeted manipulations, and novel imaging modalities. The combination of patient genomics data and the functional studies in *Xenopus* will provide necessary answers to the patients suffering from birth defects. Furthermore, it will allow for the development of better diagnostic methods to ensure early detection and intervention. Finally, with better understanding of disease pathogenesis, new treatment methods can be tailored specifically to address patient's phenotype and genotype.

1. Congenital malformations

Congenital malformations, or birth defects, occur as a result of an improper development of an embryo. These birth defects have a massive impact on pediatric health. Additionally, some congenital birth defects disrupt pregnancies leading to stillbirths and miscarriages, which is particularly devastating to parents who plan to start a family (Gregory, MacDorman, & Martin, 2014; Stillbirth Collaborative Research Network Writing Group, 2011). Every year around 8 million children are born with severe birth defects worldwide, which constitutes roughly 6% of births (Christianson, Howson, & Modell, 2006). In the United States, during the first year of life, congenital malformations are the leading cause of medical expenditure (Waitzman, Romano, & Scheffler, 1994), pediatric hospitalizations (Yoon et al., 1997), and death (Murphy, Mathews, Martin, Minkovitz, & Strobino, 2017). Moreover, birth defects remain a leading cause of death among children in different age categories. In particular, it is the number two cause of death in children between 1 and 4 years old, number three in pediatric patients from 5 to 14 years of age, and number 6 in patients from 15 to 24 years old (United States Department of Health and Human Services, 2018). Therefore, considering the tremendous impact of birth defects on pediatric patients and their families, there is an urgent need to improve both diagnostic approaches and treatment methods. Such improvements could be achieved through the combination of genetic testing and the understanding of the molecular mechanisms contributing to congenital malformations. Currently, the advancements in human genomics allow for the use of massively parallel sequencing platforms to identify *candidate* genes in patients with birth defects. These methods, however, fail to reveal the mechanisms behind disease pathogenesis. Hence, despite our best efforts to navigate the human genome, we still lack the ability to efficiently translate genomic data into diagnostic applications and treatments tailored specifically to patients' genotypes. Therefore, there is an increasing need to develop platforms for testing *candidate* genes to determine their roles in proper embryo development and how mutations in those genes could lead to birth defects. This review describes the advantages of using *Xenopus* as a model organism to study congenital malformations and connect genomics knowledge with patient phenotypes and embryological mechanisms.

2. Congenital heart disease

Congenital heart disease (CHD) is the most common and the most life-threatening birth defect. Every year CHD affects ~1% of live births and 1.3 million patients worldwide (van der Linde et al., 2011). Because much of CHD is associated with high rates of both mortality and morbidity, these heart malformations will normally need to be corrected with the use of surgery or other interventions (Xu, Murphy, Kochanek, Bastian, & Arias, 2018). Advancements in medical and surgical management allow more CHD patients to reach adulthood. Specifically, the number of adult CHD survivors is expected to rise by 5% every year (van der Bom et al., 2011; Warnes et al., 2001). This increase in survival also presents a new challenge for developing novel medical and surgical strategies to care for adult CHD patients. In particular, these adult patients have high potential to eventually develop multiple CHD associated complications, including arrhythmias, hemodynamic instability, pregnancy loss and infertility, pulmonary disease and neurodevelopmental disorders (Marino et al., 2012; Mussatto et al., 2015). Additionally, CHD patients encounter high medical care costs, which in the United States exceed 1.75 billion dollars annually (Russo & Elixhauser, 2007). Although access to health care remains one of the most important factors in the diagnosis and management of CHD, both genetic and environmental factors definitely contribute to this process as well (van der Linde et al., 2011).

Given the growing impact of CHD, it is crucial to develop platforms to test candidate genes identified through genomic sequencing. Such platforms will allow physicians to provide patients with better answers about their disease mechanisms. Further, it will improve diagnostic testing and encourage the use of individualized treatment approaches that could be specifically tailored to patient's genotype rather than just phenotype. *Xenopus* is an excellent model organism to test the function of CHD candidate genes in order to investigate their role in vertebrate development and disease pathogenesis.

3. Heterotaxy can lead to a severe form of CHD

Heterotaxy (Htx) is the abnormal position of the internal organs along the left-right axis. This can have a severe effect on the function of the heart as the right and left sides of the heart have very different functions. The right heart chambers together with associated blood vessels enable the oxygenation of blood in the lungs, while the left side of the heart delivers oxygenated blood to organs and tissues. Other organs also have a left-right asymmetry including the stomach, rotation of the gut, spleen, lungs, and even the brain. Any developmental disruptions of left-right body axis patterning could lead to abnormal positioning of internal organs and their subsequent malfunction. The arrangement of internal organs along left-right body axis can be divided into three categories: *Situs solitus*, which is characterized by normal positioning of body organs; *situs inversus*, which is a completely reversed, mirror image of the normal orientation; and *situs ambiguous*, in which there is no defined orientation of body organs with respect to left-right body axis.

Given the left-right asymmetry of the heart, heterotaxy can lead to a severe form of CHD. It occurs in 1 in 10,000 of children with CHD and it encompasses 3% of all CHD cases (Brueckner, 2007; Zhu, Belmont, & Ware, 2006). Heterotaxy in the *situs ambiguous* category is one of the most severe forms of CHD (Harden et al., 2014). Hence, patients with heterotaxy often need to undergo surgery or interventional procedures but despite advancements in care, heterotaxy has a relatively poor prognosis. For example, heterotaxy patients may suffer from multiple post-operative complications including arrhythmias and respiratory issues (Amula, Ellsworth, Bratton, Arrington, & Witte, 2014; Harden et al., 2014).

Like other congenital malformations, heterotaxy is a result of abnormal left-right embryo patterning. Left-Right embryo patterning follows a precise program that is conserved in multiple organisms. During gastrulation, the embryo establishes dorsoventral and anteroposterior body axes after the morphogenesis of the three germ layers: endoderm, mesoderm, and ectoderm. Near the end of gastrulation, the superficial dorsal mesoderm gives rise to the Left-Right Organizer (LRO) tissue, which will eventually break the bilateral symmetry of the embryo with respect to its left-right body

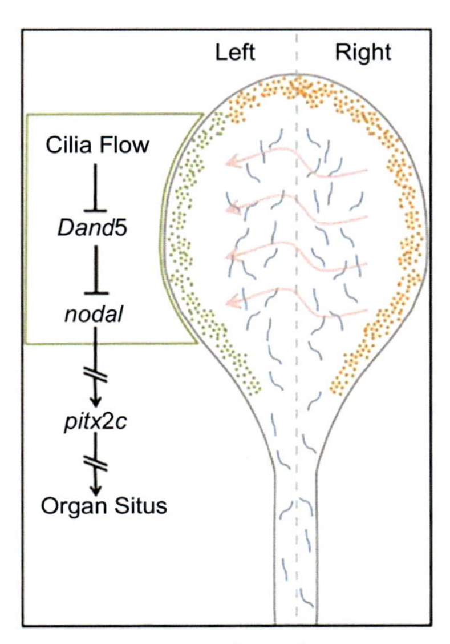

Fig. 1 Schematic of the LR signaling cascade in the LRO.

axis. The LRO is a conserved tissue found in multiple organisms: it is the node in the mouse, the Kupffer's vesicle in zebrafish, and the gastrocoel roof plate (GRP) in the frog. Left-right asymmetry is established by the leftward extracellular fluid flow generated by motile cilia in the central portion of the LRO (Fig. 1) (Nonaka, Shiratori, Saijoh, & Hamada, 2002; Okada, Takeda, Tanaka, Belmonte, & Hirokawa, 2005). This flow is detected by the peripherally located immotile cilia and translated into asymmetric gene expression (McGrath, Somlo, Makova, Tian, & Brueckner, 2003; Tabin & Vogan, 2003). Specifically, this leftward flow leads to the inhibition of *dand5*, an antagonist of *nodal* (Schweickert et al., 2010; Vonica & Brivanlou, 2007). The absence of *dand5* expression leads to the activation of nodal signaling, the phosphorylation of Smad2, and the subsequent activation of *pitx2* in the left lateral plate mesoderm (Kawasumi et al., 2011; Lee & Anderson, 2008). *Pitx2* expression is crucial for the asymmetric formation of heart, gut and lungs (Davis et al., 2008; Kurpios et al., 2008). The heart first forms from the fusion of precursor cardiac cells from lateral plate mesoderm to form a straight cardiac tube in the midline (Abu-Issa & Kirby, 2008; Stalsberg & DeHaan, 1969). The central region of the tube will give rise to the left ventricle. One end of the cardiac tube will form the outflow tracts and the atria, while the other end will form the inflow tracts and the right ventricle. Most importantly, the rightward rotation of this cardiac tube will eventually establish the left-right asymmetry of the adult heart. Multiple signaling programs contribute to proper establishment of this asymmetry (Hamada & Tam, 2014). These include cilia driven flow in the Left-Right Organizer, that leads to the suppression of dand5, a nodal antagonist, with the subsequent activation of the nodal pathway on the left side of the embryo. Additionally, BMP signaling is thought to specify the right side of the embryo. These an additional pathways integrate to specify the LR axis and subsequent cardiac looping.

4. Patient-driven gene discovery

4.1 Heterotaxy and rare copy-number variations

As genomic analysis has become more affordable, numerous efforts have been made to identify the genes associated with heterotaxy. In one study, the investigators identified rare copy-number variations (CNV) (Fakhro et al., 2011) in patients with heterotaxy. Using high-resolution SNP arrays in 262 heterotaxy patients and 991 control individuals, a total of 45 copy-number variations were found in patients with heterotaxy. In particular, 7 of 45 identified CNVs were large chromosomal deletions (6–25 Mb), while the other 38 were small in size (27–1488 kb). Examining the expression of some of these genes in *Xenopus* revealed that many were expressed in the left-right Organizer. Seven of these genes were selected for knockdown studies, and remarkably, the knockdown of five genes (NEK2, ROCK2, TGFBR2, GALNT11, and NUP188) using an antisense oligonucleotide morpholino caused significant abnormalities in the left-right patterning cascade (detected by *in situ* hybridization for *pitx2c*) with subsequent abnormal heart looping (Fakhro et al., 2011). More importantly, the tested genes are involved in diverse independent cellular processes, most of which have not had a previously established connection to cardiac development. For example, subsequent studies of the identified genes revealed a unique subcellular location of NUP188 and its binding partner NUP93 at the base of cilia (rather than the nuclear pore complex) and that their morpholino-mediated knockdown leads to the loss of cilia (Del Viso et al., 2016). Additionally, GALNT11 was found to regulate Notch signaling and establish the balance between motile and immotile cilia within the LRO tissue (Boskovski et al., 2013). As a third example, RAPGEF5 was recently found to regulate beta-catenin nuclear localization in the Wnt signaling pathway (Griffin et al., 2018). Therefore, pursuing a mechanism behind an identified candidate gene presents an outstanding opportunity to discover its novel function in cardiac development.

Another connection between copy-number variants and heterotaxy was made through the population based analysis conducted in New York State between 1998 and 2005 (Rigler et al., 2015). This study examined the DNA from dried blood spots from 77 newborns with classic heterotaxy. This investigation revealed 20 rare copy-number variants in genes contributing to various developmental pathways. Particularly, CNVs were identified in several members of the transforming growth factor-β superfamily, including bone morphogenic protein 2 (BMP2), fibroblast growth factor 12 (FGF12), and growth differentiation factor 7 (GDF7). Furthermore, two heterozygous deletions were identified in the gene encoding myeloid cell nuclear differentiation antigen (*MNDA*). Additionally, similar 20 kb duplications in the oligosaccharyltransferase complex subunit (*OSTC*) gene were identified in three separate cases. Despite the fact that multiple genes continue to be discovered, we still have a limited knowledge about the mechanisms behind candidate gene action and their role in CHD pathogenesis.

Another study examined 69 patients with classic heterotaxy identified from 1998 to 2009 in California (Hagen et al., 2016). 56 rare CNVs were identified including genes in crucial developmental pathways such as Nodal, BMP, and WNT. More importantly, none of these genes have been previously described within a context of heterotaxy pathogenesis. Moreover, CNVs in RBFOX1 and near MIR302F were found in multiple patients encouraging further investigation of their roles in heterotaxy and left-right embryo patterning. This study provides another example of how genomic methods could be used to identify candidate without known function in cardiac development, and emphasizes the importance of understanding the mechanism of disease to better serve this patient population.

4.2 *De novo* variants in CHD patients

In addition to CNVs, rare *de novo* variants have been identified in patients with congenital heart disease. CHD is generally a sporadic illness with healthy parents that have a child with CHD. Therefore, a *de novo* model of disease pathogenesis seems plausible. Using exome sequencing, 2871 probands were examined, including 2645 parent-offspring trios (Jin et al., 2017). *De novo* mutations were found in 443 genes and are thought to contribute to 8% of all cases of CHD. The candidate genes found in this study are involved in a variety of cellular processes. Interestingly, chromatin modifiers constitute a major class of novel genes in CHD patients. In particular, loss of function *de novo* mutations of chromatin modifiers were identified in

2.3% of probands. Additionally, this study also emphasized the link between CHD and neurodevelopmental disorders (NDD). 3% of patients had isolated CHD phenotype, while 28% of CHD patients had both neurodevelopmental and extra-cardiac manifestations. Specifically, patients with loss of function mutations in chromatin modifiers demonstrated an 87% higher risk of developing NDD. Interestingly, autism studies also identified *de novo* mutations in chromatin modifiers with a surprising overlap between autism and CDD. Furthermore, from a clinical perspective, many patients with CHD also suffer from NDD. NDD had been thought to be a complication of brain hypoxia operatively or peri-operatively for CHD, but these studies strongly suggested that NDD in CHD patients may be due to genes that affect both cardiac and brain development. Hence, detecting specific mutations associated with both CHD and NDD will allow for early interventions that could significantly improve outcomes in patients with neurodevelopmental abnormalities.

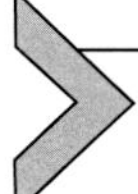

5. The challenges of patient-driven gene discovery efforts

The advancement of research methods to interrogate the human genome in CHD patients can improve our understanding of CHD and heterotaxy mechanisms. However, there are multiple challenges in defining precise disease causality using genomic data only. While the use of advanced sequencing methods enables the identification of the genes associated with heterotaxy, the establishment of their disease causality presents a challenge. Specifically, proof of disease causality requires that multiple detrimental alleles be identified in CHD patients compared to controls. However, due to high locus heterogeneity, multiple alleles are difficult in small to medium sized cohorts of CHD patients. From an embryological perspective, high locus heterogeneity is not surprising as multiple genes contribute to such complex processes as cardiac development and left-right patterning. In the future, the identification of multiple alleles will become possible with additional reductions in sequencing costs.

The next challenge is to unveil the molecular mechanisms behind the identified candidate genes and how their disruption contributes to congenital heart disease and embryo patterning pathogenesis. In order to accomplish this aim, we need an efficient high-throughput model to screen candidate genes identified by genome sequencing. While we await multiple allele identification to confirm disease causality, if we had an efficient screening platform, we could still use patient based gene discovery to identify genes that affect heart development. *Xenopus* offers a highly efficient model organism that could provide multiple benefits for patients with congenital heart disease. First, most patients and families yearn for some meaning in the patients' illness. For some, gene sequencing can identify a known cause of CHD and this provides a critical answer and the potential for genetic counseling in the future. But for many, no defined genetic cause is identified, but instead, variants of unknown significance. By testing these in *Xenopus*, we can establish that at least in this animal model the candidate gene does indeed affect heart development, a lunching platform for mechanistic studies. Many patients/families are very grateful for such discoveries as this provides some meaning (by creating a scientific research avenue) to their child's illness. In the future, understanding molecular mechanism could provide more accurate information about disease prognosis and will advance patient care. Additionally, most genetic conditions are often described in terms of patient symptoms and phenotypes. Such evaluation fails to acknowledge the differences in patients' genotypes as this information is often unavailable. Therefore, known treatment approaches are mostly targeted toward patients' phenotypes and are not ideal. A better understanding of patients' genotypes will also provide a more tailored treatment method and will improve disease outcomes.

To make this a reality, we need to address two challenges: assigning the disease causality to a specific gene, and determining mechanism of the disease. As discussed previously, the proof of disease causality presents a challenge due to high locus heterogeneity. Furthermore, a large number of identified genes may not have a known function in cardiac development or left-right patterning. In addition, some of the identified genes do not have a function previously described at all. Therefore, in the absence of massive population sequencing data, the ascertainment of molecular mechanisms should not wait until multiple alleles are identified since *Xenopus* is a perfect model organism to recapitulate human disease phenotypes and advance our understanding of plausible disease mechanisms. This screening platform has the potential to reveal the molecular mechanisms of human disease and explain complicated patient phenotypes. Therefore, *Xenopus* is a perfect animal model to test the function of candidate genes in congenital organ malformations, including cardiac abnormalities and left-right patterning defects such as heterotaxy.

6. *Xenopus* tools to study congenital disease

Xenopus is a fantastic model organism to study human disease. It enables a rapid, efficient and cost-effective screening approach of candidate genes to determine their role in left-right patterning and cardiac development (Blum et al., 2009; Wallingford, Liu, & Zheng, 2010; Warkman & Krieg, 2007). Two species of *Xenopus* are widely used in developmental research. Each species offers specific advantages. *Xenopus laevis* produces large allotetraploid embryos that are excellent for gene overexpression analysis and biochemical studies because the embryos can be cooled to allow time to inject mRNAs at the early cleavage stages. Additionally, they produce a large number of eggs *via in vitro* fertilization so abundant that biochemical experiments are greatly facilitated. *Xenopus tropicalis*, on the other hand, is a diploid organism that is a perfect match by facilitating loss of function studies. In addition, both *Xenopus* species develop externally, which allows for analysis and easy manipulation of an embryo. Furthermore, *Xenopus* early tadpoles are transparent, which provides an easy way to examine the gross morphology of internal organs under a light microscope. More significantly, *Xenopus* organogenesis is a rapid process. For example, the embryonic heart can be visualized as early as 3 days after fertilization. In addition, abnormal cardiac looping can be detected simply through the examination of the outflow tract of stage 45 embryos under light microscope. More importantly, the early development of the *Xenopus* heart does not depend on blood circulation as oxygen delivery takes place through simple diffusion. This allows for a greater variety of genomic manipulations that may not be possible in mice due to potential embryonic lethality. In addition, the *Xenopus* genome has significant conservation with human. The genome of *Xenopus tropicalis* contains orthologues for approximately 79% of genes identified in human disease (Grant et al., 2015; Hellsten et al., 2010; Khokha, 2012). Moreover, the frog genome has a high degree of synteny with the human genome, containing long equivalent regions with genes positioned in a conserved order (Blitz, 2012; Blitz, Biesinger, Xie, & Cho, 2013). Synteny is particularly useful for distinguishing orthologues from paralogs. Additionally, the anatomic structure and the development of *Xenopus* heart are characterized by a higher degree of conservation with human than other aquatic animals (Showell & Conlon, 2007). In comparison to a four-chambered human heart, a zebrafish heart has two chambers only: an atrium and a ventricle. *Xenopus*, on the other hand, has a three-chamber heart including a ventricle, and left and right atria separated by a septum (Mohun, Leong, Weninger, & Sparrow, 2000). Also, similar to human, a *Xenopus* heart contains trabeculae within the ventricular myocardium. More importantly, the background rate of cardiac malformation is very low, which allows for a robust identification of genes that could be disruptive for heart formation. Finally, the maintenance of *Xenopus* species is inexpensive compared to mammalian models.

6.1 Manipulations of gene expression in *Xenopus*

Gene expression in *Xenopus* embryos can be easily manipulated using microinjections of mRNA and antisense morpholino oligonucleotides (MO) for gain or loss of function, respectively. MOs are designed to target either the start site or the splice site of the mRNA transcript. They have been traditionally used for gene knockdown strategies due to their ability to target both maternal and zygotic transcripts. As an alternative, CRISPR-Cas9 mediated gene knockout works well even in the F0 generation (Bhattacharya, Marfo, Li, Lane, & Khokha, 2015). This strategy works best with the co-injection of Cas9 protein with the target sgRNA. The injection of Cas9 protein, rather than Cas9 mRNA, has been shown to have a higher cutting efficiency and a lower toxicity. Targeted DNA editing can be easily evaluated *via* PCR amplification and Sanger sequencing followed by subsequent mismatch analysis using a free online Interference of CRISPR Edits (ICE) tool (Hsiau et al., 2018). Importantly, successful CRISPR-Cas9 mediated genome editing of both alleles can be detected as early as 2h post injections in F0 embryos (Bhattacharya et al., 2015). Furthermore, CRISPR embryos can be raised to generate stable frog lines carrying specific mutations or indels. Therefore, CRISPR-Cas9 is an effective genome editing tool to use in *Xenopus* to knockout genes from the zygotic genome.

Another major advantage of using *Xenopus* to study organ development is the ability to efficiently generate transgenic animals (Amaya & Kroll, 2010; Hirsch et al., 2002). Successful incorporation of exogenous DNA into the genome of a one-cell embryo prior to cell division has been reported in both *Xenopus tropicalis* and *Xenopus laevis* (Marsh-Armstrong, Huang, Berry, & Brown, 1999; Offield, Hirsch, & Grainger, 2000). Moreover, a precise, targeted genomic integration using homology-directed repair (HDR) has been shown in *Xenopus* (Aslan, Tadjuidje, Zorn, & Cha, 2017). When coupled to a fluorescent reporter, the expression of a transgene can be monitored in a living, developing embryo over time. Moreover, tissue explants from transgenic animals can be transplanted into ectopic sites of a wild-type animal and examined *via* detection of a fluorescent signal combining the power of cut and paste embryology with transgenic reporter lines. Additionally, the use of different fluorescent reporters allows for detection of the interaction between multiple transgenes in the same animal. Furthermore, an inducible transgene activation has been shown to provide a controlled spatio-temporal expression (Chae, Zimmerman, & Grainger, 2002; Rankin, Zorn, & Buchholz, 2011). Together these properties make

Xenopus an effective model organism to study multiple developmental processes including kidney development (Corkins et al., 2018), lymphangiogenesis (Ny et al., 2013), signaling pathways (Tran, Sekkali, Van Imschoot, Janssens, & Vleminckx, 2010), apoptosis (Kominami et al., 2006), and epigenetic changes (Suzuki et al., 2016). More specifically, transgenic *Xenopus* animals have been used in the studies of heart chamber formation to visualize muscle fiber development (Smith et al., 2005). In this study, the expression of green fluorescent protein (GFP) was placed under control of the cardiac-specific MLC1v gene promoter, to examine the muscle formation of both the ventricle and the atria throughout tadpole development. The fluorescent signal was first detected during the heart tube looping stage, which establishes the separation between the ventricle and the outflow tract. Later, the MLC1v transgene became apparent on the atrial side of the atrioventricular boundary. Moreover, the transgenic GFP expression also showed a distinct arrangement of muscle fibers in the chamber *versus* the outflow tract myocardium. Therefore, this is yet another example how the use of transgenic animals can aid in studies of the tissue- and stage-specific expression of target genes.

6.2 Targeted injections in *Xenopus*

In addition to successful, cost-effective knockout and knockdown methods, *Xenopus* has advantages for the analysis of left-right patterning that is not shared by any other organism. For example, injection of one cell of the two-cell embryo can target either the left or right side of the embryo. Using fluorescent tracers, one can easily detected the injected side of the embryo and compare it to the un-injected side. Such manipulations are highly useful as they provide an internal control in the same animal for each experiment. In addition to two-cell stage injection, *Xenopus* embryos have a well-defined cell fate map for each organ system (Moody, 1987). This allows for targeted injections, where the expression of genes can be modified in specific organs or tissues. Additional studies of cardiac marker gene expression analysis as well as fate mapping data of cardiac progenitor cells at different stages of development are now available and could be used to better understand cardiac development (Gessert & Kühl, 2009). This study also describes an early segregation of cardiac lineages and could be used to understand the molecular mechanisms behind cardiogenesis. Therefore, *Xenopus* embryos offer straightforward manipulations to precisely track specific phenotypes.

6.3 Imaging modalities: Optical coherence tomography

Novel imaging modalities, such as optical coherence tomography (OCT), can be used to evaluate heart and craniofacial malformations in *Xenopus* embryos (Deniz et al., 2017). Similar to ultrasound, OCT uses light waves to obtain *in vivo* cross-sectional images of internal structures at micrometer resolution (Huang et al., 1991). OCT can visualize the atria, trabeculated ventricle, the outflow tract, and the atrioventricular valve and provide quantitative data. Moreover, this technology is non-destructive so live tadpole can be imaged repetitively over time so disease progression can be tracked over time in the same animal. Furthermore, subsequent imaging presents an opportunity to distinguish between the primary and secondary effects of a specific phenotype. Due to all these advantages, OCT is an effective tool to model human disease states in *Xenopus*. Recently, this imaging modality was used to recapitulate a known cause of cardiomyopathy. Mutation in myosin heavy chain 6 (*myh6*) disrupts the structure of a sarcomere, which is a functional unit of cardiac muscle. A MO-mediated knockdown of *myh6* in *Xenopus* resulted in higher end systolic diameters, a sign of systolic dysfunction from reduced contractility. Moreover, the presence of dilated atria combined with the lower excursion distance of the atrioventricular valve suggested that the reduced expression of *myh6* leads to a non-compliant ventricle with poor contractility. Additionally, the outflow tract in the depleted tadpole was much narrower and had lower excursion distance compared to control tadpoles, again reflecting a reduction in cardiac output. Thus, OCT can be effectively used in *Xenopus* to model human heart disease.

6.4 Tissue explant assays

The use of tissue explant experiments in *Xenopus* makes it an effective model to study cardiac development and congenital heart malformations. Tissue explant technique dates back to the famous experiments by Spemann and Mangold, who showed that the transplantation of the blastopore lip to an ectopic site can induce the formation of the secondary axis from the ectodermal tissue that was supposed to become epidermis (Spemann & Mangold, 1923). Subsequent work by Nieuwkoop showed that ectodermal tissue can be induced by endoderm to transform into mesoderm (Nieuwkoop, 1969). This work demonstrated the importance of signaling patterns in the establishment of the three germ layers and their derivatives in the developing embryos. The heart is a derivative of the mesodermal layer and multiple signaling pathways contribute to proper cardiac development. These signaling pathways could be directly or indirectly involved in the development of other organs as well. Additionally, development is a cascade of serial steps so signaling events at one time point could have primary as well as secondary effects on cardiac structure. Therefore, there is a need to isolate progenitor cardiac tissue at a specific time points while minimizing the effects from other signaling factors and surrounding tissues. Explant experiments provide such a framework and can be successfully used to define the mechanism of cardiac differentiation as a mesodermal derivative. Specifically, *Xenopus laevis* eggs are often used in these experiments due to their larger size as compared to *Xenopus tropicalis*. These embryos can also efficiently regenerate after a microsurgery. Importantly, because early cleavage stages are holoblastic (as opposed to meroblastic cleavages in chick or zebrafish embryos), yolk is distributed to all cells so explants do not require any supplemental nutrition as it can survive using the nutrients stored intracellularly. Dissected tissue develops and differentiates in a simple salt solution.

Heart development begins in the mesodermal region called the dorsal marginal zone (DMZ), while the ventral marginal zone (VMZ) does not contribute to the cardiac development. Cardiogenesis in *Xenopus* starts at the onset of gastrulation with the formation of precardiac mesoderm on both sides of the Spemann organizer. During gastrulation, the progenitor cells from those regions will migrate anteriorly to the ventral midline where they will undergo fusion. Later this fused crescent-like structure will split into two distinct lineages forming the first heart field and the second heart field. The first heart field contributes to the formation of a ventricle and two atria, while the second heart field will form the outflow tract.

Both DMZ and VMZ explants can be utilized to study heart morphogenesis. DMZ explants allow investigators to examine the endogenous signals and processes present in the cardiogenic dorsoanterior mesoderm that are necessary for heart formation. On the contrary, VMZ explants provide noncardiogenic tissue that could be subjected to signals capable of inducing the heart in an ectopic environment. Molecular manipulations can be easily targeted to these tissues at the four-cell stage of development by targeting the dorsal blastomeres. Additionally, during later stages of development (stages 10–10.5), the appearance of the dorsal blastopore lip marks the DMZ, from which the heart derives. Once isolated and cultured, both DMZ and VMZ can be evaluated for specific gene expression changes using a variety of methods including *in situ* hybridization, Western blotting, and differentiation into beating cardiomyocyte tissue. For instance, explant tissue experiments were used to show that Wnt antagonism initiated heart development (Schneider & Mercola, 2001). Using *Xenopus laevis* embryos at four-cell stage, the mRNAs of Wnt and BMP antagonists were injected into both ventral blastomeres. VMZ explants were removed at stage 10, kept in culture and analyzed at stage 30 using RT-PCR for the expression of cardiac-specific genes. The ectopic expression of *dkk-1*, an antagonist of Wnt signaling, in the VMZ led to the expression of cardiac-specific genes including *Nkx2.5* and *Tbx5* as well as cardiomyocyte contractile proteins *TnIc* and *MHCα*. Additionally, *TnIc* transcripts were shown to localize to the VMZ region *via in situ* hybridization. Furthermore, the expression of *crescent*, also a Wnt antagonist, led to the expression of *muscle actin*, a known marker of skeletal and cardiac muscle. Ectopic *crescent* expression also resulted in the localization of *TnIc* in the ventral explants. When examined at later stages of development (stage 41), the injection of both *dkk-1* and *crescent* resulted in the generation of a heart beat in the VMZ explants. In addition, *via* immunohistochemistry staining, the VMZ explants expressed cardiac-specific isoform of troponin-T. Besides *dkk-1* and *crescent*, the injection of GSK3β mRNA (which degrades β catenin and inhibits Wnt signaling) into ventral blastomeres was also shown to induce the expression of cardiac-specific genes, *TnIc* and *MHCα*. Together these experiments showed that the inhibition of Wnt/β-catenin signaling is sufficient for the formation of the heart in an ectopic ventral tissue.

Since the expression of Wnt antagonists in the VMZ caused the induction of cardiogenesis, the overexpression of Wnt proteins in the DMZ was expected to abolish cardiac formation. There are four Wnt genes expressed during gastrulation: *Wnt3A*, *Wnt5A*, *Wnt8*, and *Wnt11*. The injection of *Wnt3A* and *Wnt8* cDNA into dorsal blastomeres resulted in the inhibition of cardiac gene expression. The DMZ explants from injected embryos had reductions in the expression of *Nkx2.5* and *TnIc*. Interestingly, *Wnt5A* and *Wnt11* injections did not change the cardiac gene expression profile. Together, these experiments involving tissue explants established the significance of Wnt signaling in heart development.

Besides DMZ and VMZ explants, animal cap tissue can also be utilized to study heart development. The cells in the naïve animal caps initially are pluripotent but do not spontaneously differentiate into cardiac tissue. Different concentration of activin can be added to the growth media in order to induce different cell fates. A low concentration of activin will promote the formation of ventral mesoderm, while growing embryos in a medium concentration of activin will result in the formation of dorsal mesoderm.

Moreover, a high concentration of activin will stimulate the development of endodermal tissue. Since normal cardiogenesis relies on the interplay between the endodermal signal induction of mesodermal tissue, it is possible to use a specific activin concentration to enable communication between the endodermal and mesodermal layers. This approach improved our understanding of the role of canonical and non-canonical Wnt signaling in heart development (Afouda et al., 2008). Since the canonical Wnt signaling pathway (Wnt/β-catenin) is known to suppress cardiac development, and non-canonical Wnt signaling (Wnt11/JNK) promotes heart formation, it is interesting to understand the factors involved in the interplay between these two opposing pathways. Specifically, the use of tissue explants simplifies the approach as the manipulation of these crucial developmental pathways in an entire embryo will be confounding due to the multiple tissues affected. Animal cap explants from embryos injected with activin mRNA were able to initiate cardiogenesis as indicated by the expression of cardiac specific markers such as GATA4, GATA6, Nkx2.5, *etc.* When an inducible form of β-catenin was introduced, it reduced the expression of GATA transcription factors. Moreover, an inducible form of GATA transcription factors rescued the phenotypes caused by the overexpression of β-catenin. These preliminary results suggested a regulatory pathway in which Wnt/β-catenin signaling suppresses GATA gene expression to prevent cardiogenesis. These results were further confirmed when gene expression analysis was compared in DMZ explants and whole embryos. Specifically, in the whole embryo, β-catenin overexpression reduced the expression of cardiac-specific genes MLC2 and TnIc; moreover, the activation of GATA in these embryos rescued this reduction in cardiac markers. These findings demonstrated that Wnt/β-catenin signaling has a negative effect on cardiogenesis, and that GATA transcription factors are downstream effectors in this process.

In order to understand the role of GATA transcription factors in the process of heart formation and to identify their downstream targets, a MO-mediated knockdown of GATA4 and GATA6 was performed in three different tissue samples: a DMZ explant, a VMZ explant injected with a Dkk-1 mRNA to stimulate cardiogenesis, and the whole embryo tissue. In all three sample, the reduction of GATA4 and GATA6 led to the reduced expression of Wnt11, a non-canonical Wnt ligand. Additionally, this work also confirmed that Wnt11 is a direct target of GATA4 and GATA6 transcription factors. Moreover, a MO-mediated reduction in Wnt11 both in the DMZ explants and whole embryos resulted in the reduced expression of cardiac-specific genes and differentiation into beating cardiomyocytes. These findings established the link between the canonical and non-canonical Wnt signaling in the process of heart formation. Using both DMZ and VMZ explants as well as whole embryos, this work established a negative regulation of GATA transcription factors by canonical Wnt/β-catenin signaling. In turn, GATA factors were shown to directly induce the expression of Wnt11, which is part of the non-canonical Wnt11/JNK signaling pathway. More importantly, because all findings in the explants from the DMZ, VMZ and animal caps explants are consistent with the whole embryos findings, it justifies the use of explant tissue experiments to study specific organ development and to eliminate any additional spatio-temporal signaling factors present in the whole embryo.

This work in *Xenopus* was further expanded by the work done in mice focusing on the combination of factors necessary to induce cardiac tissue formation from mesoderm (Takeuchi & Bruneau, 2009). In these studies, the combination of two transcription factors, GATA4 and Tbx5, and a cardiac-specific subunit of BAF chromatin-remodeling complexes, Baf60c, induced the differentiation of cardiac myocytes from both non-cardiogenic posterior mesoderm and the extraembryonic mesoderm of the amnion. Interestingly, the combination of Gata4 and Baf60c stimulated the expression of cardiac-specific genes but failed to induce contractile tissue. The addition of Tbx5, however, resulted in the development of contracting cardiomyocytes. Moreover, Baf60c was shown to help GATA4 to bind to the loci of its cardiac target genes. These additional findings emphasized the importance of chromatin-remodeling complexes in the process of tissue-specific gene expression.

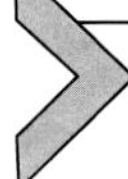

7. *Xenopus* as a system to understand human disease

7.1 The role of nucleoporins in CHD and heterotaxy

Xenopus is a rapid, cost-effective system that enables the screening of multiple candidate genes identified through large-scale genome sequencing studies in order to identify a role in cardiac development. The variety of methods available in *Xenopus* provides an opportunity to study genes with no known role in cardiac development or even genes with no defined role in development at all. For example, the use of *Xenopus* allowed us to discover a novel function of nucleoporins in development. Nucleoporin proteins are the building blocks of the nuclear pore complexes (NPCs). NPCs are massive (~100 MDa) structures embedded in the nuclear membrane, the main function of which is to regulate nucleocytoplasmic transport. Each NPC contains about 30 nucleoporins that are organized into sub-complexes and share an eightfold concentric symmetry (Rout & Aitchison, 2000). The NPC scaffold contains the outer ring complex (Nup107-160 or "Y" complex) and the inner ring complex (Nup93 complex) (Hurt & Beck, 2015). The scaffold proteins play an important role for the NPC assembly as they anchor the other nucleoporins, including the Phe-Gly (FG)-rich nucleoporins. FG nucleoporins are particularly significant as they form a size-selective barrier for passive diffusion, and they have the ability to interact with nuclear transport receptors for active transport. While nucleoporins are best characterized by their role within the NPC, there is growing evidence about additional functions of nucleoporins during development. For example, a duplication of the inner ring Nup188 was found in a patient with heterotaxy and CHD (Fakhro et al., 2011). Considering a crucial role of NPCs play in cellular processes, it is interesting to know what effect such duplication would have on the organism development. Moreover, since nucleocytoplasmic transport is important for all cells, it is fascinating that this patient could have a potential alteration to the NPC structure and assembly, and could present with a specific developmental phenotype that spares some organs and tissues in the body.

The depletion of Nup188 with a MO caused abnormal heart looping in *Xenopus* embryos at stage 45 of development. Furthermore, since Nup188 forms an inner ring complex with other nucleoporins, we hypothesized that knockdown of Nup93, another inner ring protein that binds to Nup188, would also show a cardiac looping phenotype and indeed MO depletion of Nup93 did. To investigate the effect of inner ring nucleoporins on the left-right patterning, the expression of two markers was examined: *pitx2* and *coco* (*dand5, cerl2*). The depletion of both Nup188 and Nup93 caused an abnormal, mostly absent, expression of *pitx2*. Furthermore, the loss of Nup188 and Nup93 resulted in a bilateral, symmetrical *coco* expression suggesting a ciliary dysfunction in the Left-Right Organizer tissue. To better understand this finding, MO injections were performed into one cell of the two-cell embryo. Since ciliary motility is crucial on the left side of the LRO where it represses the expression of *coco* (Vick et al., 2009), the injection of Nup188 MO into the left side had a more detrimental effect on heart looping compared to right sided targeting. This result further supported the potential cilia abnormality in the context of Nup188 and Nup93 depletion. Interestingly, the depletion of inner ring nucleoporins caused a reduction in cilia both in the frog LRO and in human retinal pigmented epithelial (RPE) cells. Moreover, the epidermal multiciliated cells were dramatically lost with Nup93 depletion. This result seemed specific to these inner ring nucleoporins as depletion of the outer ring nucleoporin Nup133 did not lead to cilia defects.

To eliminate the possibility that this phenotype could be due to a drastic disruption of nucleocytoplasmic transport, a NLS-GFP construct normally localized to the nucleus even in Nup93 or Nup188 depleted embryos. Furthermore, the density of the NPCs remained the same for injected and un-injected embryos. Additionally, the overexpression of Nup188, as seen in the patient, also led to abnormal heart looping and the loss of cilia in the LRO and the multiciliated cells. These findings suggested that changes in Nup188 levels neither affected nuclear transport, nor alter the number of NPC complexes per cell. While subtle changes could not be eliminated in NPC function, the investigators began looking for alternative functions for these inner ring nucleoporins.

Surprisingly, both Nup188 and Nup93 were also found at the cilium base where they co-localize with centrioles in *Xenopus* embryos and human RPE cells. Interestingly, these nucleoporins do not form a "ciliary pore complex" at the base of the cilium. First, these nucleoporins were found at the centriole rather than the transition zone which is where the ciliary diffusion barrier is thought to function. Next, using super-resolution PALM imaging, the size of the Nup188 puncta was much smaller when compared to the NPC dimensions and had no central pore. Finally, the Nup188 clusters were shown to form a structure consisting of two barrels that are perpendicular to each other in the pericentriolar material. This is a unique finding since it describes a novel subcellular localization of Nup188 outside of NPC. Moreover, the presence of Nup188 at the cilium base provides insight about why the depletion or overexpression of Nup188 has such a detrimental effect on cilium, left-right patterning and heart formation. Based on this work, the presence of two distinct cellular pools of Nup188 was subsequently confirmed (Vishnoi et al., 2020). Specifically, centrosomal accumulation of Nup188 in the pericentriolar material results from newly synthesized Nup188 protein. This Nup188 pool is distinct from the one present at the NPC, which is not exchanged or incorporated into the centrosomal Nup188 structures. Even after mitosis, nuclear Nup188 remains in the NPCs and is not redistributed into the centrosomal pool.

The presence and turnover of Nup188 at the centrosomes are regulated by proteasome activity. Interestingly, the turnover of centriolar Nup188 is more dynamic when compared with the NPC pool. This observation provides a plausible hypothesis why patients with either a duplication or depletion of Nup188 may have tissue-specific defects that spare some of the organs within a body. Since the NPC pool of Nup188 is more stable, the more dynamic centriolar pool of Nup188 will be more affected by the changes in Nup188 levels. Such changes in Nup188 expression will directly affect the structure and function of cilia, which is crucial for the establishment of left-right asymmetry of human body. Together these findings provide insight into the role of Nup188 in cilia structure and function, left-right patterning and cardiac formation.

In addition to Nup188, the role of other nucleoporins continues to be discovered in the context of development and congenital heart disease. Like Nup188, nucleoporin 205 (Nup205) is also a component of the inner ring complex of the NPC. Moreover, it is a paralogue of Nup188 (Beck & Hurt, 2017). Mutations in Nup205 have also been implicated as a cause of CHD (Chen et al., 2019). The depletion of Nup205 in *Xenopus* embryos caused abnormal heart looping and disrupted the expression of *pitx2* and *dand5* similar to Nup188 and Nup93 depletion (Marquez, Bhattacharya, Lusk, & Khokha, 2021). Moreover, *nup205* depletion also resulted in the abnormal formation of the pronepros and suggested the role of Nup205 in kidney development as well. Due to the combination of cardiac looping and kidney phenotype, a ciliary defect was suspected in the Nup205-depleted embryos. Indeed, the loss of Nup205 resulted in the loss of cilia in the LRO tissue, epidermal MCCs and pronephroi. More importantly, Nup205 was also found to localize both to the NPC and the ciliary base. Furthermore, the depletion of Nup205 caused an abnormal docking of basal bodies to the apical cell surface. Similarly, the depletion of Nup93 and Nup188 also led to the mislocalization of the basal bodies. To better understand this connection between inner ring nucleoporins in the ciliary function, MO-mediated depletion of Nup205, Nup93 and Nup188 was performed and resulted in the loss of cilia. Interestingly, the ciliary loss caused by the depletion of Nup188 and Nup205 was rescued by the expression of either Nup188 or Nup205. On the other hand, ciliary defect from the depletion of Nup93 was rescued by Nup93 expression only. Therefore, this suggests a redundant function of Nup188 and Nup205 at the ciliary base. This work describes a novel subcellular localization of nucleoporins and their new function within the context of ciliary biology and human disease.

7.2 Patient-driven gene discovery and insights into developmental pathways with *Xenopus*

7.2.1 Wnt/β-catenin pathway

Mutations in guanine nucleotide exchange factor (RAPGEF5) have been reported in patients with CHD and heterotaxy (Fakhro et al., 2011). This factor has recently been shown to contribute to the Wnt signaling pathway *via* regulating nuclear translocation of β-catenin (Griffin et al., 2018). The Wnt signaling pathway plays an important role in the development of embryonic tissue and its dysregulation contributes to cancer. A key step in this pathway is the nuclear import of β-catenin. In the absence of Wnt ligand binding, β-catenin remains in the cytoplasm, where is undergoes phosphorylation by the β-catenin degradation complex and subsequent degradation by the proteasome. However, when the Wnt ligand is bound to its receptor, the β-catenin degradation complex is inhibited, and β-catenin can translocate into the nucleus, bind to the TCF/LEF and induce the expression of Wnt-responsive genes. Although crucial, the mechanism of β-catenin nuclear import remains unknown. The nuclear transport receptor for β-catenin import has not been identified yet. However, these is evidence that the nuclear import of β-catenin requires energy and a GTPase activity (Fagotto, 2013).

To understand the contribution of RAPGEF5 to CHD and heterotaxy, its effect on left-right patterning was examined first. The depletion of RAPGEF5 resulted in an abnormal cardiac *situs* and abnormal expression of *pitx2* (a marker of left-right patterning) in the lateral plate mesoderm. Interestingly, the expression of *coco* in the LRO tissue was reduced even before extracellular flow was initiated to break the left-right symmetry. Moreover, the expression of other LRO markers, *xnr1* and *gdf3*, was also significantly reduced when RAPGEF5 was depleted. These findings suggested that the reduction of RAPGEF5 expression prevented the proper patterning of the LRO tissue. Since the LRO is derived from the superficial dorsal mesoderm and is formed near the end of gastrulation, the expression of dorsal mesodermal markers was examined. Specifically, the depletion of RAPGEF5 caused the reduction of *foxj1* and *xnr3* expression. These two genes are known direct targets of the Wnt signaling pathway. Interestingly, reduced expression of RAPGEF5 caused changes in both total β-catenin levels as well as in stabilized (unphosphorylated) β-catenin levels. To understand such an effect on β-catenin levels and what roles RAPGEF5 could play in this pathway, the investigators used the *Xenopus* secondary axis assay. The overexpression of β-catenin is known to induce secondary axes in *Xenopus* embryos (Funayama, Fagotto, McCrea, & Gumbiner, 1995). This strategy is extremely useful as the resulting secondary axes can be easily examined in live embryos under the light microscope. Interestingly, the depletion of RAPGEF5 significantly reduced the formation of secondary axis caused by the injection of both wildtype and surprisingly stabilized β-catenin mRNAs. This result was also consistent with the findings from a TOPFlash luciferase assay, in which the expression of luciferase is under control of TCF/LEF binding. In this assay, the depletion of Rapgef5 significantly reduced the luciferase signal. These findings suggested a negative regulation of Wnt/β-catenin signaling pathway by the depletion of RAPGEF5.

Of note, the depletion of Rapgef5 reduced nuclear accumulation of β-catenin. More importantly, when a β-catenin fusion to an N-terminal classic nuclear localization signal from SV40 large T antigen was expressed, RAPGEF5 depletion had no effect. This NLS is known to interact with the Ran/importin-β1 nuclear transport pathway. These findings indicated that nuclear localization of β-catenin *via* RAPGEF5 was independent of Ran/ImportinB1. Moreover, this study also confirmed the nuclear localization of Rapgef5 and its connection to the nuclear Raps, which are known to interact with β-catenin. This work is yet another example how *Xenopus* could be used to elucidate the mechanisms behind human disease and identify new genes involved in conserved developmental pathways. Moreover, since nuclear translocation of β-catenin contributes to the development of colorectal cancer, these findings could be used in therapeutic applications.

7.2.2 Notch pathway

A heterotaxy gene GALNT11, encoding for a polypeptide *N*-acetylgalactosaminyltransferas
e, has been shown to play a crucial role in the Notch signaling pathway (Boskovski et al., 2013). GALNT11 controls the initiation of GalNAc-type O-glycosylation and had no known function in cardiac development or heterotaxy. Hence, this example further emphasizes why we need an efficient system, such as *Xenopus*, to test these candidate genes. In order to understand the role of GALNT11 in cardiac development, the process of heart looping was examined in GALNT11-depleted embryos. In addition to abnormal heart lopping, these embryos also had abnormal expression of *pitx2* and *coco*, which suggested a defect in the cilia-mediated leftward flow of extracellular fluid in the LRO. Given the ease at which epidermal MCCs can be assayed for cilia since they are external and abundant, the examination of the epidermal MCCs revealed that GALNT11 knockdown increased the density of MCCs on the epidermis, while the overexpression of GALNT11 caused a decreased density of MCCs. This phenotype mimicked Notch signaling effects on MCC formation suggesting that GALNT11 may play a role in this pathway. Notch signaling consists of multiple factors. Specifically, it requires the binding of the *Delta* or *Jagged* ligand to the transmembrane Notch receptor, and the CBF1/Su(H)/Lag-1 (CSL) transcription factor

complex. The binding of the ligand induces multiple cleavage events that result in the release of the Notch intracellular domain (NICD), its translocation into the nucleus to drive the expression of Notch-responsive genes. In order to test the contribution of Galnt11 to the Notch signaling pathway, a rescue experiment was carried out using Notch pathway members to attempt a rescue of the Galnt11 knockdown phenotype in the *pitx2* assay. *Delta* gain-of-function did not rescue the abnormal *pitx2* expression in the Galnt11 morphants. However, both *nicd* and a constitutively active CSL protein rescued the phenotype. This suggested that Galnt11 functions downstream of the Notch receptor binding by the Delta ligand. Considering that Notch receptor is subject to different O-glycosylations, mass spectrometry analysis was used to identify whether the Notch receptor can also be a target of the GalNAc-type-O-glycosylation. It was found that there is a glycosylation site right next to the ADAM protease cleavage site within the Notch receptor. This glycosylation appears to enhance the cleavage of the Notch receptor and increase the release of the NICD from the membrane enhancing Notch signaling.

In order to better understand the connection between ciliary function, Notch signaling pathway and Galnt11 function, the LRO cilia was examined using live imaging. Based on the two cilia model (McGrath et al., 2003; Tabin & Vogan, 2003), there are two types of cilia in the LRO: motile cilia that generate extracellular fluid flow to break the left-right symmetry, and immotile cilia located in the periphery of the LRO that mechanically sense the extracellular fluid flow and drive downstream signaling. A major question in the field was how are two cilia types specified in the LRO: motile and immotile cilia. When notch1 was overexpressed, the LR phenotypes mimicked cilia immotility while when galnt11 or notch1 were depleted, the LR phenotypes suggested that there was a loss of cilia signaling. One possible explanation is that with gain of Notch signaling, immotile cilia were preferentially formed while with loss of Notch signaling, motile cilia formed. Indeed, by live confocal imaging cilia in the *Xenopus* LRO, the overexpression of *nicd* resulted in a reduced ratio of the motile to immotile cilia. In contrast, the knockdown of *notch1* or *galnt11* demonstrated a reverse phenotype with an increased ratio of motile to immotile cilia. Consistent with this finding, the overexpression of *nicd* or Galnt11 reduced the expression of Foxj1 and Rfx2, critical transcription factors that regulate motile ciliogenesis. Together these results indicate that the identity of the cilia was changed from immotile to motile. This work is a great example of how *Xenopus* can be used to understand the role of novel candidate genes in signaling pathways affecting both

Notch signaling and ciliary function. More importantly, it reveals a plausible mechanism of human disease, even when there is just a small cohorts of patients lacking the second allele to define disease causality.

7.3 CHD and Down syndrome

Xenopus can be used as a model organism to study diseases with rather complex phenotypes. For example, patients with Down syndrome often present with multiple cardiac malformations including tetralogy of Fallot, patent ductus arteriosus, atrial and ventricular septal defects. Therefore, it is rather interesting to discover genes responsible for such complex phenotypes. One such candidate gene is congenital heart disease protein 5 (CHD5), or tryptophan-rich basic protein (WRB). This gene is located within the restricted region of chromosome 21 in patients with both Down syndrome and heart defects. It is unknown whether CHD5 has a distinct function in heart formation that is independent of Trisomy 21. The analysis of CHD5 mRNA expression in the Medaka fish (*Oryzias latipes*) *via in situ* hybridization revealed signal in the differentiating ventricle in the endocardial cells at stage 28 (Murata, Degmetich, Kinoshita, & Shimada, 2009). Later, at stages 35 and 39, CHD5 mRNA and protein were found in the looped ventricle and atrium. The depletion of CHD5 using a morpholino resulted in abnormal heart looping, atrium elongation and abnormally large ventricles. Besides cardiac abnormalities, these morphants also had defects in ocular development characterized by narrowed interpupillary distances and cyclops. Together these findings support a contribution of CHD5 to heart development; however, they do not demonstrate a distinct role of CHD5 in cardiac formation that is independent from Trisomy 21.

In addition to CHD5, the zinc-finger transcription factor CASTOR (CASZ1) was implicated in vertebrate heart development. Specifically, the reduction in CASZ1 in *Xenopus* embryos prohibited cardiac progenitor cells from differentiating into cardiac myocytes, which resulted in cardia bifida and abnormal cardiac morphogenesis (Christine & Conlon, 2008). A link between CASZ1 and CHD5 was established using a yeast two-hybrid system, where full-length CASZ1 was evaluated for its interaction with a library generated from the stage 28 *Xenopus* cardiac enriched tissue (Sojka et al., 2014). Moreover, *in vivo* interaction between these two proteins was confirmed using co-immunoprecipitation in *Xenopus* embryos. Additionally, both CASZ1 and CHD5 had expression in the developing myocardium and co-localized to the nuclei of cardiomyocytes in *Xenopus* embryos.

Interestingly, MO-mediated depletion of *chd5* in *Xenopus* did not demonstrate any cardiac specification defects at early tadpole stages of development. However, reduced levels of *chd5* had a detrimental effect on later cardiac development at stage 37. Specifically, the reduction of CHD5 did not affect initial differentiation as the expression of tropomyosin, a cardiac marker, remained intact. Low CHD5 levels, however, abolished the processes of cardiac looping and chamber formation, and resulted in a thicker myocardial layer. Moreover, the depletion of CHD5 prevented cell movements toward the midline and fusion of the cardiac fields. Parts of the heart also had disrupted cell movements even prior to chamber formation. Scanning electron microscopy (SEM) also revealed that reduction in CHD5 levels caused a failure in cell shape changes that are normally required for maturation of the linear cardiac tube. Additionally, similar phenotypes were observed when CASZ1 was depleted; however, CHD5 depletion caused more severe phenotypes.

In terms of the mechanism behind these cellular changes, CHD5-depleted embryos had decreased cardiomyocyte numbers and a lower mitotic index rather than an increased level of apoptosis. Moreover, the depletion of both proteins caused a looser association of the cells within cardiac tissue indicating that these proteins may play a role in the cell-to-cell adhesion processes. Indeed, the expression of zonula occludens-1 (ZO-1), a tight junction marker, was either diffuse or absent in the myocardial tissue with reduced levels of CHD5 and CASZ1. Additional transmission electron microscopy (TEM) analysis of CHD5- and CASZ1-depleted *Xenopus* embryos revealed increased gaps between cardiomyocytes, breaks in the basement membrane of the myocardium, and the ectopic deposition of laminin in the deeper myocardial layers. Together these findings indicate that these proteins have an important function regulating the integrity of the myocardial tissue. Additionally, the expression of ZO-1 and claudin-5 were reduced and demonstrated a diffuse pattern of deposition in CHD5- and CASZ1-depleted embryos. In contrast, no changes were observed in the expression of β1-integrin. These findings suggested that the depletion of these two proteins specifically change the structure of the tight junctions rather than globally changing the extracellular matrix-cytoskeletal junctions.

This work also examined the nature of interaction between CASZ1 and CHD5. Specifically, the depletion of CHD5 did not affect the expression or nuclear localization of CASZ1. However, the activity of CASZ1 was found to depend on its interaction with CHD5. While the CASZ1 depletion phenotype was partially rescued by the full-length CASZ1 mRNA injection, the CASZ1 mRNA lacking the CHD5-interacting domain (CID) did not rescue the CASZ1 depletion phenotype. This finding suggests that while CHD5 does not influence the expression or nuclear localization of CASZ1, its interaction is necessary for the activity of CASZ1 in cardiac development. Additionally, the overexpression of either full-length of CASZ1 or its CHD5-interacting domain (CID) resulted in the abnormalities of the cardiac looping and morphogenesis defects. Therefore, a balanced interaction between CASZ1 and CHD5 is necessary for proper cardiac development.

7.4 CHD and chromatin regulators

Recent large scale studies of CHD have shown an enrichment of *de novo* mutations in chromatin modifiers. Hence, there has been substantial interest in understanding the role of these genes in heart development. Novel functions of chromatin regulators in heart development have been discovered using *Xenopus* as a model organism. A *de novo* mutation in *WDR5* was identified in patients with heterotaxy and CHD (Jin et al., 2017). The role of *WDR5* in chromatin remodeling is well-defined: it is a scaffolding protein in the H3K4 methyltransferase complex. Methylation of lysine 4 in histone 3 (H3K4) plays an important role in the gene expression regulation. In particular, the tri-methylation of this residue is often found within the promoter of active genes, while the mono- and di-methylation of H3K4 is often present within enhancer regions. However, the role of *WDR5* in development or heart formation had been unknown. Since the patient carrying a *de novo* missense *WDR5* mutation presented with a right (rather than a normal left) aortic arch, a defect in Left-Right patterning was suspected. The depletion of *WDR5* using a morpholino caused both LR patterning markers, *pitx2* and *dand5*, to have an abnormal expression (Kulkarni & Khokha, 2018). Since the disruption of cilia on the left side of the LRO has more detrimental effects on LR patterning (Vick et al., 2009), the injections of *WDR5* MO were performed into one cell of the two-cell embryo. The depletion of *WDR5* on the left side of the embryo resulted in a higher percentage of abnormal *pitx2* and *dand5* expression as compared to the right-side injections. This suggested a role of *WDR5* in ciliary function. Indeed, using confocal microscopy, the depletion of *WDR5* resulted in shorter cilia that were reduced in number in the LRO.

Given the loss of cilia, the expression of cilia-specific genes *dnah9, rfx2* and *foxj1* were examined in the LRO. The expression of *dnah9* and *rfx2* remained intact. However, the depletion of *WDR5* caused reduced levels of *foxj*, which is a known target of *WDR5* methyltransferase activity. The injection of *foxj1* mRNA into *WDR5*-depleted embryos rescued the phenotype confirming *foxj1* as a downstream effector of the *WDR5* action providing an explanation for the loss of cilia.

Interestingly, during the course of these experiments, *WDR5* was found to have an H3K4-independent role in the LR patterning, which had not been appreciated before. In addition to abnormal cilia in the LRO, WDR5 is also essential for the cilia in MCCs on the epidermal surface. Depletion of WDR5 dramatically eliminates the cilia in the MCCs which are responsible for a brisk extracellular fluid flow much like the mucociliary clearance in mammalian lungs. As expected, depletion of WDR5 dramatically eliminated the extracellular fluid flow across the epidermal surface and the addition of human wildtype WDR5 rescued this loss of flow. Biochemically, WDR5 binds different subunits of the methyltransferase that is essential for methyltransferase activity. An S91K WDR5 mutant completely abolishes this binding as well as methyltransferase activity. Unexpectedly, the S91K WDR5 mutant rescued the extracellular fluid flow and cilia of the MCCs in WDR5 depleted embryos suggesting a non-chromatin role for WDR5. Consistent with this finding, the patient had a K7Q mutation. For methyltransferase activity, the first 26 amino acids of WDR5 are not necessary which has been established functionally, biochemically, and by crystallographic structure and therefore the relevance of a K7Q mutation in the patient was far from certain. Using *Xenopus*, the K7Q mutation could not rescue the WDR5 depletion phenotype in the MCC indicating that it was critical for function in this context. While investigating non-chromatin molecular mechanism of WDR5, WDR5 was localized to the base of the cilia and appeared essential for the formation of the actin network in which the basal bodies of cilia are embedded at the apical surface of a MCC. In fact, the K7Q mutation appeared to affect the ciliary localization of WDR5 providing an explanation of its pathogenicity. Taken together, these results suggest that the examination of genes in the context of patient phenotypes and the investigation of pathogenic molecular mechanism can reveal new functions for proteins that had previously been considered only in a single context. Patient-driven gene discovery and molecular mechanism analysis can identify new functions for proteins thought to be already carefully characterized.

Other chromatin modifications have been shown to contribute to cardiac development by controlling ciliary motility rather than ciliogenesis (Robson et al., 2019). Specifically, *de novo* mutations affecting genes that affect the monoubiquitination of histone H2B on K120 (H2Bub1) were identified in patients with CHD. H2Bub1 is catalyzed by the RNF20-RNF40 complex together with the ubiquitin-conjugating enzyme UBE2B. MO-mediated depletion of *rnf20* resulted in abnormal expression of both *pitx2* and *dand5*, especially when *rnf20* was depleted on the left side of the embryo. Additionally, *rnf20* was found to be expressed in the LRO and multiciliated tissues in *Xenopus* and mouse including oviduct, trachea, and brain. Together these findings suggested a role of *rnf20* in the function of the cilia. The depletion of *rnf20* using a MO did not affect the cilia length, number, or the orientation in the LRO. However, it reduced the ciliary rotational frequency. Moreover, a similar motility defect was observed in the multiciliated cells of the *Xenopus* epidermis, while the morphology and distribution of the cilia remained intact. Furthermore, the depletion of *rnf20* resulted in the reduction of the epidermal fluid flow. Additionally, TEM analysis revealed shortened or absent inner dynein arm in the epidermal cilia of the *rnf20*-depleted embryos. Since *rnf20* is part of a complex, the contribution of other members to the ciliary function was examined. The loss of other components, including *rnf40* and *ube2b*, in combination with *rnf20* resulted in the reduced velocity of the epidermal cilia. Therefore, other members of the complex contribute to the proper motility of the cilium.

To identify the possible mechanism behind RNF20 complex activity, the levels of H2Bub1 were examined throughout development and were found to be upregulated during the formation of both the LRO and the epidermal cilia. Additionally, when *rnf20* is depleted, the levels of H2Bub1 were reduced suggesting a developmental regulation of H2B modification, which is dependent on the Rnf20 activity and correlates with the formation of the LRO and epidermal cilia. Using ChIP-Seq, a ciliary transcription factor *Rfx3* was identified as one of the genes enriched with H2Bub1 marks. To determine the downstream targets of *rnf20*, the expression levels of ChIP-Seq target genes was examined in the LRO. Using qPCR analysis of the LRO tissue from the *rnf20*-depleted embryos, reduced levels of *dnah7* and *rfx3* mRNAs were detected. Additional analysis using *in situ* hybridization examining the localization of these transcripts and RNA-Seq experiments confirmed these results. Moreover, overexpression of *rfx3* rescued the defects in ciliary motility, epidermal flow and abnormal *pitx2* expression in the *rnf20*-depleted embryos. This result established that monoubiquitination of histone H2B by RNF20 complex activates *rfx3* transcription, which acts an important transcription factor for cilia genes, including dyneins. Therefore, RNF20 mutations identified in the CHD patients have a direct detrimental effect on the ciliary function and subsequent left-right patterning and cardiac development.

Additional chromatin modifiers have been implicated in cardiac development and disease. Mutations in the H3K4 methyltransferase KMT2D have been identified in patients with Kabuki syndrome (Schwenty-Lara, Nürnberger, & Borchers, 2019). Kabuki syndrome is a rare, autosomal dominant disorder characterized by malformations of multiple organ systems. Multiple abnormalities include distinct facial features (cleft palate and abnormally small jaw), short stature, intellectual disability and skeletal abnormalities. The two most common cardiac manifestations in children with Kabuki syndrome include the narrowing of the main artery in the body (coarctation of the aorta) and the presence of holes between heart chambers (atrial and ventricular septal defects). Other malformations include renal dysplasia or hypoplasia, hydronephrosis and the fusion of two kidneys (a horseshoe kidney).

KMT2D is a large protein capable of catalyzing mono-, di- and trimethylation reactions. Moreover, the C-terminus of this peptide contains the SET domain that is necessary for its methyltransferase activity. Using *in situ* hybridization, it was shown that KMT2D is ubiquitously expressed in *Xenopus* at stages of cardiac development, with enrichment in the anterior region including cardiac precursor cells. To understand the impact of KMT2D mutations on heart formation, a MO-mediated depletion of this protein was performed in the cardiac mesoderm and the expression of α-myosin heavy chain (MHCα) was examined. Control embryos maintained the expression of MHCα in the atria, the ventricle, and the jaw. In contrast, KMT2D morphants developed cardiac abnormalities and a misplacement/reduction of jaw muscles. Heart defects included a hypoplastic morphology including malformed, laterally displaced hearts as well as the formation of tube-like structures that lacked three-chamber structural organization. These findings were also consistent with the histological sections demonstrating a narrow lumen with no separation of the heart tube into atria and a ventricle. In addition, the depletion of KMT2D affected cardiac differentiation. Specifically, it resulted in an abnormal expression of MHCα at the onset of cardiac differentiation and cardiac Troponin I, a marker of terminal differentiation.

To evaluate the impact of KMT2D depletion on the development of the first and second heart fields, the expression of Tbx20 and Isl1 were examined, respectively. KMT2D morphants had reduced levels of Tbx20 and the cells expressing Tbx20 were positioned more posterior than in un-injected embryos. The expression of Isl1 was also reduced in the KMT2D morphants, however, there was no cell misplacement. This result confirmed an important role of KMT2D in the cardiac differentiation and the establishment of the first and second heart fields. Additionally, the expression of Nkx2.5 was examined since it contributes to the development of both first and second cardiac fields. The depletion of KMT2D resulted in a reduced number of cells expressing of Nkx2.5, with some of cells demonstrating a complete loss of Nkx2.5 expression. These results suggested that KMT2D depletion has a global effect on cardiac specification early on in development. The analysis of gene expression in the precardiac precursor cells showed that KMT2D did not have an effect on this process. Together these findings show that KMT2D does not influence early cardiac development but becomes crucial during the separation of first and second heart fields.

8. Conclusion

The significant burden of congenital birth defects on child health requires more efficient, accessible and affordable diagnostic and treatment methods. To develop such practices, we must better understand the mechanisms leading to the development of birth defects. The advances in genome sequencing combined with the decline in its costs allow for identification of multiple genes in patients with CHD and heterotaxy, a few of which are described in this article. However, sequencing by itself does not provide answers to patients about the pathogenesis of their conditions. Considering the broad spectrum of molecular pathways these genes have been found to contribute to, it is important to appreciate how complex these diseases are in terms of physiology and embryo development. Moreover, we also need to consider utilizing a proper model organism that will allow for the exploration of patient phenotype and underlying mechanisms. Based on multiple examples described in this article, *Xenopus* does serve as an efficient, rapid, cost-effective system that offers a variety of methods to study congenital malformations. The ability to model human disease in *Xenopus* will allow us to understand disease pathogenesis. In turn, this will not only provide answers to patients and their families, but will also allow for better counseling regarding future risks based on the mechanism of pathogenesis of the disease. Moreover, crucial laboratory findings can be brought back to patient care *via* improvements in diagnostic testing, tailored treatment methods, and specific outcomes. Importantly, since most genetic syndromes are grouped based on shared phenotypes, specific genotype information will allow for a more precise management and outcome predictions in patient that may explain clinical outcome differences even though some of the phenotypes are shared.

Therefore, as sequencing becomes more rapid and less expensive, the need for high-throughput screening platforms to model patient disease will increase, and *Xenopus* provides all the tools that can be used for this process. Finally, this patient-driven gene discovery process will establish the necessary collaboration between different experts in the fields of medicine and biological sciences. Specifically, clinicians can provide care to patients suffering from birth defects, while medical geneticists can identify the candidate genes mutated in those patients. Furthermore, biomedical scientists in the variety of disciplines including developmental, cellular and molecular biology can combine their efforts to explore the complex mechanisms and signaling pathways that could be dysregulated in patients affected by CHD and heterotaxy. This collaborative effort in combination with the advantages of proper model organisms, such as *Xenopus*, will lead to the improvements in care for patients with CHD, heterotaxy, and birth defects in general.

Author conflict of interest

None.

References

Abu-Issa, R., & Kirby, M. L. (2008). Patterning of the heart field in the chick. *Developmental Biology*, *319*(2), 223–233. https://doi.org/10.1016/j.ydbio.2008.04.014.

Afouda, B. A., Martin, J., Liu, F., Ciau-Uitz, A., Patient, R., & Hoppler, S. (2008). GATA transcription factors integrate Wnt signalling during heart development. *Development*, *135*(19), 3185–3190. https://doi.org/10.1242/dev.026443.

Amaya, E., & Kroll, K. (2010). Production of transgenic *Xenopus laevis* by restriction enzyme mediated integration and nuclear transplantation. *Journal of Visualized Experiments*, (42). https://doi.org/10.3791/2010.

Amula, V., Ellsworth, G. L., Bratton, S. L., Arrington, C. B., & Witte, M. K. (2014). Heterotaxy syndrome: Impact of ventricular morphology on resource utilization. *Pediatric Cardiology*, *35*(1), 38–46. https://doi.org/10.1007/s00246-013-0736-y.

Aslan, Y., Tadjuidje, E., Zorn, A. M., & Cha, S. W. (2017). High-efficiency non-mosaic CRISPR-mediated knock-in and indel mutation in F0. *Development*, *144*(15), 2852–2858. https://doi.org/10.1242/dev.152967.

Beck, M., & Hurt, E. (2017). The nuclear pore complex: Understanding its function through structural insight. *Nature Reviews. Molecular Cell Biology*, *18*(2), 73–89. https://doi.org/10.1038/nrm.2016.147.

Bhattacharya, D., Marfo, C. A., Li, D., Lane, M., & Khokha, M. K. (2015). CRISPR/Cas9: An inexpensive, efficient loss of function tool to screen human disease genes in Xenopus. *Developmental Biology*, *408*(2), 196–204. https://doi.org/10.1016/j.ydbio.2015.11.003.

Blitz, I. L. (2012). Navigating the Xenopus tropicalis genome. *Methods in Molecular Biology*, *917*, 43–65. https://doi.org/10.1007/978-1-61779-992-1_4.

Blitz, I. L., Biesinger, J., Xie, X., & Cho, K. W. (2013). Biallelic genome modification in F(0) Xenopus tropicalis embryos using the CRISPR/Cas system. *Genesis*, *51*(12), 827–834. https://doi.org/10.1002/dvg.22719.

Blum, M., Beyer, T., Weber, T., Vick, P., Andre, P., Bitzer, E., et al. (2009). Xenopus, an ideal model system to study vertebrate left-right asymmetry. *Developmental Dynamics*, *238*(6), 1215–1225. https://doi.org/10.1002/dvdy.21855.

Boskovski, M. T., Yuan, S., Pedersen N. B., Goth, C. K., Makova, S., Clausen, H., et al. (2013). The heterotaxy gene GALNT11 glycosylates Notch to orchestrate cilia type and laterality. *Nature*, *504*(7480), 456–459. https://doi.org/10.1038/nature12723.

Brueckner, M. (2007). Heterotaxia, congenital heart disease, and primary ciliary dyskinesia. *Circulation*, *115*(22), 2793–2795. https://doi.org/10.1161/CIRCULATIONAHA.107.699256.

Chae, J., Zimmerman, L. B., & Grainger, R. M. (2002). Inducible control of tissue-specific transgene expression in Xenopus tropicalis transgenic lines. *Mechanisms of Development*, *117*(1–2), 235–241. https://doi.org/10.1016/s0925-4773(02)00219-8.

Chen, W., Zhang, Y., Yang, S., Shi, Z., Zeng, W., Lu, Z., et al. (2019). Bi-allelic mutations in NUP205 and NUP210 are associated with abnormal cardiac left-right patterning. *Circulation. Genomic and Precision Medicine*, *12*(7), e002492. https://doi.org/10.1161/CIRCGEN.119.002492.

Christianson, A., Howson, C., & Model, B. (2006). Global report on birth defects. The hidden toll of dying and disabled children. In *White plains*. New York: March of Dimes Birth Defects Foundation.

Christine, K. S., & Conlon, F. L. (2008). Vertebrate CASTOR is required for differentiation of cardiac precursor cells at the ventral midline. *Developmental Cell*, *14*(4), 616–623. https://doi.org/10.1016/j.devcel.2008.01.009.

Corkins, M. E., Hanania, H. L., Krneta-Stankic, V., DeLay, B. D., Pearl, E. J., Lee, M., et al. (2018). Transgenic *Xenopus laevis* line for in vivo labeling of nephrons within the kidney. *Genes (Basel)*, *9*(4), 197. https://doi.org/10.3390/genes9040197.

Davis, N. M., Kurpios, N. A., Sun, X., Gros, J., Martin, J. F., & Tabin, C. J. (2008). The chirality of gut rotation derives from left-right asymmetric changes in the architecture of the dorsal mesentery. *Developmental Cell*, *15*(1), 134–145. https://doi.org/10.1016/j.devcel.2008.05.001.

Del Viso, F., Huang, F., Myers, J., Chalfant, M., Zhang, Y., Reza, N., et al. (2016). Congenital heart disease genetics uncovers context-dependent organization and function of nucleoporins at cilia. *Developmental Cell*, *38*(5), 478–492. https://doi.org/10.1016/j.devcel.2016.08.002.

Deniz, E., Jonas, S., Hooper, M. N., Griffin, J., Choma, M. A., & Khokha, M. K. (2017). Analysis of craniocardiac malformations in Xenopus using optical coherence tomography. *Scientific Reports*, 7, 42506. https://doi.org/10.1038/srep42506.

Fagotto, F. (2013). Looking beyond the Wnt pathway for the deep nature of β-catenin. *EMBO Reports*, *14*(5), 422–433. https://doi.org/10.1038/embor.2013.45.

Fakhro, K. A., Choi, M., Ware, S. M., Belmont, J. W., Towbin, J. A., Lifton, R. P., et al. (2011). Rare copy number variations in congenital heart disease patients identify unique genes in left-right patterning. *Proceedings of the National Academy of Sciences of the United States of America*, *108*(7), 2915–2920. https://doi.org/10.1073/pnas.1019645108.

Funayama, N., Fagotto, F., McCrea, P., & Gumbiner, B. M. (1995). Embryonic axis induction by the armadillo repeat domain of beta-catenin: Evidence for intracellular signaling. *The Journal of Cell Biology*, *128*(5), 959–968. https://doi.org/10.1083/jcb.128.5.959.

Gessert, S., & Kühl, M. (2009). Comparative gene expression analysis and fate mapping studies suggest an early segregation of cardiogenic lineages in *Xenopus laevis*. *Developmental Biology*, *334*(2), 395–408. https://doi.org/10.1016/j.ydbio.2009.07.037.

Grant, I. M., Balcha, D., Hao, T., Shen, Y., Trivedi, P., Patrushev, I., et al. (2015). The Xenopus ORFeome: A resource that enables functional genomics. *Developmental Biology*, *408*(2), 345–357. https://doi.org/10.1016/j.ydbio.2015.09.004.

Gregory, E. C., MacDorman, M. F., & Martin, J. A. (2014). Trends in fetal and perinatal mortality in the United States, 2006–2012. *NCHS Data Brief, 169*, 1–8.
Griffin, J. N., Del Viso, F., Duncan, A. R., Robson, A., Hwang, W., Kulkarni, S., et al. (2018). RAPGEF5 regulates nuclear translocation of β-catenin. *Developmental Cell, 44*(2), 248–260.e244. https://doi.org/10.1016/j.devcel.2017.12.001.
Hagen, E. M., Sicko, R. J., Kay, D. M., Rigler, S. L., Dimopoulos, A., Ahmad, S., et al. (2016). Copy-number variant analysis of classic heterotaxy highlights the importance of body patterning pathways. *Human Genetics, 135*(12), 1355–1364. https://doi.org/10.1007/s00439-016-1727-x.
Hamada, H., & Tam, P. P. (2014). Mechanisms of left-right asymmetry and patterning: Driver, mediator and responder. *F1000Prime Reports, 6*, 110. https://doi.org/10.12703/P6-110.
Harden, B., Tian, X., Giese, R., Nakhleh, N., Kureshi, S., Francis, R., et al. (2014). Increased postoperative respiratory complications in heterotaxy congenital heart disease patients with respiratory ciliary dysfunction. *The Journal of Thoracic and Cardiovascular Surgery, 147*(4), 1291–1298.e1292. https://doi.org/10.1016/j.jtcvs.2013.06.018.
Hellsten, U., Harland, R. M., Gilchrist, M. J., Hendrix, D., Jurka, J., Kapitonov, V., et al. (2010). The genome of the Western clawed frog Xenopus tropicalis. *Science, 328*(5978), 633–636. https://doi.org/10.1126/science.1183670.
Hirsch, N., Zimmerman, L. B., Gray, J., Chae, J., Curran, K. L., Fisher, M., et al. (2002). Xenopus tropicalis transgenic lines and their use in the study of embryonic induction. *Developmental Dynamics, 225*(4), 522–535. https://doi.org/10.1002/dvdy.10188.
Hsiau, T., Conant, D., Rossi, N., Maures, T., Waite, K., Yang, J., et al. (2018). *Inference of CRISPR edits from sanger trace data*. bioRxiv.
Huang, D., Swanson, E. A., Lin, C. P., Schuman, J. S., Stinson, W. G., Chang, W., et al. (1991). Optical coherence tomography. *Science, 254*(5035), 1178–1181. https://doi.org/10.1126/science.1957169.
Hurt, E., & Beck, M. (2015). Towards understanding nuclear pore complex architecture and dynamics in the age of integrative structural analysis. *Current Opinion in Cell Biology, 34*, 31–38. https://doi.org/10.1016/j.ceb.2015.04.009.
Jin, S. C., Homsy, J., Zaidi, S., Lu, Q., Morton, S., DePalma, S. R., et al. (2017). Contribution of rare inherited and de novo variants in 2,871 congenital heart disease probands. *Nature Genetics, 49*(11), 1593–1601. https://doi.org/10.1038/ng.3970.
Kawasumi, A., Nakamura, T., Iwai, N., Yashiro, K., Saijoh, Y., Belo, J. A., et al. (2011). Left-right asymmetry in the level of active Nodal protein produced in the node is translated into left-right asymmetry in the lateral plate of mouse embryos. *Developmental Biology, 353*(2), 321–330. https://doi.org/10.1016/j.ydbio.2011.03.009.
Khokha, M. K. (2012). Xenopus white papers and resources: Folding functional genomics and genetics into the frog. *Genesis, 50*(3), 133–142. https://doi.org/10.1002/dvg.22015.
Kominami, K., Takagi, C., Kurata, T., Kitayama, A., Nozaki, M., Sawasaki, T., et al. (2006). The initiator caspase, caspase-10beta, and the BH-3-only molecule, Bid, demonstrate evolutionary conservation in Xenopus of their pro-apoptotic activities in the extrinsic and intrinsic pathways. *Genes to Cells, 11*(7), 701–717. https://doi.org/10.1111/j.1365-2443.2006.00983.x.
Kulkarni, S. S., & Khokha, M. K. (2018). WDR5 regulates left-right patterning via chromatin-dependent and -independent functions. *Development, 145*(23). https://doi.org/10.1242/dev.159889.
Kurpios, N. A., Ibañes, M., Davis, N. M., Lui, W., Katz, T., Martin, J. F., et al. (2008). The direction of gut looping is established by changes in the extracellular matrix and in cell: cell adhesion. *Proceedings of the National Academy of Sciences of the United States of America, 105*(25), 8499–8506. https://doi.org/10.1073/pnas.0803578105.

Lee, J. D., & Anderson, K. V. (2008). Morphogenesis of the node and notochord: The cellular basis for the establishment and maintenance of left-right asymmetry in the mouse. *Developmental Dynamics*, *237*(12), 3464–3476. https://doi.org/10.1002/dvdy.21598.

Marino, B. S., Lipkin, P. H., Newburger, J. W., Peacock, G., Gerdes, M., Gaynor, J. W., et al. (2012). Neurodevelopmental outcomes in children with congenital heart disease: Evaluation and management: A scientific statement from the American Heart Association. *Circulation*, *126*(9), 1143–1172. https://doi.org/10.1161/CIR.0b013e318265ee8a.

Marquez, J., Bhattacharya, D., Lusk, C. P., & Khokha, M. K. (2021). Nucleoporin NUP205 plays a critical role in cilia and congenital disease. *Developmental Biology*, *469*, 46–53. https://doi.org/10.1016/j.ydbio.2020.10.001.

Marsh-Armstrong, N., Huang, H., Berry, D. L., & Brown, D. D. (1999). Germ-line transmission of transgenes in *Xenopus laevis*. *Proceedings of the National Academy of Sciences of the United States of America*, *96*(25), 14389–14393. https://doi.org/10.1073/pnas.96.25.14389.

McGrath, J., Somlo, S., Makova, S., Tian, X., & Brueckner, M. (2003). Two populations of node monocilia initiate left-right asymmetry in the mouse. *Cell*, *114*(1), 61–73. https://doi.org/10.1016/s0092-8674(03)00511-7.

Mohun, T. J., Leong, L. M., Weninger, W. J., & Sparrow, D. B. (2000). The morphology of heart development in *Xenopus laevis*. *Developmental Biology*, *218*(1), 74–88. https://doi.org/10.1006/dbio.1999.9559.

Moody, S. A. (1987). Fates of the blastomeres of the 32-cell-stage Xenopus embryo. *Developmental Biology*, *122*(2), 300–319. https://doi.org/10.1016/0012-1606(87)90296-x.

Murata, K., Degmetich, S., Kinoshita, M., & Shimada, E. (2009). Expression of the congenital heart disease 5/tryptophan rich basic protein homologue gene during heart development in medaka fish, *Oryzias latipes*. *Development, Growth & Differentiation*, *51*(2), 95–107. https://doi.org/10.1111/j.1440-169X.2008.01084.x.

Murphy, S. L., Mathews, T. J., Martin, J. A., Minkovitz, C. S., & Strobino, D. M. (2017). Annual summary of vital statistics: 2013–2014. *Pediatrics*, *139*(6). https://doi.org/10.1542/peds.2016-3239.

Mussatto, K. A., Hoffmann, R., Hoffman, G., Tweddell, J. S., Bear, L., Cao, Y., et al. (2015). Risk factors for abnormal developmental trajectories in young children with congenital heart disease. *Circulation*, *132*(8), 755–761. https://doi.org/10.1161/CIRCULATIONAHA.114.014521.

Nieuwkoop, P. D. (1969). The formation of the mesoderm in urodelean amphibians: I. Induction by the endoderm. *Wilhelm Roux' Archiv für Entwicklungsmechanik der Organismen*, *162*(4), 341–373. https://doi.org/10.1007/BF00578701.

Nonaka, S., Shiratori, H., Saijoh, Y., & Hamada, H. (2002). Determination of left-right patterning of the mouse embryo by artificial nodal flow. *Nature*, *418*(6893), 96–99. https://doi.org/10.1038/nature00849.

Ny, A., Vandevelde, W., Hohensinner, P., Beerens, M., Geudens, I., Diez-Juan, A., et al. (2013). A transgenic Xenopus laevis reporter model to study lymphangiogenesis. *Biology Open*, *2*(9), 882–890. https://doi.org/10.1242/bio.20134739.

Offield, M. F., Hirsch, N., & Grainger, R. M. (2000). The development of Xenopus tropicalis transgenic lines and their use in studying lens developmental timing in living embryos. *Development*, *127*(9), 1789–1797.

Okada, Y., Takeda, S., Tanaka, Y., Belmonte, J. I., & Hirokawa, N. (2005). Mechanism of nodal flow: A conserved symmetry breaking event in left-right axis determination. *Cell*, *121*(4), 633–644. https://doi.org/10.1016/j.cell.2005.04.008.

Rankin, S. A., Zorn, A. M., & Buchholz, D. R. (2011). New doxycycline-inducible transgenic lines in Xenopus. *Developmental Dynamics*, *240*(6), 1467–1474. https://doi.org/10.1002/dvdy.22642.

Rigler, S. L., Kay, D. M., Sicko, R. J., Fan, R., Liu, A., Caggana, M., et al. (2015). Novel copy-number variants in a population-based investigation of classic heterotaxy. *Genetics in Medicine, 17*(5), 348–357. https://doi.org/10.1038/gim.2014.112.

Robson, A., Makova, S. Z., Barish, S., Zaidi, S., Mehta, S., Drozd, J., et al. (2019). Histone H2B monoubiquitination regulates heart development via epigenetic control of cilia motility. *Proceedings of the National Academy of Sciences of the United States of America, 116*(28), 14049–14054. https://doi.org/10.1073/pnas.1808341116.

Rout, M. P., & Aitchison, J. D. (2000). Pore relations: Nuclear pore complexes and nucleocytoplasmic exchange. *Essays in Biochemistry, 36*, 75–88. https://doi.org/10.1042/bse0360075.

Russo, C. A., & Elixhauser, A. (2007). Hospitalizations for birth defects, 2004. In *Healthcare Cost and Utilization Project (HCUP) Statistical Briefs [Internet]*. Agency for Healthcare Research and Quality (US).

Schneider, V. A., & Mercola, M. (2001). Wnt antagonism initiates cardiogenesis in *Xenopus laevis*. *Genes & Development, 15*(3), 304–315. https://doi.org/10.1101/gad.855601.

Schweickert, A., Vick, P., Getwan, M., Weber, T., Schneider, I., Eberhardt, M., et al. (2010). The nodal inhibitor coco is a critical target of leftward flow in Xenopus. *Current Biology, 20*(8), 738–743. https://doi.org/10.1016/j.cub.2010.02.061.

Schwenty-Lara, J., Nürnberger, A., & Borchers, A. (2019). Loss of function of Kmt2d, a gene mutated in Kabuki syndrome, affects heart development in *Xenopus laevis*. *Developmental Dynamics, 248*(6), 465–476. https://doi.org/10.1002/dvdy.39.

Showell, C., & Conlon, F. L. (2007). Decoding development in Xenopus tropicalis. *Genesis, 45*(6), 418–426. https://doi.org/10.1002/dvg.20286.

Smith, S. J., Ataliotis, P., Kotecha, S., Towers, N., Sparrow, D. B., & Mohun, T. J. (2005). The MLC1v gene provides a transgenic marker of myocardium formation within developing chambers of the Xenopus heart. *Developmental Dynamics, 232*(4), 1003–1012. https://doi.org/10.1002/dvdy.20274.

Sojka, S., Amin, N. M., Gibbs, D., Christine, K. S., Charpentier, M. S., & Conlon, F. L. (2014). Congenital heart disease protein 5 associates with CASZ1 to maintain myocardial tissue integrity. *Development, 141*(15), 3040–3049. https://doi.org/10.1242/dev.106518.

Spemann, H., & Mangold, H. (1923). Induction of embryonic primordia by implantation of organizers from a different species. *The International Journal of Developmental Biology, 45*, 13–38.

Stalsberg, H., & DeHaan, R. L. (1969). The precardiac areas and formation of the tubular heart in the chick embryo. *Developmental Biology, 19*(2), 128–159. https://doi.org/10.1016/0012-1606(69)90052-9.

Stillbirth Collaborative Research Network Writing Group. (2011). Causes of death among stillbirths. *JAMA, 306*(22), 2459–2468. https://doi.org/10.1001/jama.2011.1823.

Suzuki, M., Takagi, C., Miura, S., Sakane, Y., Sakuma, T., Sakamoto, N., et al. (2016). In vivo tracking of histone H3 lysine 9 acetylation in *Xenopus laevis* during tail regeneration. *Genes to Cells, 21*(4), 358–369. https://doi.org/10.1111/gtc.12349.

Tabin, C. J., & Vogan, K. J. (2003). A two-cilia model for vertebrate left-right axis specification. *Genes & Development, 17*(1), 1–6. https://doi.org/10.1101/gad.1053803.

Takeuchi, J. K., & Bruneau, B. G. (2009). Directed transdifferentiation of mouse mesoderm to heart tissue by defined factors. *Nature, 459*(7247), 708–711. https://doi.org/10.1038/nature08039.

Tran, H. T., Sekkali, B., Van Imschoot, G., Janssens, S., & Vleminckx, K. (2010). Wnt/beta-catenin signaling is involved in the induction and maintenance of primitive hematopoiesis in the vertebrate embryo. *Proceedings of the National Academy of Sciences of the United States of America, 107*(37), 16160–16165. https://doi.org/10.1073/pnas.1007725107.

United States Department of Health and Human Services. (2018). *National vital statistics system—Mortality data.* Retrieved from https://www.cdc.gov/nchs/fastats/child-health.htm.

van der Bom, T., Zomer, A. C., Zwinderman, A. H., Meijboom, F. J., Bouma, B. J., & Mulder, B. J. (2011). The changing epidemiology of congenital heart disease. *Nature Reviews. Cardiology*, *8*(1), 50–60. https://doi.org/10.1038/nrcardio.2010.166.

van der Linde, D., Konings, E. E., Slager, M. A., Witsenburg, M., Helbing, W. A., Takkenberg, J. J., et al. (2011). Birth prevalence of congenital heart disease worldwide: A systematic review and meta-analysis. *Journal of the American College of Cardiology*, *58*(21), 2241–2247. https://doi.org/10.1016/j.jacc.2011.08.025.

Vick, P., Schweickert, A., Weber, T., Eberhardt, M., Mencl, S., Shcherbakov, D., et al. (2009). Flow on the right side of the gastrocoel roof plate is dispensable for symmetry breakage in the frog *Xenopus laevis*. *Developmental Biology*, *331*(2), 281–291. https://doi.org/10.1016/j.ydbio.2009.05.547.

Vishnoi, N., Dhanasekeran, K., Chalfant, M., Surovstev, I., Khokha, M. K., & Lusk, C. P. (2020). Differential turnover of Nup188 controls its levels at centrosomes and role in centriole duplication. *The Journal of Cell Biology*, *219*(3). https://doi.org/10.1083/jcb.201906031.

Vonica, A., & Brivanlou, A. H. (2007). The left-right axis is regulated by the interplay of Coco, Xnr1 and derrière in Xenopus embryos. *Developmental Biology*, *303*(1), 281–294. https://doi.org/10.1016/j.ydbio.2006.09.039.

Waitzman, N. J., Romano, P. S., & Scheffler, R. M. (1994). Estimates of the economic costs of birth defects. *Inquiry*, *31*(2), 188–205.

Wallingford, J. B., Liu, K. J., & Zheng, Y. (2010). Xenopus. *Current Biology*, *20*(6), R263–R264. https://doi.org/10.1016/j.cub.2010.01.012.

Warkman, A. S., & Krieg, P. A. (2007). Xenopus as a model system for vertebrate heart development. *Seminars in Cell & Developmental Biology*, *18*(1), 46–53. https://doi.org/10.1016/j.semcdb.2006.11.010.

Warnes, C. A., Liberthson, R., Danielson, G. K., Dore, A., Harris, L., Hoffman, J. I., et al. (2001). Task force 1: The changing profile of congenital heart disease in adult life. *Journal of the American College of Cardiology*, *37*(5), 1170–1175. https://doi.org/10.1016/s0735-1097(01)01272-4.

Xu, J., Murphy, S. L., Kochanek, K. D., Bastian, B., & Arias, E. (2018). Deaths: Final data for 2016. *National Vital Statistics Reports*, *67*(5), 1–76.

Yoon, P. W., Olney, R. S., Khoury, M. J., Sappenfield, W. M., Chavez, G. F., & Taylor, D. (1997). Contribution of birth defects and genetic diseases to pediatric hospitalizations. A population-based study. *Archives of Pediatrics & Adolescent Medicine*, *151*(11), 1096–1103. https://doi.org/10.1001/archpedi.1997.02170480026004.

Zhu, L., Belmont, J. W., & Ware, S. M. (2006). Genetics of human heterotaxias. *European Journal of Human Genetics*, *14*(1), 17–25. https://doi.org/10.1038/sj.ejhg.5201506.

CHAPTER TEN

Xenopus, an emerging model for studying pathologies of the neural crest

Laura Medina-Cuadra[a,b] and Anne H. Monsoro-Burq[a,b,c,*]

[a]Université Paris-Saclay, Faculté des Sciences d'Orsay, CNRS UMR3347, Inserm U1021, Signalisation radiobiologie et cancer, Orsay, France
[b]Institut Curie, Université PSL, CNRS UMR3347, Inserm U1021, Signalisation radiobiologie et cancer, Orsay, France
[c]Institut Universitaire de France, Paris, France
*Corresponding author: e-mail address: anne-helene.monsoro-burq@curie.fr

Contents

Abstract

Neural crest cells are a multipotent embryonic stem cell population that emerges from the lateral border of the neural plate after an epithelium-to-mesenchyme transition. These cells then migrate extensively in the embryo and generate a large variety of differentiated cell types and tissues. Alterations in almost any of the processes involved in neural crest development can cause severe congenital defects in humans. Moreover, the malignant transformation of one of the many neural crest derivatives, during

Current Topics in Developmental Biology, Volume 145
ISSN 0070-2153
https://doi.org/10.1016/bs.ctdb.2021.03.002

childhood or in adults, can cause the development of aggressive tumors prone to metastasis such as melanoma and neuroblastoma. Collectively these diseases are called neurocristopathies. Here we review how a variety of approaches implemented using the amphibian *Xenopus* as an experimental model have shed light on the molecular basis of numerous neurocristopathies, and how this versatile yet underused vertebrate animal model could help accelerate discoveries in the field. Using the current framework of the neural crest gene regulatory network, we review the pathologies linked to defects at each step of neural crest formation and highlight studies that have used the *Xenopus* model to decipher the cellular and molecular aspects of neurocristopathies.

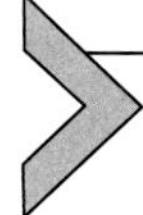

1. Introduction

1.1 The neural crest cells

The neural crest (NC) cells form a multipotent cell population that arises from the ectoderm germ layer during early vertebrate development. They give rise to an exceptionally rich variety of cell types forming diverse tissues and participating to the development of many organs. This includes typical ectodermal and neuro-ectodermal derivatives (e.g., neurons, glia), pigment cells, as well as ectomesenchymal derivatives, which are cell types formed by the cranial NC cells in the head and neck (e.g., smooth muscle, cartilage, bone) while these cell types typically arise from the mesoderm germ layer in the trunk (Alkobtawi & Monsoro-Burq, 2020; Le Douarin & Kalcheim, 1999). At the end of neurulation, when the neural tube is about to close, NC cells undergo an important cellular remodeling process known as the epithelium-to-mesenchyme transition (EMT). During gastrulation and neurulation, NC progenitors are induced in the epithelium adjacent to the neural plate ectoderm, so that in the head, as the cephalic neural folds elevate, NC precursors lie on both sides of the future brain while they are located at the tip of the dorsal neural folds in the trunk. EMT allows NC cells to exit the ectoderm epithelium, to acquire motility, to migrate toward internal or superficial tissues and to finally differentiate into more than 30 cell types throughout the body. The cranial NC cells initiate emigration collectively while the trunk NC cells tend to travel individually or in very small groups toward the target tissues, guided by anatomical or chemical cues provided by the environment (Theveneau & Linker, 2017). NC cells navigate along several stereotyped pathways according to their prospective differentiation into neurons and glia of the peripheral nervous system (sensory, autonomous, or enteric nervous system), pigment cells of the skin and internal organs, ectomesenchyme in the head, cardiac and pulmonary outflow tract, and secretory chromaffin cells in adrenal medulla (Fig. 1).

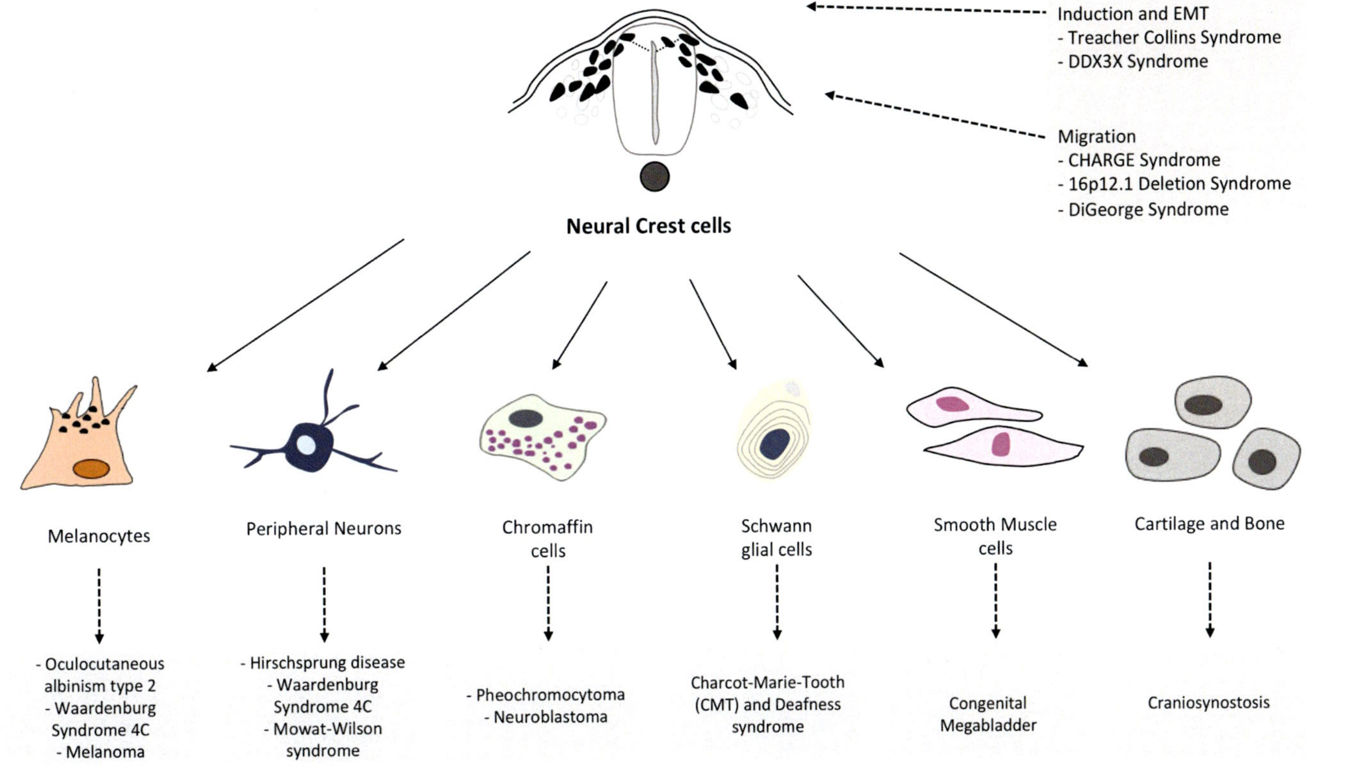

Fig. 1 After undergoing a stereotypical epithelium-to-mesenchyme transition, the motile neural crest cells delaminate from the neuroepithelium, migrate along several specific routes across the body, and reach diverse destinations where they differentiate into a wide spectrum of cells types, the main ones being illustrated here. Severe or milder alterations during the delamination and migration processes, in the cell guidance parameters, or during the differentiation of each lineage can result into a broad range of human pathologies and complex syndromes. A subset of those diseases, classified by the cell process affected or a neural crest-derived cell type affected for each pathology, is presented here and detailed in Table 1.

1.2 Neural crest-linked congenital defects and other pathologies of the neural crest derivatives

During embryonic development, any mild temporal or functional alteration of the molecules and signals involved in NC cells induction, EMT, migration or differentiation is likely to produce a significant developmental disruption. These defects lead either to the failure of embryonic development (embryonic lethality) or to one of the diseases and syndromes collectively named neurocristopathies (NCPs; Bolande, 1974). Therefore, understanding the details of NC cells development and function has immense medical implications. NCPs encompass a broad spectrum of congenital malformations and diseases affecting individual or multiple NC derivatives, as well as the tumors arising from the malignant transformation of one of the NC-derived cells. It is estimated that 1/4th to 1/3rd of the human congenital defects involve NC cells derivatives. With tumors formed by NC-derived cells during childhood or adult life, collectively these diseases constitute a heavy societal burden. However, individual NCPs are clinically classified by their specific phenotypes which can cover a range from relatively frequent to very rare pathologies: for instance, Hirschsprung disease is described in 1/5000 live birth in human (Amiel & Lyonnet, 2001); Charge syndrome in 0.1–1.2/10000 human live birth (Bajpai et al., 2010) and Waardenburg syndrome is found in 1/42000 infants at birth (Song et al., 2016).

NCPs were subdivided into four main categories by Bolande: (1) isolated tumors, (2) tumor syndromes, (3) congenital malformations, (4) other NC-linked pathologies (Bolande, 1974). Recently, a new classification has proposed to group NCPs according to the stage of NC development affected at the onset of the disease (Vega-Lopez et al., 2018). This proposal takes advantage of the tremendous progress made in understanding the sequential steps of the NC gene regulatory network (GRN) during the last 15 years, using a variety of vertebrate animal models *in vivo*, including the clawed frogs *Xenopus laevis* and *tropicalis*, the chick *Gallus gallus*, the zebrafish *Danio rerio*, and the mouse *Mus musculus*. To this list of models suitable to explore NC biology, research of the last decade has added human cells derived from induced pluripotent stem cells (iPSCs) reprogrammed into NC cells (Barrell et al., 2019; Leung et al., 2016). In this chapter, we will use this novel classification for NCPs and will thus define four main groups of diseases: NCPs arising from defects primarily occurring during NC induction, specification and EMT (group 1), NC migration (group 2), NC differentiation (group 3) and tumorigenesis (group 4). Table 1 references selected

Table 1 Selected neurocristopathies with studies using *Xenopus* frogs as a model organism.

Disease name	OMIM reference	Mutations identified in patients	References using *Xenopus* as a model
Group 1: NC induction and EMT			
Branchio-Oto-Renal (BOR) syndrome	#113650	EYA1	Li, Manaligod, and Weeks (2010) and Moody, Neilson, Kenyon, Alfandari, and Pignoni (2015)
Chromosome 16p12.1 deletion syndrome	#136570	–	Lasser et al. (2020)
DDX3 syndrome	#300160	DDX3X	Perfetto et al. (2021)
Diamond Blackfan Anemia	#105650	RPS19, RPS24, RPS17, RPL35A, RPL5, RPL11, RPS7, RPS10, RPS26, RPL26, RPL15, RPS29, TSR2, RPS28, RPL27, RPS27, RPL18, RPL35, RPS15A	Griffin, Sondalle, Del Viso, Baserga, and Khokha (2015) and Robson, Owens, Baserga, Khokha, and Griffin (2016)
Mowat-Wilson syndrome	#235730	ZEB2	Verstappen et al. (2008)
Treacher Collins syndrome	#154500	TCOF1, POLR1C, POLR1D, POLR1B, DDX21	Gonzales, Yang, Henning, and Valdez (2005), Robson et al. (2016) and Calo et al. (2018)
Group 2: NC migration defects			
Auriculo condylar syndrome	#602483	GNAI3	Marivin et al. (2016)
Charge syndrome	#214800	SEMA3E, CHD7	Schulz et al. (2014), Ufartes et al. (2018) and Bajpai et al. (2010)
Chromosome 16p12.1 deletion syndrome	#136570	–	Lasser et al. (2020)
DiGeorge syndrome	#188400	TBX1	Tazumi, Yabe, and Uchiyama (2010) and Alharatani et al. (2020)
Hamamy syndrome	#611174	IRX5	Bonnard et al. (2012)

Continued

Table 1 Selected neurocristopathies with studies using *Xenopus* frogs as a model organism.—cont'd

Disease name	OMIM reference	Mutations identified in patients	References using *Xenopus* as a model
Hypohidrotic ectodermal dysplasia	#305100	EDA	Essenfelder et al. (2004)
Kabuki syndrome	#147920	KMT2D	Schwenty-Lara, Nehl, and Borchers (2020)
Noonan syndrome	#163950	PTPN11, KRAS, SOS1, RAF1, NRAS, BRAF, RIT1, SOS2, LZTR1, MRAS, RRAS2, SHOC2, PPP1CB, CBL	Popov et al. (2019)
Wolf-Hirschhorn syndrome	#194190	–	Mills et al. (2019)
Group 3: Differentiation defects			
Axenfeld-Rieger syndrome	#180500; #602482	PITX2, FOXC1	Cha et al. (2007)
Bamforth-Lazarus syndrome	#241850	FOXE1	El-Hodiri et al. (2005)
Blepharocheilodontic syndrome 2	#617681	CTNND1	Alharatani et al. (2020)
Charcot-Marie-Tooth (CMT) and deafness syndrome (Schwann cell differentiation)	#118300	PMP22, PRPS1, GJB1	Tae, Rahman, and Park (2015) and Bae et al. (2014)
Craniosynostosis (cartilage and bone differentiation)	#123100 #604757 #600775	MSX2, TWIST1, FGF receptors	Twigg et al. (2015)
Dyschromatosis Symmetrica Hereditaria (melanocyte differentiation)	#127400	DSRAD	Wagner, Smith, Cooperman, and Nishikura (1989)

Hermansky-Pudlak syndrome 6	#614075	HPS6	Nakayama et al. (2017)
Hirschsprung disease (enteric nervous system)	#142623	RET, EDNRB	Li et al. (2018)
Klippel-Feil syndrome	#118100 #214300 #613702	GDF6, MEOX1, GDF3	Martens et al. (2020) and Tassabehji et al. (2008)
Megabladder, congenital (cardiac NC cell differentiation defects)	#618719	Myocardin	Wang et al. (2001)
Oculocutaneous albinism II (melanocyte differentiation)	#203200	OCA2, MC1R, TYR	Shi et al. (2019) and Barsh, Gunn, He, Schlossman, and Duke-Cohan (2000)
Waardenburg syndrome, type 2E, type 4C (pigment cell, enteric neurons)	#277580	SOX10, EDNRB, EDN3	McGary et al. (2010)
Group 4: Tumors			
Melanoma	#155600	Multiple	Kuznetsov et al. (2019), Kuznetsoff et al. (2021), Fürst et al. (2019) and Plouhinec et al. (2017)
Neuroblastoma	#256700	KIF1B, PHOX2B, ALK	Corallo et al. (2020) and Gonzalez Malagon et al. (2018)

Diseases were selected using both the OMIM database (search with "neural crest" and "*Xenopus*" as key words, 39 hits were further sub-selected if they matched with a NC-linked defect) and the neurocristopathies listed in Vega-Lopez, Cerrizuela, Tribulo, and Aybar (2018) further sub-selected for the availability of functional studies in *Xenopus* (using both Xenbase and PubMed websites). When such studies were not available, some gene expression studies of particular interest for neurocristopathies were mentioned. When the molecular and cellular mechanisms underlying the pathology were available, the disease was listed under one of the four following groups: (1) "induction and EMT," (2) "migration," (3) "differentiation" and (4) "tumors." Within each group, alphabetical order is used to classify diseases. The genes identified in human patients are listed in column 2 and the main references using *Xenopus* biology to understand the disease are listed column 3. We apologize to the authors that could not be listed here due to space constraints and hope that the indicated references will point toward their work.

NCPs which have been explored using *Xenopus* frogs as animal models, either by a functional study or, at minimum, by the expression analysis of a gene involved in the human pathology and has shown expression in the developing NC. This table indicates the main gene(s) involved in humans when this is known. References for studies conducted in other animal models can be found elsewhere (Vega-Lopez et al., 2018; Watt & Trainor, 2014). Using the sequential steps of the current NC-GRN is very useful to establish a hierarchical sequence of molecular causes for each disease. However, this new classification does not fully describe the complexity of these pathologies, as most NCPs are not exclusive members of a single category. The frequent overlap of multiple defects occurring at distinct steps of NC formation renders NCP description and understanding very complex. This facet of neural crest biology is probably one of the main scientific challenges of the next decade.

1.3 Conservation of the neural crest gene regulatory network in vertebrates

In this context, the multiplicity of vertebrate animal models used in developmental biology brings many options for genetic or experimental approaches to decipher the biology of NCPs. Cellular models further increase the range of possibilities to explore the details of NC cells behavior *in vitro* and in controlled environments (Scarpa et al., 2015). Collectively, animal models and human-derived NC cells present complementary advantages to tackle almost any question in NC and NCPs biology (Leung et al., 2016; Tchieu et al., 2017). Importantly, the recent comparison of the NC-GRN across various vertebrate animal models, both amniotes and non-amniotes, highlights the high degree of conservation of the global architecture of this gene network and of the associated signaling pathways, starting from the base of the vertebrate evolutionary lineage. The main steps and gene families identified in the animal models classically used for NC studies such as bird (amniotes, tetrapod; chick, quail), mouse (amniote, tetrapod, mammal) or frogs (non-amniote, tetrapod) are found conserved in lamprey (cyclostome; non-amniote, non-tetrapod; Hockman et al., 2019; Nikitina, Sauka-Spengler, & Bronner-Fraser, 2008; Sauka-Spengler, Meulemans, Jones, & Bronner-Fraser, 2007) and in bony fishes (jawed vertebrates, non-amniote, non-tetrapod) (Fig. 2). Additionally, specific features of the NC, such as its high multipotency, jaw formation and the recruitment of specific signaling pathways, have emerged at various evolutionary time points, supported by genetic innovations unleashed by the whole genome

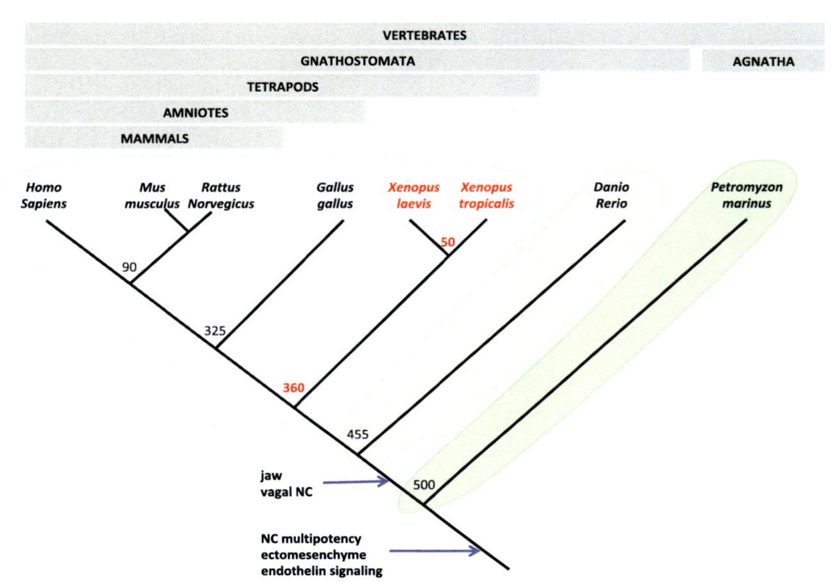

Fig. 2 A simplified vertebrate evolutionary tree presents the main groups, with a highlight on the clawed frogs *Xenopus laevis* and *Xenopus tropicalis* among the Tetrapod superclass. The early evolutionary acquisition of characteristic features of the neural crest, at the base of the vertebrate tree (multipotency, endothelin signaling, ectomesenchyme potential) or just after early-derived vertebrates (jaws and vagal neural crest derivatives), is reflected in the high conservation of the gene regulatory network governing neural crest development. The main scaffold of this network is highly comparable across vertebrates, yet subtle species/class-specific variations also modulate the fine-tuning of neural crest developmental molecular regulations. Tetrapods are grouped together and highlighted in blue, to emphasize the close evolutionary position of *Xenopus* frogs compared to mammals. Jawed fishes are highlighted in beige and the jawless vertebrates in green. Median divergence times are indicated in million years.

duplications that occurred in vertebrate evolution. These genetic novelties have allowed a range of molecular modulations on top the main NC-GRN scaffold. Hence the acquisition of multipotency, of the ectomesenchymal potential, and endothelin signaling are found at the base of the vertebrate tree (Scerbo & Monsoro-Burq, 2020; Square et al., 2020), but the acquisition of jaws, odontoblasts and vagal neural crest derivatives likely appeared in gnathostomes (Martik et al., 2019; Fig. 2). This high degree of conservation of the NC-GRN, taken with the caution of controlling for species-specific modulations, allows the efficient use of several animal models to explore the mechanisms of human NCPs. For instance, a survey of the steps leading human iPSCs from pluripotency to premigratory NC specification has highlighted the recapitulation of the key elements defined for the frog NC-GRN (Kobayashi et al., 2020; Pla & Monsoro-Burq, 2018). Moreover, during later stages of NC development, craniofacial malformations are obtained when gene expression causing NCPs in human are experimentally altered in frog embryos (e.g., Bajpai et al., 2010).

In this chapter, we will present some regulatory mechanisms affected in human NCPs that have been explored using the clawed frog *Xenopus laevis* or *tropicalis* models. We will focus on several examples illustrating the advances made using this versatile amphibian tetrapod model, ideally suited for studying the earliest stages of NC development, NC migration and various aspects of NC differentiation, either *in vivo* or *in vitro*. We will further describe the use of this animal model to understand NCPs and NC-derived tumors, using high throughput screening strategies *in vivo*.

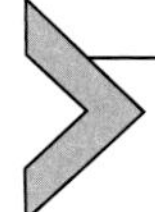

2. General considerations about *Xenopus* frog embryos as models for neural crest developmental studies

Xenopus, the African clawed frog, has been widely used in cell and developmental biology since early 1950s. The advantages of *Xenopus laevis* and *tropicalis* for experimental and genetic research, and the high conservation of developmental genes and molecular mechanisms from amphibians to human have promoted this important model for biomedical research. However, perhaps because it has been used for the best classical embryological studies for a long time, researchers using other models sometimes do not fully appreciate the amazing possibilities of the frog models to understand complex disease-related molecular networks. With modern genetics, epigenetics, single cell approaches, imaging and other tools, combined to the rich heritage of experimental embryology, *Xenopus* research must be highlighted for its important contributions to the better understanding of human pathologies, through the detailed molecular studies of candidate genes associated with human diseases (Fortriede et al., 2020). Similar to other aquatic laboratory models such as the zebrafish *Danio rerio*, *Xenopus* also respects the ethical needs for limiting the experimentation on mammalian models. Moreover, *Xenopus* frogs are tetrapods with lungs and a three-chambered heart, presenting a closer evolutionary relationship with humans than bony fishes for example (Fig. 2). The *Xenopus* genome shares approximately 80% of the genes known to cause human diseases, listed under an easily searchable repository linked to the OMIM database (Nenni et al., 2019; www.xenbase.org). From an experimental point of view, *Xenopus* embryos are robust and generated in large numbers by a simple hormone injection and these externally fertilized embryos form large cohorts of siblings allowing sound statistical analyses. Wildtype, inbred and transgenic frog lines are available at Resource Centers in Europe and the United States of America (EXRC, NRC, listed at www.xenbase.org). NC-related reporter transgenic lines have been developed (Alkobtawi et al., 2018; Li et al., 2019). Relatively large size eggs can be injected at early stages of development and embryos remain accessible at all subsequent stages, including gastrulation and neurulation, the critical stages of neural crest induction, which are poorly accessible in mammalian models. Gene expression can be easily disrupted with morpholino-mediated knockdown or CRISPR/cas9-based approaches. The spatial and temporal control of gene expression is also possible using hormone-inducible constructs, targeted injections or tissue grafting. The studies of frog NC cell migration have extensively explored the mechanisms of collective cell migration and cell motility parameters *in vitro* and *in vivo*. At later stages, *Xenopus* tadpoles are transparent and suited for imaging similar to zebrafish larvae, allowing a follow-up of neural crest-derived differentiated cells. Given all these advantages, the *Xenopus* animal model counts among the current key models to approach neural crest-derived diseases both for basic and biomedical research.

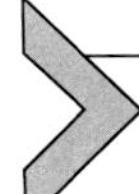

3. Congenital disorders caused by defects in early formation of the neural crest

It is very unlikely that genetic alterations affecting deeply the induction and specification steps of the neural crest cells be found in live human births. As they would affect a large set of NC derived cells with critical functions during organogenesis, such defects would most often result in early embryonic lethality. For example, *pax3* and its paralog *pax7* have been identified as critical regulators of neural crest induction at the neural border of frog and chick embryos during gastrulation and neurulation (Basch, Bronner-Fraser, & García-Castro, 2006; Milet, Maczkowiak, Roche, & Monsoro-Burq, 2013; Monsoro-Burq, Wang, & Harland, 2005). The lack of *pax3* gene expression in mouse embryos severely affects cardiac and trunk neural crest cell formation, resulting in failure of heart development and death of the mutant mouse embryos (Morgan et al., 2008). Interestingly, cranial NC cells seem to form normally in those mouse mutants, whereas cephalic NC depleted for *pax3* is virtually not induced in the frog animal model (Monsoro-Burq, 2015; Monsoro-Burq et al., 2005). There are however, multiple alleles for *Pax3* mutation in mouse: in the spontaneous *Pax3* mutant *Splotch*, homozygotes present an open neural tube, severe neural crest defects and limb muscle anomalies, resulting in embryonic lethality at various time points (Epstein, Vekemans, & Gros, 1991). Interestingly in mouse, *pax7* is also expressed in the neural crest cell lineages and the mouse *pax7* mutants exhibit craniofacial phenotypes of varying severity in different genetic backgrounds, suggesting complex and still incompletely understood specific and redundant functions for *pax3/7* in mice (Mansouri, Stoykova, Torres, & Gruss, 1996; Murdoch, DelConte, & García-Castro, 2012; Zalc, Rattenbach, Auradé, Cadot, & Relaix, 2015). In humans, *PAX3* mutation (*PAX3* (2q36.1)) causes Waardenburg syndrome type I and III, with type III being homozygous or heterozygous, and type I being heterozygous (Pingault et al., 2010). The main features of Waardenburg syndrome include congenital hearing loss, mild craniofacial anomalies and pigmentation defects in the skin, hair and eyes. Waardenburg syndrome Type III or Klein-Waardenburg syndrome is exceedingly rare, suggesting that in humans too, the complete lack of *PAX3* gene function causes embryonic lethality. Interestingly, Klein-Waardenburg syndrome also involves upper limb anomalies due to muscle defects, as observed in the *splotch* mice, consistent with the development of mesoderm-derived limb muscle cells, which depend on early *PAX3* gene functions. This exemplifies the complexity of NCPs syndromes, due to multiple functions of a given gene during development. Moreover, in the same cell lineage, here the neural crest lineage, a same gene may display successive functions at different steps of the NC-GRN. In the case of *PAX3*, heterozygous mutations in Waardenburg syndrome type I (or Shah-Waardenburg syndrome) cause defects in melanocyte biology, indicating anomalies in the later control of NC-derived melanoblast differentiation by Pax3.

The example of the Waardenburg syndrome illustrates three key points in NCPs analyses.

- Early regulators of the NC-GRN are essential for the global development of all NC cell populations throughout the embryo. In consequence, their null mutations will likely cause embryonic lethality. Only milder alterations of such genes would be found in human live births, with more subtle phenotypes than the ones usually studied in animal models. Those alterations can include several situations with reduced (but not absent) protein levels, such as found with heterozygous gene mutations, mutations resulting in a partially active protein or affecting protein stability, or when the epigenetic regulation of elements controlling the gene expression levels is altered. Finally, and more complex to understand, human syndromes may result from the alteration of the protein's co-factors, not the gene itself.

- Animal models are useful to understand NCPs, but a careful analysis of paralog genes activity has to be validated in each model, as a developmental usage swapping between genes with closely related functions has been described: for example *pax3* and *pax7* display both similar and distinct functions in frogs and chick (Basch et al., 2006; Maczkowiak et al., 2010; Monsoro-Burq et al., 2005). According to the genetic background, *Pax3* mutations have various outcomes in the mouse. Moreover, genetic duplications in vertebrates may create redundant activities masking the role of individual genes, but providing with interesting indications for safeguard modalities created during evolution, as it is the case for the four paralogs *pax3a/3b*, *pax7a/7b* in zebrafish (Minchin & Hughes, 2008).
- Altered development of the NC-derived cells is often accompanied by alterations in cell types formed from other germ layers. This can be due to the use of a given gene/protein—in particular transcription factors—for the development of unrelated cell types. In this instance, the developmental trees established from whole embryos single cell transcriptomes allow tracing the expression of a given gene in each cell type and better understand phenotypic alterations observed after mutation of this gene in human. Such a comprehensive resource is available during gastrulation and neurulation in frog embryos (Briggs et al., 2018).

To further illustrate these points, another example can be taken from the TFAP2 family of transcription factors. Of the five paralogs of this family, four (*tfap2a, b, c* and *e*) are expressed during ectoderm and neural crest patterning. TFAP2a has been involved in non-neural ectoderm formation and in early NC induction in mouse and frog embryos (Luo, Lee, Saint-Jeannet, & Sargent, 2003; Luo, Matsuo-Takasaki, Thomas, Weeks, & Sargent, 2002; Schorle, Meier, Buchert, Jaenisch, & Mitchell, 1996). Moreover, in *Xenopus laevis tfap2a* displays several successive functions in the NC-GRN: TFAP2a first initiates neural border patterning in response to canonical WNT signaling, then TFAP2a collaborates with the key transcription factors Pax3 and Zic1 to specify neural crest, and TFAP2a is then necessary during NC cell EMT and migration (De Crozé, Maczkowiak, & Monsoro-Burq, 2011). TFAP2 factors act as dimers on the target DNA motifs. A key switch between TFAP2a/c and TFAP2a/b dimerization controls the onset of NC specification at the neural border in chick neurulas (Rothstein & Simoes-Costa, 2020). In zebrafish, gene redundancy between *tfap2a* and *tfap2c* ensures robust NC induction, but their individual mutation also reveals which target genes are most sensitive to *tfap2* gene dosage (Dooley et al., 2019; Li & Cornell, 2007). In humans however, the recorded NCPs involving TFAP2 transcription factors comprise Branchioculofacial syndrome (TFAP2a) and Char syndrome (TFAP2b) which are found in live births, indicating that NC cells did develop in those patients, albeit with severe alterations of their derivatives, including various craniofacial and heart anomalies associated with malformation of tissues unrelated with the NC lineage (e.g., limbs).

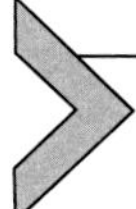

4. Congenital disorders caused by defects in neural crest EMT and migration

Embryonic development requires dramatic morphogenetic changes, at tissue-level and cell-level during gastrulation, neurulation and organogenesis. Among the most notorious changes in tissue architecture, EMT is a process by which an epithelial cell acquires a mesenchymal phenotype, a state often accompanied by the acquisition of cell motility and the ability to migrate along specific paths of the body. The reverse process, mesenchyme-to-epithelium transition (MET) is involved during diverse tissues and organs formation, and allows migratory cells to settle into distant epithelial derivatives. For example, sequential rounds of EMT and MET control gastrulation (in chick embryos), somite formation, and craniofacial development. While EMT and MET are tightly orchestrated in the embryo, the uncontrolled reactivation of EMT in adults is exploited by cancer cells to disseminate and establish secondary tumors in distant regions of the body (Brabletz, Kalluri, Nieto, & Weinberg, 2018). EMT is also involved in tissue healing and fibrosis in adults.

At the end of neurulation (in the head) and during organogenesis (in the trunk), the NC cells emigrate from the neural border area by a stereotypical EMT with a strictly controlled spatiotemporal onset along the anterior-posterior body axis. Chick and *Xenopus* embryos have been instrumental for the identification of the molecular actors driving EMT and the precise dissection of NC EMT controls. Although much remains to be understood in this complex process, NC development provides one of the best settings to understand EMT *in vitro* and *in vivo*.

4.1 The epithelium-to-mesenchyme transition in *Xenopus* neural crest cells

4.1.1 General cellular features of EMT

As described in other vertebrates, frog premigratory NC cells initiate EMT by the loss of epithelial markers such as Cadherin 1 (cdh1, Ecadherin), a transmembrane adhesion molecule localized in adherens junctions of the cell basolateral membrane (Fig. 3). In parallel, NC cells gain expression of mesenchymal markers such as the cell-cell adhesion molecule Cadherin 2 (cdh2, N-cadherin), the intermediate filament Vimentin and the extracellular matrix adhesion molecule Fibronectin (Kalluri & Weinberg, 2009). As cells lose their E-cadherin-based complexes and their apical-basal polarity, the stable epithelial-type cytoskeleton is reorganized into the dynamic actin and microtubule system of motile cells. Further changes in the repertoire of cadherins and adhesion molecules accompany the graded steps of EMT and the onset of NC cell migration.

4.1.2 Direct molecular control of EMT

The cellular transdifferentiation from an epithelial to a mesenchymal phenotype is activated by a core set of transcription factors (TFs) that regulate cell-cell adhesion, cell polarity and the onset of cell migration. Collectively, EMT-inducing TFs repress the expression of the genes and the function of the proteins maintaining the epithelial shape and activate mesenchymal ones. The three main groups of EMT-TFs comprise the zinc-finger transcription factors Snail 1 and Snail 2, the basic helix-loop-helix (bHLH) factor Twist 1, the zinc-finger E-box-binding homeobox factors Zeb 1 and Zeb 2, and the Prrx 1 and Prrx 2 factors. Most of these factors have been discovered in chick and *Xenopus* embryos during the 1990s by degenerate cloning aimed at finding vertebrate orthologs of *Drosophila melanogaster* genes. Their expression in the premigratory NC immediately caught researcher's attention: *snail 1* (*snai1*, initially named *xsna*), *snail 2* (*snai2*, *xslu*), *twist 1* (*xtwi*), *prrx 1* (*prx-1*) and *zeb 1/zeb 2* (initially named *deltaef1* and *sip-1*, respectively) (Hopwood, Pluck, & Gurdon, 1989; Mayor, Morgan, & Sargent, 1995; Nieto, Bennett, Sargent, & Wilkinson, 1992; Nieto, Sargent, Wilkinson, & Cooke, 1994; Takahashi et al., 1998; Van Grunsven et al., 2006). These factors display non-redundant roles and their interactions are still not fully understood (Fazilaty et al., 2019; Stemmler, Eccles, Brabletz, & Brabletz, 2019). NC model still remains essential to identify

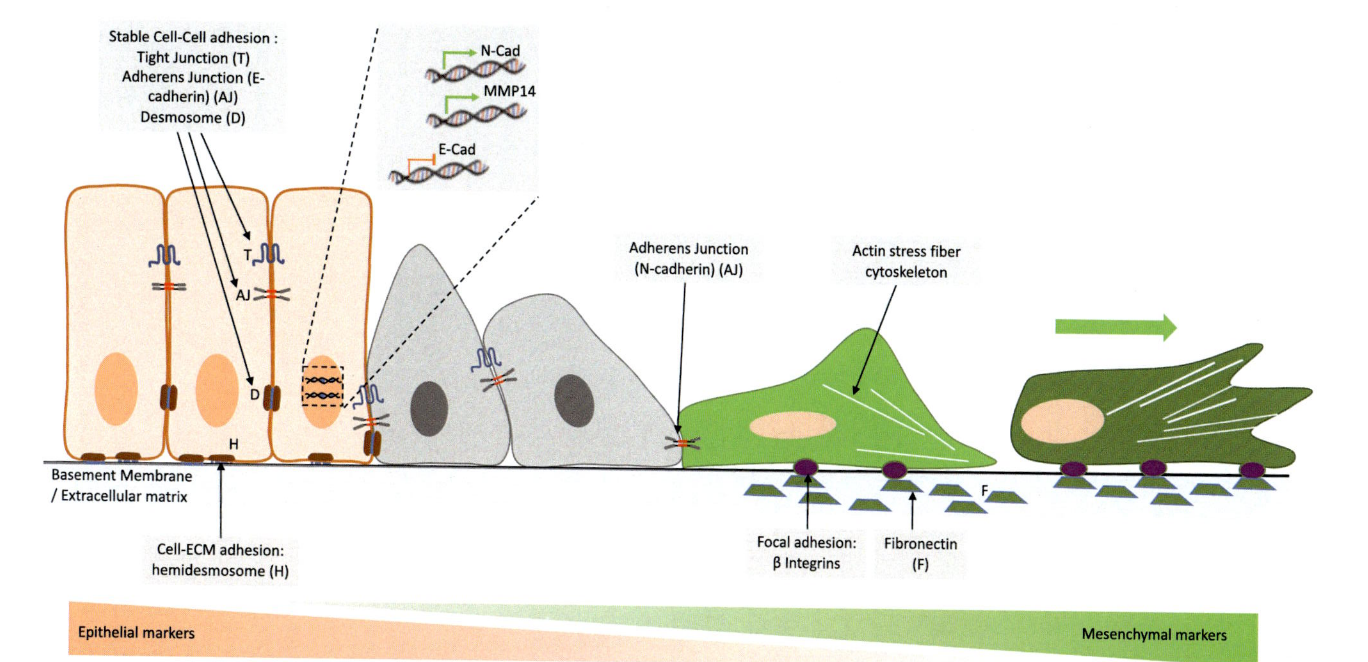

Fig. 3 The epithelium-to-mesenchyme transition (EMT). When located in the neural border ectoderm prior to EMT, the neural crest cells display an epithelial phenotype. They are closely attached to one another by tight and adherens junctions, and display stable adhesion complexes with the extracellular matrix. During the process of EMT, the epithelial cells experience a series of transcriptional and post-transcriptional changes that progressively convert them into cells with a mesenchymal phenotype. Premigratory NC cells loose the expression epithelial markers such as E-cadherin (Cdh1), encoding a transmembrane adhesion molecule localized in adherens junctions and gain the expression mesenchymal markers, such as N-Cadherin (Cdh2) and fibronectin. An extensive remodeling of the cytoskeleton accompanies these changes in cell-cell and cell-matrix adhesion and yields to a motile cell phenotype.

novel features of EMT-TFs nowadays. For example, the cooperation between EMT-TFs *in vivo* has been studied in frog NC cells, showing that Twist 1 controls post-translational modifications of Snail1/2 proteins, thereby modulating their activity (Lander et al., 2013). Further general interactions between the EMT-TFs at different stages of premigratory and migratory NC development remain largely unknown. In addition to the transcriptional regulations directly triggered by the EMT-TFs, complex epigenetic regulations participate to the repression/activation of either epithelial or mesenchymal genes, respectively, involving polycomb repressor complexes (PRCs), histone deacetylases (HDACs), and histone lysine demethylases (KDMs) (reviewed in Tam & Weinberg, 2013). The details of these regulations, on each of the genes forming the NC-GRN remain to be understood, as are the details of the direct and indirect transcriptional regulations of the network. In sum, although tremendous progress has been done in the last 10 years to establish the EMT-GRN in development and cancer, there is still a vast horizon to be explored.

4.1.3 Control of EMT-TF gene expression: Deciphering the early NC-GRN

Snail 1 and Snail 2 are the earliest core EMT-TFs expressed during NC formation *in vivo*. Their expression is first detected in NC progenitors at the end of gastrulation (stage 11.5–12 in *Xenopus laevis*) (Mayor et al., 1995; Plouhinec et al., 2017). This means that the discovery of the upstream regulators of EMT-TFs has been conducted very efficiently in frog embryos, as they are highly suitable for experimental and genetic manipulations during gastrulation. Hence, the early NC-GRN was first drafted in 2005 using combined gain and loss of function both in *Xenopus laevis* embryos *in vivo* and in dissected animal caps grown *in vitro* (a well-established frog spheroid model of pluripotent blastula-stage stem cell ectoderm, extensively used for the last 30 years to discover new genes and new genetic regulation). This first NC-GRN displayed the epistasis relationships between the secreted BMP, FGF8 and canonical Wnt signals, the transcription factors acting at the neural border Pax3, Zic1 and Msx1, and the EMT-TF Snail 2 (Monsoro-Burq et al., 2005). In this case, the robustness of the experimental assays established in *Xenopus* has allowed the simultaneous gain or depletion of several genes *in vivo*. The short development time of this embryo has then allowed the analysis of hundreds of conditions including multiple gene combinations at different doses and different developmental time points. Combined, these unique features of the frog model were the key, used by many since, to successfully build a complex vertebrate GRN during early development

(Buitrago-Delgado, Nordin, Rao, Geary, & LaBonne, 2015; De Crozé et al., 2011; Gutkovich et al., 2010; Macrì et al., 2016; Milet et al., 2013; Scerbo & Monsoro-Burq, 2020). For a review of the current NC-GRN upstream of EMT-TFs expression with focus on the regulations identified using the frog model, please see Pla and Monsoro-Burq (2018).

4.1.4 Mutation of EMT-TFs in human disease

As noted for their upstream regulators in the neural border, Pax3 and TFAP2a, the complete loss by mutations in the core EMT-TFs would affect either gastrulation or whole NC formation. The human syndromes related to mutations in Snail 2 (Waardenburg syndrome type 2D) or Twist 1 (Saethre-Chotzen syndrome) found in live births, thus correspond to later functions of these genes for example in the differentiation NC-derived of skull bones (Saethre-Chotzen syndrome) or pigmentation phenotypes (WS2D).

4.1.5 Mutations causing altered NC survival upon EMT

An important feature of EMT is the acquisition of enhanced cell survival properties, resistance to apoptosis and to various cellular stresses. In consequence, several genes with a "housekeeping" function, in addition to be important for all cell types of the embryo, display a particularly acute phenotype when they are altered in the premigratory neural crest. This is the case in particular for factors monitoring the global cellular homeostasis, which trigger cell death when cell parameters are altered.

A first example is the AKT signaling pathway, a central hub of the cell homeostasis acting in almost all cell types. Disrupted AKT signaling was recently described to alter ectoderm patterning and NC induction (Pegoraro et al., 2015) as well as NC EMT and migration (Bahm et al., 2017; Figueiredo et al., 2017). In both cases, upstream regulators of the NC-GRN, PFKFB4 and PDGFRa, respectively, stimulated the appropriate levels of AKT signaling in premigratory NC cells. In particular, the PFKFB4/AKT signaling deficiency prevented *ncad/cdh2* activation, leading to EMT failure. Interestingly, DDX3, the ortholog of the human gene DDX3X which, when mutated, is involved in the intellectual disability and craniofacial dysmorphogenesis, was studied in *Xenopus tropicalis* (Perfetto et al., 2021). Similar to PDGFRa and PFKFB4, DDX3 is important for NC induction, by sustaining high AKT levels in NC progenitors.

Another example is the Treacher Collins syndrome (TCS), an autosomal dominant disorder causing severe craniofacial anomalies (Dixon et al., 2006). In human, TCS is caused by a mutation in TCOF1, POLR1C or

PLOR1D, encoding proteins involved in ribosome biogenesis. Altered ribosome biogenesis causes nucleolar stress and triggers apoptosis. Mouse models reproduce the human phenotype and have shown that mutant NC cells undergo extensive p53-dependent apoptosis in the neural folds upon EMT (Dixon et al., 2006). In frogs, the genes causing Treacher-Collins syndrome have not yet been studied in NC, although their function in ribosome genesis is conserved (Gonzales et al., 2005). Interestingly, the detailed analysis of 13 genes mutated in various ribosomopathies conducted in *X. tropicalis* revealed a surprising diversity of those genes expression patterns including several enriched in NC cells (Robson et al., 2016). This study suggests a fine-tuned and dynamic control of ribosome biogenesis in early NC formation that may be conserved with other vertebrate models and creates an opportunity to develop new models of ribosomopathies in frog (Griffin et al., 2015).

4.2 *Xenopus* neural crest cell migration

4.2.1 Mechanisms of cranial neural crest collective migration

In frog, cranial NC (CNC) cells are the most accessible and most studied NC subpopulation during neurulation and organogenesis, because trunk NC cell delaminate at later tailbud stages in smaller numbers (Fig. 4). During delamination, CNC cells move collectively as cohesive population and undergo a progressive dissociation into smaller cell clusters then separate as individual cells as their migration progresses further toward target tissues (Fig. 4A). The complex dynamics of CNC cell migration is controlled by several complementary mechanisms, which have been deciphered mainly using *Xenopus laevis* as a model (reviewed in Shellard & Mayor, 2019). These include individual cell-cell interactions, cell adhesion properties within a moving tissue and cell-environment interactions precisely guiding the migration. Interactions between two migrating CNC cells involve contact inhibition of locomotion (CIL) and co-attraction. Cells located in the larger sheet of migrating CNC regulate the adhesion with their neighbors in a dynamic fashion, involving LPAR2 and N-cadherin-dependent remodeling of the cell-cell junctions. Collectively, the population of migrating cells efficiently moves by creating a contractile supracellular ring of actomyosin at the rear of the cellular group creating a net forward force. Lastly, the environment presents chemical and physical cues guiding the migration, involving chemotaxis toward specific neighboring tissues (e.g., CXCR4-SDF-1 signaling with adjacent placode ectoderm) and repulsive cues such as Ephrin signaling, semaphorins, and extracellular matrix components which shape and constrain the stream of migrating CNC cells. Although

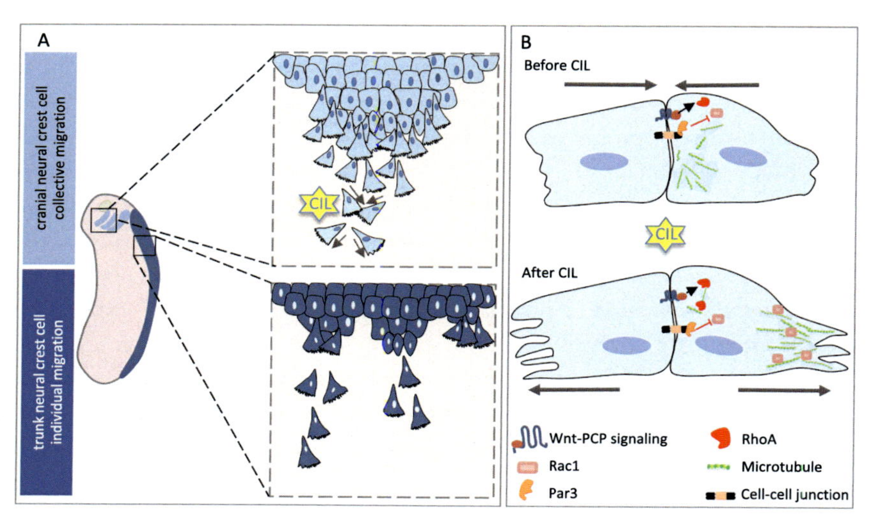

Fig. 4 (A) Cranial (light blue) and trunk (dark blue) neural crest cell migration in *Xenopus* embryo. (A) Cranial neural crest cells migrate collectively, as a cell population with dynamic cell-cell and cell-environment relationships. On the free edge of the migrating population, forward movement is promoted by Rac1 expression in the protrusion side, and forces resulting from a supracellular contractile actomyosin ring formed at the back end of the group of cells. Active Rac1 stabilizes the microtubule cytoskeleton and therefore promotes lamellipodia formation and cell migration in the forward direction. (B) When two cells collide Rac1 activity is inhibited at the site of collision by the activation of RhoA, which is recruited by the WNT/PCP signaling pathway. After collision, Par3 is activated at the cell-cell adhesion complexes and inhibits Rac1 activity resulting into a microtubule catastrophe. This generates a new gradient of RhoA-Rac1 activity in the cell and repolarizes the cytoskeleton. The cells form new protrusions away from the contact site and migrate in opposite directions.

molecular mechanisms controlling CNC migration remain incompletely understood, tremendous advances were brought by studies using frog CNC cells migrating *in vitro* on a fibronectin substratum or using elegant micro-channels approaches (Scarpa et al., 2015; Szabó et al., 2016).

As an example, CIL is the mechanism by which a cell-cell collision causes the arrest of migration, cell repolarization and a change in the direction of migration (Carmona-Fontaine et al., 2008; Scarpa et al., 2015) (Fig. 4B). Importantly, this process is also involved in metastatic dissemination and the mechanisms discovered in NC studies have highlighted several important mechanisms also activated during cancer progression. The non-canonical Wnt/PCP signaling pathway controls both CIL and the directionality of cell movements during migration (Carmona-Fontaine et al., 2008; De Calisto, Araya, Marchant, Riaz, & Mayor, 2005). Disheveled, a central cytoplasmic element of the non-canonical Wnt/PCP signaling is recruited at the plasma membrane at the level of the cell-cell contacts. This activates the small Rho GTPase RhoA and recruits Par3, a protein associated with cell-cell adhesion complexes which promotes the collapse of the microtubule at the level of the cell-cell contacts sites through the inactivation of Rac1/TRIO interaction (Moore et al., 2013). At the opposite end of the cell, the small Rho GTPase Rac1 is active and promotes the formation of membrane protrusions and focal adhesion contacts with the extracellular matrix. Several parameters, including the E-to-N cadherin switch upon EMT and PDGFR/Akt signaling favor active Rac1 polarization in the new direction of migration (Bahm et al., 2017; Scarpa et al., 2015). In sum, by controlling the gradient of RhoA and Rac1 activity at two opposite poles of the cell, CIL leads to the dispersion of the CNC cells. In addition, Cadherin-11 initiates the formation of filopodia and lamellipodia *in vivo* (Kashef et al., 2009). Inhibiting Cadherin 11 or the processing of Cadherin 11 by the ADAM13 metalloprotease affects CIL and the directionality of migration (Abbruzzese, Becker, Kashef, & Alfandari, 2016; Becker, Mayor, & Kashef, 2013). Importantly, in chick NC, Cadherin-11 also controls p53-dependent cell death, suggesting complex roles of such adhesion molecules at the onset of NC cell emigration as well (Manohar, Camacho-Magallanes, Echeverria Jr, & Rogers, 2020). As our understanding of CNC cell migration deepens, the new molecular actors will help understanding how human mutations create NCPs caused by defective NC migration.

4.2.2 Human syndromes affecting neural crest cell EMT and migration

In vivo and *in vitro* analyses of NC migration are thus a strong point of studies using frog models. The consequences of altered NC migration can be detected morphologically at differentiation stage, for example at head levels by the abnormal/reduced shape of the craniofacial cartilages or at trunk levels by the altered repartition of melanoblasts/melanocyte. Two examples that benefited from studies in frog, CHARGE syndrome and 16p12.1 deletion syndrome are detailed below, but other examples can be found in Table 1.

CHARGE syndrome is a rare congenital syndrome characterized by a variable combination of multiple anomalies, mainly eye coloboma; choanal atresia; peripheral nervous system defective cranial nerves; characteristic ear anomalies (known as the major 4C's), combined with conotruncal heart defects and craniofacial anomalies. Several other syndromes share close similarities with CHARGE, such as the KABUKI syndrome. CHARGE is caused by the heterozygous loss of function mutation in the chromodomain helicase DNA-binding CHD7, a chromatin remodeler (Bajpai et al., 2010). Chd7 frog ortholog are highly similar to the human protein (Linder, Cabot, Schwickert, & Rupp, 2004). Knockdown studies in *Xenopus* phenocopy the main features of the pathology and show that NC specification is affected (defective *sox9, snail2* and *twist1* expression) together with NC cell migration (Bajpai et al., 2010). Moreover, comparison with human iPS cells has helped identifying the action of CHD7 on regulatory elements controlling *sox9* and *twist1* (Bajpai et al., 2010). Putative CHD7 targets, including *ephrin* and *semaphorin* genes, were identified in a screen of mutant mouse embryos. Those targets were validated in *Xenopus* embryos *in vivo*. Semaphorin 3a (*sema3a*) expression was altered in Chd7 morphant frog embryos (Schulz et al., 2014). The functional and evolutionary conservation of the molecular mechanisms causing the CHARGE disease were validated by the phenotypic rescue of the frog morphants after co-injection with human Chd7 mRNA (Ufartes et al., 2018). Interestingly, SEMA3f is mutated in KABUKI syndrome (Schwenty-Lara et al., 2020). Collectively, these results uncover that defective CNC cell migration contributes to the pleiotropic alterations seen in CHARGE patients, and involves guidance molecules of the Semaphorin family. Moreover, other syndromes related to CHARGE have involved other causative mutations, such as in the p120 catenin gene CTNND1 (Alharatani et al., 2020). It will be interesting to explore further these candidates as targets of CDH7. Lastly, all these studies gather excellent examples of the combined use of *Xenopus* assays and mammalian models to explore the molecular mechanisms of NCPs.

Complex or multigenic syndromes can also be very effectively explored using *Xenopus* assays. For example, the hemizygous deletion of the p12.1 region of Chromosome 16 causes a series of neurodevelopmental defects, including severe craniofacial malformations. The four genes involved in this deletion encode proteins with very different biological functions, *polr3e* (encodes RNA Polymerase IIIE; synthesis of small RNAs), *mosmo* (encodes a modulator of Smoothened; Sonic Hedgehog signaling), *uqcrc2* (encodes a component of mitochondrial respiratory chain; ATP biosynthesis), and *cdr2* (encodes an oncoprotein) were systematically studied at the different steps of the NC-GRN *in vivo* (Lasser et al., 2020). This study found different roles for each candidate genes in NC specification (*mosmo, polr3e*), NC motility *in vitro* (*polr3e, uqcrc2*, cdr2), NC migration (all) altogether resulting in craniofacial anomalies similar to the 16p12.1 syndrome malformations in human. Importantly, the availability of a large number of stage-matched sibling embryos to conduct parallel analyses for the different genes has allows a direct and systematical comparison of these complex phenotypes, at several sequential steps of the NC-GRN.

4.3 Pathologies linked to defective differentiation of neural crest derivatives or their tumoral transformation

4.3.1 Studies of neural crest differentiation in NCPs

As the development of frog NC cells in the cranial area is better understood than in the trunk, it is not surprising that few studies focus on the defects in the trunk NC derivatives related to human NCPs (Table 1). For instance, the expression of *pmp22*, the gene encoding a myelin protein causing defective Schwann myelin protein function in the Charcot-Marie-Tooth syndrome, is controlled by Pax3 and Zic1 in the frog NC-GRN during early organogenesis (Bae et al., 2014; Tae et al., 2015). The expression of other genes causing altered differentiation of glial cells, such as in the Congenital Central Hypoventilation syndrome (altered sympathetic glia) was described in frog embryos, but not with a focus on glial differentiation (*phox2a/b*, Talikka, Stefani, Brivanlou, & Zimmerman, 2004; *gdnf*, Kyuno & Jones, 2007; *ret*, Carroll, Wallingford, Seufert, & Vize, 1999; *edn3/ednr*, Square, Jandzik, Cattell, Hansen, & Medeiros, 2016). This means that some molecular tools have been developed and could be directly used in NCP studies. Moreover, studies which independently explore the developmental signaling driving the migration and the differentiation of derivatives such as melanoblasts and melanocytes or trunk sensory neurons as readouts will be valuable to further understanding NCPs (e.g., Nakano et al., 2003; Valluet et al., 2012).

However, although this remains seldom used, *Xenopus* tadpoles are fully suitable for the analysis of trunk NC derivatives formation. For example, Hirschsprung disease is caused by mutations in RET, GDNF, and several other genes, resulting in the defective colonization of the caudal parts of the digestive tract by NC progenitors of the enteric nervous system. Li and colleagues have studied the impact of retinoic acid (RA) signaling and CREB-binding protein on the differentiation of the enteric nervous system in *Xenopus laevis* tadpoles (Li et al., 2018). They show that RA signaling is important for NC induction globally and affects ENS formation in particular. Although RA signaling has been studied in-depth in many studies of early NC patterning, this analysis includes an interesting late differentiation step.

In contrast, the differentiation of the cranial skeleton has received high attention in the frog model, as size, morphology and differentiation of the cartilage elements forming the visceral skeleton are readouts easily examined in *Xenopus* tadpoles. Most alterations in the NC-GRN affecting CNC cells are tested to alter ectomesenchyme formation (e.g., Monsoro-Burq et al., 2005; Scerbo & Monsoro-Burq, 2020). Moreover, frog craniofacial morphogenesis was studied in relationship with cleft palate and other common human craniofacial malformations (reviewed in Dickinson, 2016). These studies involved for instance p120-catenin (Alharatani et al., 2020), Zic1/Zic4 transcription factors (Twigg et al., 2015) and Msx2 (Kennedy & Dickinson, 2012). While MSX2 is also involved in craniosynostosis in human, the exact molecular mechanisms involving the same gene in the two species remain to precisely be compared, since in slightly different cell contexts or different steps of development, the same transcription factor may act with different partners for a different function. Two of the genes causing DiGeorge syndrome, *tbx1* and *tbx6*, are another example where frog craniofacial cartilage biology uncovered detailed mechanistic information (Ataliotis, Ivins, Mohun, & Scambler, 2005; Tazumi et al., 2010). While the initial induction of NC appeared normal, and the NC cell migration only slightly delayed, the chondrocyte progenitors failed to maintain *sox9* gene expression and consequently did not differentiate appropriately. Thus, an important aspect of DiGeorge pathology was explored, while other NC derivatives such as cardiac NC affected in the pathology await further analysis. In sum, the analysis of craniofacial cartilage differentiation is one of the major assets of the frog model. Since recently, it includes the possibility to develop quantitative approaches in a more systematic manner with Optical Coherence Tomography (OCT), which will be invaluable for the robust statistical analyses of the frog models of human craniofacial malformations (Deniz et al., 2017).

4.3.2 Studies of tumorigenic transformation of neural crest derivatives tadpoles or adult frogs

How frog studies have advanced our understanding of NC cell migration is highlighted above. These studies display an intricate crosstalk with studies on human cancer metastasis, as the molecules involved in NC migration have later been found in cancer and *vice-versa*. The biology of matrix metalloproteases of the MMP family brings a series of such examples. MMP14 is a transmembrane protein triggering tumor cells invasion of the extracellular matrix (Sato et al., 1994). MMP14 is also important for NC migration and EMT in cancer and organ fibrosis (Gonzalez-Molina et al., 2019; Szabova, Chrysovergis, Yamada, & Holmbeck, 2008; Xiong et al., 2017). Details of MMP14 activity, such as its control of the levels of cadherin expression, have been explored in frog (Garmon, Wittling, & Nie, 2018). Moreover, the use of transplanted NC cells into host embryos allowed to test which tissue relies on MMP14 expression or on other MMPs such as MMP2, and showed that NC cells depend on MMP14 while MMP2 is needed in the surrounding craniofacial mesenchyme (Garmon et al., 2018).

As far as tumor formation is concerned, NC derivatives form aggressive tumors such as melanoma, neuroblastoma or schwannoma. About 220 studies have used frog models as part of the analysis of mechanisms driving melanoma or neuroblastoma. For instance, BRCA-1-associated protein BAP-1 is found mutated in uveal melanoma, a pigment cell type derived from neural crest as are cutaneous melanoma cells. In several cancers, BAP-1 displays tumor suppressor functions. Partly because of its closer relationship with the human protein (*Xenopus* Bap1 protein shares 92% similarity and 71% identity with human BAP1, compared to 85% and 66%, respectively, for zebrafish), Kuznetsov and colleagues conducted a thorough study of *bap1* gene expression, regulation and function in *Xenopus* development (Kuznetsov et al., 2019). They show NC-restricted expression and identify its function as a regulator of the histone deacetylase HDAC4. They further confirm BAP1 function on HDAC4 nucleo-cytoplasmic shuttling in a human uveal melanoma cell line. Moreover, a following study from the same authors, described below, exemplifies how the frog model can both be used to decipher complex mechanisms of tumor biology, and for screening chemical therapeutic compounds (Kuznetsoff et al., 2021).

4.3.3 Drug screening for cancer studies

Both *Xenopus laevis* and *tropicalis* have proven excellent models for the screening of pharmacological compounds targeting various signaling pathways and developmental mechanisms. Those screens take advantage of the availability of large embryos number, of their fast external development, and of their easy manipulation in multi-well plates. Hence, pigment cell development readout was used for a screen looking for chemicals involved in melanocyte migration and differentiation (Tomlinson, Rejzek, Fidock, Field, & Wheeler, 2009), very similarly to the strategies used in zebrafish studies (White et al., 2011). Moreover, gastrulation and NC migration in *Xenopus* embryos were used to screen 100 novel chemical compounds and identify those inhibiting cell migration. This *in vivo* screening identified nine novel molecules that inhibited cell migration without embryonic lethality. When those molecules were tested on cancer cell line invasion, two significantly suppressed tumor cell invasion in 3D-collagen/matrigel gels and reduced melanoma and glioma tumor size in nude mice. Both components disrupted microtubule assembly and caused a dose-dependent increase in apoptosis (Tanaka et al., 2016). Lastly, *Xenopus* was recently used to identify non-toxic compounds rescuing the loss of the tumor suppressor BAP-1 *in vivo*. A two-step strategy has found a HDAC inhibitor selectively rescuing BAP-1 loss in uveal melanoma (Kuznetsoff et al., 2021).

5. Conclusion

In this review, we have attempted to highlight several fields in which the study of *Xenopus* frogs can contribute to understand neural crest cell biology and disease. With a selected series of examples, we have illustrated how biomedical research has tremendously benefited from the in-depth analyses that can be conducted in these models, both *in vivo* and *in vitro* using a variety of assays for cell migration and differentiation. Our understanding of NCPs, such as CHARGE and Hirschsprung syndromes, but also of neurodevelopmental disorders more generally such as the Bainbridge-Ropers syndrome and of other pathologies (Blackburn et al., 2019; Lichtig et al., 2020; Singh et al., 2020) will certainly include results obtained using this animal model more and more in the near future. With increasing usage of CRISPR/Cas9-based mutagenesis in *Xenopus tropicalis* (and to a lesser extent in *Xenopus laevis*), many novel mutants and variants can be studied faster than in mammalian models and in complement to studies using human or mouse approaches (Marquez et al., 2020). Novel tools including NC-centered genomic resources (Plouhinec et al., 2017) and transgenic frog lines are developed for NC reporter genes (Alkobtawi et al., 2018; Li et al., 2019). Complex syndromes, involving several gene alterations can be explored in a systematical manner (Lasser et al., 2020; Lasser, Pratt, Monahan, Kim, & Lowery, 2019). Lastly, cancer studies are increasingly aware of the advantages of this animal model (Hardwick & Philpott, 2018; Kuznetsoff et al., 2021), opening avenues to explore the neural crest-derived tumorigenesis as well as developmental neurocristopathies.

Acknowledgments

The authors are grateful to all members of the Monsoro-Burq team for their insightful comments on the manuscript, especially V. Kappès, S. Seal and A. Kotov. L. M.-C. is supported by the Institut Curie EURECA Ph.D. Program. Funding to A. H. M.-B.

laboratory is provided by Université Paris Saclay (Orsay, France), Institut Universitaire de France (Paris, France), European Union's Horizon 2020 Research and Innovation Program under Marie Skłodowska-Curie (grant agreement No 860635, ITN NEUcrest), Centre National de la Recherche Scientifique (CNRS), and Agence Nationale pour la Recherche (ANR-15-CE13-0012-01-CRESTNETMETABO).

Authors contribution

L. M.-C. and A. H. M.-B. wrote the manuscript and designed the original figures. Both authors contributed to the article and approved the submitted version.

Declaration of interest

The authors declare no competing interests.

References

Abbruzzese, G., Becker, S. F., Kashef, J., & Alfandari, D. (2016). ADAM13 cleavage of cadherin-11 promotes CNC migration independently of the homophilic binding site. *Developmental Biology*, *415*, 383–390. https://doi.org/10.1016/j.ydbio.2015.07.018.

Alharatani, R., Ververi, A., Beleza-Meireles, A., Ji, W., Mis, E., Patterson, Q. T., et al. (2020). Novel truncating mutations in CTNND1 cause a dominant craniofacial and cardiac syndrome. *Human Molecular Genetics*, *29*, 1900–1921. https://doi.org/10.1093/hmg/ddaa050.

Alkobtawi, M., & Monsoro-Burq, A. H. (2020). Chapter 1: The neural crest, a vertebrate invention. In B. F. Eames, D. M. Medeiros, & I. Adameyko (Eds.), *Evolving neural crest cells* (pp. 5–66). Boca Raton: CRC Press.

Alkobtawi, M., Ray, H., Barriga, E. H., Moreno, M., Kerney, R., Monsoro-Burq, A.-H., et al. (2018). Characterization of Pax3 and Sox10 transgenic Xenopus laevis embryos as tools to study neural crest development. *Developmental Biology*, *444*(Suppl. 1), S202–S208. https://doi.org/10.1016/j.ydbio.2018.02.020.

Amiel, J., & Lyonnet, S. (2001). Hirschsprung disease, associated syndromes, and genetics: A review. *Journal of Medical Genetics*, *38*, 729–739. https://doi.org/10.1136/jmg.38.11.729.

Ataliotis, P., Ivins, S., Mohun, T. J., & Scambler, P. J. (2005). XTbx1 is a transcriptional activator involved in head and pharyngeal arch development in Xenopus laevis. *Developmental Dynamics*, *232*(4), 979–991. https://doi.org/10.1002/dvdy.2027611.

Bae, C.-J., Park, B.-Y., Lee, Y.-H., Tobias, J. W., Hong, C.-S., & Saint-Jeannet, J.-P. (2014). Identification of Pax3 and Zic1 targets in the developing neural crest. *Developmental Biology*, *386*, 473–483. https://doi.org/10.1016/j.ydbio.2013.12.011.

Bahm, I., Barriga, E. H., Frolov, A., Theveneau, E., Frankel, P., & Mayor, R. (2017). PDGF controls contact inhibition of locomotion by regulating N-cadherin during neural crest migration. *Development (Cambridge, England)*, *144*, 2456–2468. https://doi.org/10.1242/dev.147926.

Bajpai, R., Chen, D. A., Rada-Iglesias, A., Zhang, J., Xiong, Y., Helms, J., et al. (2010). CHD7 cooperates with PBAF to control multipotent neural crest formation. *Nature*, *463*, 958–962. https://doi.org/10.1038/nature08733.

Barrell, W. B., Griffin, J. N., Harvey, J.-L., HipSci Consortium, Danovi, D., Beales, P., et al. (2019). Induction of neural crest stem cells from Bardet-Biedl syndrome patient derived hiPSCs. *Frontiers in Molecular Neuroscience*, *12*, 139. https://doi.org/10.3389/fnmol.2019.00139.

Barsh, G., Gunn, T., He, L., Schlossman, S., & Duke-Cohan, J. (2000). Biochemical and genetic studies of pigment-type switching. *Pigment Cell Research*, *13*(Suppl. 8), 48–53. https://doi.org/10.1034/j.1600-0749.13.s8.10.x.

Basch, M. L., Bronner-Fraser, M., & García-Castro, M. I. (2006). Specification of the neural crest occurs during gastrulation and requires Pax7. *Nature*, *441*, 218–222. https://doi.org/10.1038/nature04684.

Becker, S. F. S., Mayor, R., & Kashef, J. (2013). Cadherin-11 mediates contact inhibition of locomotion during Xenopus neural crest cell migration. *PLoS One*, *8*, e85717. https://doi.org/10.1371/journal.pone.0085717.

Blackburn, A. T. M., Bekheirnia, N., Uma, V. C., Corkins, M. E., Xu, Y., Rosenfeld, J. A., et al. (2019). DYRK1A-related intellectual disability: A syndrome associated with congenital anomalies of the kidney and urinary tract. *Genetics in Medicine: Official Journal of the American College of Medical Genetics*, *21*, 2755–2764. https://doi.org/10.1038/s41436-019-0576-0.

Bolande, R. P. (1974). The neurocristopathies: A unifying concept of disease arising in neural crest maldevelopment. *Human Pathology*, *5*, 409–429. https://doi.org/10.1016/S0046-8177(74)80021-3.

Bonnard, C., Strobl, A. C., Shboul, M., Lee, H., Merriman, B., Nelson, S. F., et al. (2012). Mutations in IRX5 impair craniofacial development and germ cell migration via SDF1. *Nature Genetics*, *44*, 709–713. https://doi.org/10.1038/ng.2259.

Brabletz, T., Kalluri, R., Nieto, M. A., & Weinberg, R. A. (2018). EMT in cancer. *Nature Reviews. Cancer*, *18*, 128–134. https://doi.org/10.1038/nrc.2017.118.

Briggs, J. A., Weinreb, C., Wagner, D. E., Megason, S., Peshkin, L., Kirschner, M. W., et al. (2018). The dynamics of gene expression in vertebrate embryogenesis at single-cell resolution. *Science*, *360*(6392), eaar5780. https://doi.org/10.1126/science.aar5780.

Buitrago-Delgado, E., Nordin, K., Rao, A., Geary, L., & LaBonne, C. (2015). Neurodevelopment. Shared regulatory programs suggest retention of blastula-stage potential in neural crest cells. *Science*, *348*, 1332–1335. https://doi.org/10.1126/science.aaa3655.

Calo, E., Gu, B., Bowen, M. E., Aryan, F., Zalc, A., Liang, J., et al. (2018). Tissue-selective effects of nucleolar stress and rDNA damage in developmental disorders. *Nature*, *554*, 112–117. https://doi.org/10.1038/nature25449.

Carmona-Fontaine, C., Matthews, H. K., Kuriyama, S., Moreno, M., Dunn, G. A., Parsons, M., et al. (2008). Contact inhibition of locomotion in vivo controls neural crest directional migration. *Nature*, *456*, 957–961. https://doi.org/10.1038/nature07441.

Carroll, T., Wallingford, J., Seufert, D., & Vize, P. D. (1999). Molecular regulation of pronephric development. *Current Topics in Developmental Biology*, *44*, 67–100. https://doi.org/10.1016/s0070-2153(08)60467-6.

Cha, J. Y., Birsoy, B., Kofron, M., Mahoney, E., Lang, S., Wylie, C., et al. (2007). The role of FoxC1 in early Xenopus development. *Developmental Dynamics: An Official Publication of the American Association of the Anatomists*, *236*, 2731–2741. https://doi.org/10.1002/dvdy.21240.

Corallo, D., Donadon, M., Pantile, M., Sidarovich, V., Cocchi, S., Ori, M., et al. (2020). LIN28B increases neural crest cell migration and leads to transformation of trunk sympathoadrenal precursors. *Cell Death and Differentiation*, *27*, 1225–1242. https://doi.org/10.1038/s41418-019-0425-3.

De Calisto, J., Araya, C., Marchant, L., Riaz, C. F., & Mayor, R. (2005). Essential role of non-canonical Wnt signalling in neural crest migration. *Development (Cambridge, England)*, *132*, 2587–2597. https://doi.org/10.1242/dev.01857.

De Crozé, N., Maczkowiak, F., & Monsoro-Burq, A. H. (2011). Reiterative AP2a activity controls sequential steps in the neural crest gene regulatory network. *Proceedings of the National Academy of Sciences of the United States of America*, *108*, 155–160. https://doi.org/10.1073/pnas.1010740107.

Deniz, E., Jonas, S., Hooper, M., N Griffin, J., Choma, M. A., & Khokha, M. K. (2017). Analysis of craniocardiac malformations in Xenopus using optical coherence tomography. *Science Reports*, *7*, 42506. https://dci.org/10.1038/srep42506.

Dickinson, A. J. G. (2016). Using frogs faces to dissect the mechanisms underlying human orofacial defects. *Seminars in Cell & Developmental Biology*, *51*, 54–63. https://doi.org/10.1016/j.semcdb.2016.01.016.

Dixon, J., Jones, N. C., Sandell, L. L., Jayasinghe, S. M., Crane, J., Rey, J.-P., et al. (2006). Tcof1/Treacle is required for neural crest cell formation and proliferation deficiencies that cause craniofacial abnormalities. *Proceedings of the National Academy of Sciences of the United States of America*, *103*, 13403–13408. https://doi.org/10.1073/pnas.0603730103.

Dooley, C. M., Wali, N., Sealy, I. M., White, R. J., Stemple, D. L., Collins, J. E., et al. (2019). The gene regulatory basis of genetic compensation during neural crest induction. *PLoS Genetics*, *15*, e1008213. https://doi.org/10.1371/journal.pgen.1008213.

El-Hodiri, H. M., Seufert, D. W., Nekkalapudi, S., Prescott, N. L., Kelly, L. E., & Jamrich, M. (2005). Xenopus laevis FoxE1 is primarily expressed in the developing pituitary and thyroid. *The International Journal of Developmental Biology*, *49*, 881–884. https://doi.org/10.1387/ijdb.052011he.

Epstein, D. J., Vekemans, M., & Gros, P. (1991). Splotch (Sp2H), a mutation affecting development of the mouse neural tube, shows a deletion within the paired homeodomain of Pax-3. *Cell*, *67*, 767–774. https://doi.org/10.1016/0092-8674(91)90071-6.

Essenfelder, G. M., Bruzzone, R., Lamartine, J., Charollais, A., Blanchet-Bardon, C., Barbe, M. T., et al. (2004). Connexin30 mutations responsible for hidrotic ectodermal dysplasia cause abnormal hemichannel activity. *Human Molecular Genetics*, *13*, 1703–1714. https://doi.org/10.1093/hmg/ddh191.

Fazilaty, H., Rago, L., Kass Youssef, K., Ocaña, O. H., Garcia-Asencio, F., Arcas, A., et al. (2019). A gene regulatory network to control EMT programs in development and disease. *Nature Communications*, *10*, 5115. https://doi.org/10.1038/s41467-019-13091-8.

Figueiredo, A. L., Maczkowiak, F., Borday, C., Pla, P., Sittewelle, M., Pegoraro, C., et al. (2017). PFKFB4 control of AKT signaling is essential for premigratory and migratory neural crest formation. *Development (Cambridge, England)*, *144*, 4183–4194. https://doi.org/10.1242/dev.157644.

Fortriede, J. D., Pells, T. J., Chu, S., Chaturvedi, P., Wang, D., Fisher, M. E., et al. (2020). Xenbase: Deep integration of GEO & SRA RNA-seq and ChIP-seq data in a model organism database. *Nucleic Acids Research*, *48*, D776–D782. https://doi.org/10.1093/nar/gkz933.

Fürst, K., Steder, M., Logotheti, S., Angerilli, A., Spitschak, A., Marquardt, S., et al. (2019). DNp73-induced degradation of tyrosinase links depigmentation with EMT-driven melanoma progression. *Cancer Letters*, *442*, 299–309. https://doi.org/10.1016/j.canlet.2018.11.009.

Garmon, T., Wittling, M., & Nie, S. (2018). MMP14 regulates cranial neural crest epithelial-to-mesenchymal transition and migration. *Developmental Dynamics: An Official Publication of the American Association of the Anatomists*, *247*, 1083–1092. https://doi.org/10.1002/dvdy.24661.

Gonzales, B., Yang, H., Henning, D., & Valdez, B. C. (2005). Cloning and functional characterization of the Xenopus orthologue of the Treacher Collins syndrome (TCOF1) gene product. *Gene*, *359*, 73–80. https://doi.org/10.1016/j.gene.2005.04.042.

Gonzalez Malagon, S. G., Lopez Muñoz, A. M., Doro, D., Bolger, T. G., Poon, E., Tucker, E. R., et al. (2018). Glycogen synthase kinase 3 controls migration of the neural crest lineage in mouse and Xenopus. *Nature Communications*, *9*, 1126. https://doi.org/10.1038/s41467-018-03512-5.

Gonzalez-Molina, J., Gramolelli, S., Liao, Z., Carlson, J. W., Ojala, P. M., & Lehti, K. (2019). MMP14 in sarcoma: A regulator of tumor microenvironment communication in connective tissues. *Cell*, *8*(9), 991. https://doi.org/10.3390/cells8090991.

Griffin, J. N., Sondalle, S. B., Del Viso, F., Baserga, S. J., & Khokha, M. K. (2015). The ribosome biogenesis factor Nol11 is required for optimal rDNA transcription and craniofacial development in Xenopus. *PLoS Genetics*, *11*, e1005018. https://doi.org/10.1371/journal.pgen.1005018.

Gutkovich, Y. E., Ofir, R., Elkouby, Y. M., Dibner, C., Gefen, A., Elias, S., et al. (2010). Xenopus Meis3 protein lies at a nexus downstream to Zic1 and Pax3 proteins, regulating multiple cell-fates during early nervous system development. *Developmental Biology*, *338*, 50–62. https://doi.org/10.1016/j.ydbio.2009.11.024.

Hardwick, L. J. A., & Philpott, A. (2018). Xenopus models of cancer: Expanding the oncologist's toolbox. *Frontiers in Physiology*, *9*, 1660. https://doi.org/10.3389/fphys.2018.01660.

Hockman, D., Chong-Morrison, V., Green, S. A., Gavriouchkina, D., Candido-Ferreira, I., Ling, I. T. C., et al. (2019). A genome-wide assessment of the ancestral neural crest gene regulatory network. *Nature Communications*, *10*, 4689. https://doi.org/10.1038/s41467-019-12687-4.

Hopwood, N. D., Pluck, A., & Gurdon, J. B. (1989). A Xenopus mRNA related to Drosophila twist is expressed in response to induction in the mesoderm and the neural crest. *Cell*, *59*, 893–903. https://doi.org/10.1016/0092-8674(89)90612-0.

Kalluri, R., & Weinberg, R. A. (2009). The basics of epithelial-mesenchymal transition. *The Journal of Clinical Investigation*, *119*, 1420–1428. https://doi.org/10.1172/JCI39104.

Kashef, J., Köhler, A., Kuriyama, S., Alfandari, D., Mayor, R., & Wedlich, D. (2009). Cadherin-11 regulates protrusive activity in Xenopus cranial neural crest cells upstream of Trio and the small GTPases. *Genes & Development*, *23*, 1393–1398. https://doi.org/10.1101/gad.519409.

Kennedy, A. E., & Dickinson, A. J. G. (2012). Median facial clefts in Xenopus laevis: Roles of retinoic acid signaling and homeobox genes. *Developmental Biology*, *365*, 229–240. https://doi.org/10.1016/j.ydbio.2012 02.033.

Kobayashi, G. S., Musso, C. M., Moreira D.d. P., Pontillo-Guimarães, G., Hsia, G. S. P., Caires-Júnior, L. C., et al. (2020). Recapitulation of neural crest specification and EMT via induction from neural plate border-like cells. *Stem Cell Reports*, *15*, 776–788. https://doi.org/10.1016/j.stemcr.2020.07.023.

Kuznetsoff, J. N., Owens, D. A., Lopez, A., Rodriguez, D. A., Chee, N. T., Kurtenbach, S., et al. (2021). Dual screen for efficacy and toxicity identifies HDAC inhibitor with distinctive activity spectrum for BAP1-mutant uveal melanoma. *Molecular Cancer Research*, *19*(2), 215–222. https://doi.org/10.1158/1541-7786.MCR-20-0434.

Kuznetsov, J. N., Aguero, T. H., Owens, D. A., Kurtenbach, S., Field, M. G., Durante, M. A., et al. (2019). BAP1 regulates epigenetic switch from pluripotency to differentiation in developmental lineages giving rise to BAP1-mutant cancers. *Science Advances*, *5*, eaax1738. https://doi.org/10.1126/sciadv.aax1738.

Kyuno, J., & Jones, E. A. (2007). GDNF expression during Xenopus development. *Gene Expression Patterns*, *7*, 313–317. https://doi.org/10.1016/j.modgep.2006.08.005.

Lander, R., Nasr, T., Ochoa, S. D., Nordin, K., Prasad, M. S., & Labonne, C. (2013). Interactions between twist and other core epithelial-mesenchymal transition factors are controlled by GSK3-mediated phosphorylation. *Nature Communications*, *4*, 1542. https://doi.org/10.1038/ncomms2543.

Lasser, M., Bolduc, J., Murphy, L., O'Brien, C., Lee, S., Girirajan, S., et al. (2020). 16p12.1 deletion orthologs are expressed in motile neural crest cells and are important for regulating craniofacial development in Xenopus laevis. *bioRxiv*. 2020.12.11.421347. https://doi.org/10.1101/2020.12.11.421347.

Lasser, M., Pratt, B., Monahan, C., Kim, S. W., & Lowery, L. A. (2019). The many faces of Xenopus: Xenopus laevis as a model system to study Wolf-Hirschhorn syndrome. *Frontiers in Physiology*, *10*, 817. https://doi.org/10.3389/fphys.2019.00817.

Le Douarin, N. M., & Kalcheim, C. (1999). The neural crest. *Developmental and cell biology series* (2nd ed.). Cambridge: Cambridge University Press.

Leung, A. W., Murdoch, B., Salem, A. F., Prasad, M. S., Gomez, G. A., & García-Castro, M. I. (2016). WNT/β-catenin signaling mediates human neural crest induction via a pre-neural border intermediate. *Development (Cambridge, England)*, *143*, 398–410. https://doi.org/10.1242/dev.130849.

Li, W., & Cornell, R. A. (2007). Redundant activities of Tfap2a and Tfap2c are required for neural crest induction and development of other non-neural ectoderm derivatives in zebrafish embryos. *Developmental Biology*, *304*, 338–354. https://doi.org/10.1016/j.ydbio.2006.12.042.

Li, C., Hu, R., Hou, N., Wang, Y., Wang, Z., Yang, T., et al. (2018). Alteration of the retinoid acid-CBP signaling pathway in neural crest induction contributes to enteric nervous system disorder. *Frontiers in Pediatrics*, *6*, 382. https://doi.org/10.3389/fped.2018.00382.

Li, Y., Manaligod, J. M., & Weeks, D. L. (2010). EYA1 mutations associated with the branchio-oto-renal syndrome result in defective otic development in Xenopus laevis. *Biology of the Cell*, *102*, 277–292. https://doi.org/10.1042/BC20090098.

Li, J., Perfetto, M., Materna, C., Li, R., Thi Tran, H., Vleminckx, K., et al. (2019). A new transgenic reporter line reveals Wnt-dependent Snai2 re-expression and cranial neural crest differentiation in Xenopus. *Scientific Reports*, *9*, 11191. https://doi.org/10.1038/s41598-019-47665-9.

Lichtig, H., Artamonov, A., Polevoy, H., Reid, C. D., Bielas, S. L., & Frank, D. (2020). Modeling Bainbridge-Ropers syndrome in Xenopus laevis embryos. *Frontiers in Physiology*, *11*, 75. https://doi.org/10.3389/fphys.2020.00075.

Linder, B., Cabot, R. A., Schwickert, T., & Rupp, R. A. W. (2004). The SNF2 domain protein family in higher vertebrates displays dynamic expression patterns in Xenopus laevis embryos. *Gene*, *326*, 59–66. https://doi.org/10.1016/j.gene.2003.09.053.

Luo, T., Lee, Y.-H., Saint-Jeannet, J.-P., & Sargent, T. D. (2003). Induction of neural crest in Xenopus by transcription factor AP2alpha. *Proceedings of the National Academy of Sciences of the United States of America*, *100*, 532–537. https://doi.org/10.1073/pnas.0237226100.

Luo, T., Matsuo-Takasaki, M., Thomas, M. L., Weeks, D. L., & Sargent, T. D. (2002). Transcription factor AP-2 is an essential and direct regulator of epidermal development in Xenopus. *Developmental Biology*, *245*, 136–144. https://doi.org/10.1006/dbio.2002.0621.

Macrì, S., Simula, L., Pellarin, I., Pegoraro, S., Onorati, M., Sgarra, R., et al. (2016). Hmga2 is required for neural crest cell specification in Xenopus laevis. *Developmental Biology*, *411*, 25–37. https://doi.org/10.1016/j.ydbio.2016.01.014.

Maczkowiak, F., Matéos, S., Wang, E., Roche, D., Harland, R., & Monsoro-Burq, A. H. (2010). The Pax3 and Pax7 paralogs cooperate in neural and neural crest patterning using distinct molecular mechanisms, in Xenopus laevis embryos. *Developmental Biology*, *340*, 381–396. https://doi.org/10.1016/j.ydbio.2010.01.022.

Manohar, S., Camacho-Magallanes, A., Echeverria, C., Jr., & Rogers, C. D. (2020). Cadherin-11 Is required for neural crest specification and survival. *Frontiers in Physiology*, *11*, 563372. https://doi.org/10.3389/fphys.2020.563372.

Mansouri, A., Stoykova, A., Torres, M., & Gruss, P. (1996). Dysgenesis of cephalic neural crest derivatives in Pax7 −/− mutant mice. *Development (Cambridge, England)*, *122*, 831–838.

Marivin, A., Leyme, A., Parag-Sharma, K., DiGiacomo, V., Cheung, A. Y., Nguyen, L. T., et al. (2016). Dominant-negative Gα subunits are a mechanism of dysregulated heterotrimeric G protein signaling in human disease. *Science Signaling*, *9*, ra37. https://doi.org/10.1126/scisignal.aad2429.

Marquez, J., Criscione, J., Charney, R. M., Prasad, M. S., Hwang, W. Y., Mis, E. K., et al. (2020). Disrupted ER membrane protein complex-mediated topogenesis drives congenital neural crest defects. *The Journal of Clinical Investigation*, *130*, 813–826. https://doi.org/10.1172/JCI129308.

Martens, H., Hennies, I., Getwan, M., Christians, A., Weiss, A.-C., Brand, F., et al. (2020). Rare heterozygous GDF6 variants in patients with renal anomalies. *European Journal of Human Genetics*, *28*, 1681–1693. https://doi.org/10.1038/s41431-020-0678-9.

Martik, M. L., Gandhi, S., Uy, B. R., Gillis, J. A., Green, S. A., Simoes-Costa, M., et al. (2019). Evolution of the new head by gradual acquisition of neural crest regulatory circuits. *Nature*, *574*, 675–678. https://doi.org/10.1038/s41586-019-1691-4.

Mayor, R., Morgan, R., & Sargent, M. G. (1995). Induction of the prospective neural crest of Xenopus. *Development (Cambridge, England)*, *121*, 767–777.

McGary, K. L., Park, T. J., Woods, J. O , Cha, H. J., Wallingford, J. B., & Marcotte, E. M. (2010). Systematic discovery of nonobvious human disease models through orthologous phenotypes. *Proceedings of the National Academy of Sciences of the United States of America*, *107*, 6544–6549. https://doi.org/10.1073/pnas.0910200107.

Milet, C., Maczkowiak, F., Roche, D. D., & Monsoro-Burq, A. H. (2013). Pax3 and Zic1 drive induction and differentiation of multipotent, migratory, and functional neural crest in Xenopus embryos. *Proceedings of the National Academy of Sciences of the United States of America*, *110*, 5528–5533. https://doi.org/10.1073/pnas.1219124110.

Mills, A., Bearce, E., Cella, R., Kim, S. W., Selig, M., Lee, S., et al. (2019). Wolf-Hirschhorn syndrome-associated genes are enriched in motile neural crest cells and affect craniofacial development in Xenopus laevis. *Frontiers in Physiology*, *10*, 431. https://doi.org/10.3389/fphys.2019.00431.

Minchin, J. E. N., & Hughes, S. M. (2008). Sequential actions of Pax3 and Pax7 drive xanthophore development in zebrafish neural crest. *Developmental Biology*, *317*, 508–522. https://doi.org/10.1016/j.ydbio.2008.02.058.

Monsoro-Burq, A. H. (2015). PAX transcription factors in neural crest development. *Seminars in Cell & Developmental Biology*, *44*, 87–96. https://doi.org/10.1016/j.semcdb.2015.09.015.

Monsoro-Burq, A.-H., Wang, E., & Harland, R. (2005). Msx1 and Pax3 cooperate to mediate FGF8 and WNT signals during Xenopus neural crest induction. *Developmental Cell*, *8*, 167–178. https://doi.org/10.1016/j.devcel.2004.12.017.

Moody, S. A., Neilson, K. M., Kenyon, K. L., Alfandari, D., & Pignoni, F. (2015). Using Xenopus to discover new genes involved in branchiootorenal spectrum disorders. *Comparative Biochemistry and Physiology. Toxicology & Pharmacology*, *178*, 16–24. https://doi.org/10.1016/j.cbpc.2015.06.007.

Moore, R., Theveneau, E., Pozzi, S., Alexandre, P., Richardson, J., Merks, A., et al. (2013). Par3 controls neural crest migration by promoting microtubule catastrophe during contact inhibition of locomotion. *Development (Cambridge, England)*, *140*, 4763–4775. https://doi.org/10.1242/dev.098509.

Morgan, S. C., Lee, H.-Y., Relaix, F., Sandell, L. L., Levorse, J. M., & Loeken, M. R. (2008). Cardiac outflow tract septation failure in Pax3-deficient embryos is due to p53-dependent regulation of migrating cardiac neural crest. *Mechanisms of Development*, *125*, 757–767. https://doi.org/10.1016/j.mod.2008.07.003.

Murdoch, B., DelConte, C., & García-Castro, M. I. (2012). Pax7 lineage contributions to the mammalian neural crest. *PLoS One*, *7*, e41089. https://doi.org/10.1371/journal.pone.0041089.

Nakano, M., Kishida, R., Funakoshi, K., Tsukagoshi, M., Goris, R. C., Kadota, T., et al. (2003). Central projections of thoracic splanchnic and somatic nerves and the location of sympathetic preganglionic neurons in Xenopus laevis. *The Journal of Comparative Neurology*, *456*, 321–337. https://doi.org/10.1002/cne.10514.

Nakayama, T., Nakajima, K., Cox, A., Fisher, M., Howell, M., Fish, M. B., et al. (2017). No privacy, a Xenopus tropicalis mutant, is a model of human Hermansky-Pudlak syndrome and allows visualization of internal organogenesis during tadpole development. *Developmental Biology*, *426*, 472–486. https://doi.org/10.1016/j.ydbio.2016.08.020.

Nenni, M. J., Fisher, M. E., James-Zorn, C., Pells, T. J., Ponferrada, V., Chu, S., et al. (2019). Xenbase: Facilitating the use of Xenopus to model human disease. *Frontiers in Physiology*, *10*, 154. https://doi.org/10.3389/fphys.2019.00154.

Nieto, M. A., Bennett, M. F., Sargent, M. G., & Wilkinson, D. G. (1992). Cloning and developmental expression of Sna, a murine homologue of the Drosophila snail gene. *Development (Cambridge, England)*, *116*, 227–237.

Nieto, M. A., Sargent, M. G., Wilkinson, D. G., & Cooke, J. (1994). Control of cell behavior during vertebrate development by Slug, a zinc finger gene. *Science*, *264*, 835–839. https://doi.org/10.1126/science.7513443.

Nikitina, N., Sauka-Spengler, T., & Bronner-Fraser, M. (2008). Dissecting early regulatory relationships in the lamprey neural crest gene network. *Proceedings of the National Academy of Sciences of the United States of America*, *105*, 20083–20088. https://doi.org/10.1073/pnas.0806009105.

Pegoraro, C., Figueiredo, A. L., Maczkowiak, F., Pouponnot, C., Eychène, A., & Monsoro-Burq, A. H. (2015). PFKFB4 controls embryonic patterning via Akt signalling independently of glycolysis. *Nature Communications*, *6*, 5953. https://doi.org/10.1038/ncomms6953.

Perfetto, M., Xu, X., Lu, C., Shi, Y., Yousaf, N., Li, J., et al. (2021). The RNA helicase DDX3 induces neural crest by promoting AKT activity. *Development (Cambridge, England)*, *148*(2), dev184341. https://doi.org/10.1242/dev.184341.

Pingault, V., Ente, D., Dastot-Le Moal, F., Goossens, M., Marlin, S., & Bondurand, N. (2010). Review and update of mutations causing Waardenburg syndrome. *Human Mutation*, *31*, 391–406. https://doi.org/10.1002/humu.21211.

Pla, P., & Monsoro-Burq, A. H. (2018). The neural border: Induction, specification and maturation of the territory that generates neural crest cells. *Developmental Biology*, *444*(Suppl. 1), S36–S46. https://doi.org/10.1016/j.ydbio.2018.05.018.

Plouhinec, J.-L., Medina-Ruiz, S., Borday, C., Bernard, E., Vert, J.-P., Eisen, M. B., et al. (2017). A molecular atlas of the developing ectoderm defines neural, neural crest, placode, and nonneural progenitor identity in vertebrates. *PLoS Biology*, *15*, e2004045. https://doi.org/10.1371/journal.pbio.2004045.

Popov, I. K., Hiatt, S. M., Whalen, S., Keren, B., Ruivenkamp, C., van Haeringen, A., et al. (2019). A YWHAZ variant associated with cardiofaciocutaneous syndrome activates the RAF-ERK pathway. *Frontiers in Physiology*, *10*, 388. https://doi.org/10.3389/fphys.2019.00388.

Robson, A., Owens, N. D. L., Baserga, S. J., Khokha, M. K., & Griffin, J. N. (2016). Expression of ribosomopathy genes during Xenopus tropicalis embryogenesis. *BMC Developmental Biology*, *16*, 38. https://doi.org/10.1186/s12861-016-0138-5.

Rothstein, M., & Simoes-Costa, M. (2020). Heterodimerization of TFAP2 pioneer factors drives epigenomic remodeling during neural crest specification. *Genome Research*, *30*, 35–48. https://doi.org/10.1101/gr.249680.119.

Sato, H., Takino, T., Okada, Y., Cao, J., Shinagawa, A., Yamamoto, E., et al. (1994). A matrix metalloproteinase expressed on the surface of invasive tumour cells. *Nature*, *370*, 61–65. https://doi.org/10.1038/370061a0.

Sauka-Spengler, T., Meulemans, D., Jones, M., & Bronner-Fraser, M. (2007). Ancient evolutionary origin of the neural crest gene regulatory network. *Developmental Cell*, *13*, 405–420. https://doi.org/10.1016/j.devcel.2007.08.005.

Scarpa, E., Szabó, A., Bibonne, A., Theveneau, E., Parsons, M., & Mayor, R. (2015). Cadherin switch during EMT in neural crest cells leads to contact inhibition of locomotion via repolarization of forces. *Developmental Cell*, *34*, 421–434. https://doi.org/10.1016/j.devcel.2015.06.012.

Scerbo, P., & Monsoro-Burq, A. H. (2020). The vertebrate-specific VENTX/NANOG gene empowers neural crest with ectomesenchyme potential. *Science Advances*, *6*, eaaz1469. https://doi.org/10.1126/sciadv.aaz1469.

Schorle, H., Meier, P., Buchert, M., Jaenisch, R., & Mitchell, P. J. (1996). Transcription factor AP-2 essential for cranial closure and craniofacial development. *Nature, 381*, 235–238. https://doi.org/10.1038/381235a0.

Schulz, Y., Wehner, P., Opitz, L., Salinas-Riester, G., Bongers, E. M. H. F., van Ravenswaaij-Arts, C. M. A., et al. (2014). CHD7, the gene mutated in CHARGE syndrome, regulates genes involved in neural crest cell guidance. *Human Genetics, 133*, 997–1009. https://doi.org/10.1007/s00439-014-1444-2.

Schwenty-Lara, J., Nehl, D., & Borchers, A. (2020). The histone methyltransferase KMT2D, mutated in Kabuki syndrome patients, is required for neural crest cell formation and migration. *Human Molecular Genetics, 29*, 305–319. https://doi.org/10.1093/hmg/ddz284.

Shellard, A., & Mayor, R. (2019). Integrating chemical and mechanical signals in neural crest cell migration. *Current Opinion in Genetics & Development, 57*, 16–24. https://doi.org/10.1016/j.gde.2019.06.004.

Shi, Z., Xin, H., Tian, D., Lian, J., Wang, J., Liu, G., et al. (2019). Modeling human point mutation diseases in Xenopus tropicalis with a modified CRISPR/Cas9 system. *FASEB Journal: Official Publication of the Federation of American Societies for Experimental Biology, 33*, 6962–6968. https://doi.org/10.1096/fj.201802661R.

Singh, M. D., Jensen, M., Lasser, M., Huber, E., Yusuff, T., Pizzo, L., et al. (2020). NCBP2 modulates neurodevelopmental defects of the 3q29 deletion in Drosophila and Xenopus laevis models. *PLoS Genetics, 16*, e1008590. https://doi.org/10.1371/journal.pgen.1008590.

Song, J., Feng, Y., Acke, F. R., Coucke, P., Vleminckx, K., & Dhooge, I. J. (2016). Hearing loss in Waardenburg syndrome: A systematic review. *Clinical Genetics, 89*, 416–425. https://doi.org/10.1111/cge.12631.

Square, T., Jandzik, D., Cattell, M., Hansen, A., & Medeiros, D. M. (2016). Embryonic expression of endothelins and their receptors in lamprey and frog reveals stem vertebrate origins of complex endothelin signaling. *Scientific Reports, 6*, 34282. https://doi.org/10.1038/srep34282.

Square, T. A., Jandzik, D., Massey, J. L., Romášek, M., Stein, H. P., Hansen, A. W., et al. (2020). Evolution of the endothelin pathway drove neural crest cell diversification. *Nature, 585*, 563–568. https://doi.org/10.1038/s41586-020-2720-z.

Stemmler, M. P., Eccles, R. L., Brabletz, S., & Brabletz, T. (2019). Non-redundant functions of EMT transcription factors. *Nature Cell Biology, 21*, 102–112. https://doi.org/10.1038/s41556-018-0196-y.

Szabó, A., Melchionda, M., Nastasi, G., Woods, M. L., Campo, S., Perris, R., et al. (2016). In vivo confinement promotes collective migration of neural crest cells. *The Journal of Cell Biology, 213*, 543–555. https://doi.org/10.1083/jcb.201602083.

Szabova, L., Chrysovergis, K., Yamada, S. S., & Holmbeck, K. (2008). MT1-MMP is required for efficient tumor dissemination in experimental metastatic disease. *Oncogene, 27*, 3274–3281. https://doi.org/10.1038/sj.onc.1210982.

Tae, H.-J., Rahman, M. M., & Park, B.-Y. (2015). Temporal and spatial expression analysis of peripheral myelin protein 22 (Pmp22) in developing Xenopus. *Gene Expression Patterns, 17*, 26–30. https://doi.org/10.1016/j.gep.2015.01.001.

Takahashi, S., Uochi, T., Kawakami, Y., Nohno, T., Yokota, C., Kinoshita, K., et al. (1998). Cloning and expression pattern of Xenopus prx-1 (Xprx-1) during embryonic development. *Development, Growth & Differentiation, 40*, 97–104. https://doi.org/10.1046/j.1440-169x.1998.t01-6-00011.x.

Talikka, M., Stefani, G., Brivanlou, A. H., & Zimmerman, K. (2004). Characterization of Xenopus Phox2a and Phox2b defines expression domains within the embryonic nervous system and early heart field. *Gene Expression Patterns, 4*, 601–607. https://doi.org/10.1016/j.modgep.2004.01.012.

Tam, W. L., & Weinberg, R. A. (2013). The epigenetics of epithelial-mesenchymal plasticity in cancer. *Nature Medicine*, *19*, 1438–1449. https://doi.org/10.1038/nm.3336.

Tanaka, M., Kuriyama, S., Itoh, G., Kohyama, A., Iwabuchi, Y., Shibata, H., et al. (2016). Identification of anti-cancer chemical compounds using Xenopus embryos. *Cancer Science*, *107*, 803–811. https://doi.org/10.1111/cas.12940.

Tassabehji, M., Fang, Z. M., Hilton, E. N., McGaughran, J., Zhao, Z., de Bock, C. E., et al. (2008). Mutations in GDF6 are associated with vertebral segmentation defects in Klippel-Feil syndrome. *Human Mutation*, *29*, 1017–1027. https://doi.org/10.1002/humu.20741.

Tazumi, S., Yabe, S., & Uchiyama, H. (2010). Paraxial T-box genes, Tbx6 and Tbx1, are required for cranial chondrogenesis and myogenesis. *Developmental Biology*, *346*, 170–180. https://doi.org/10.1016/j.ydbio.2010.07.028.

Tchieu, J., Zimmer, B., Fattahi, F., Amin, S., Zeltner, N., Chen, S., et al. (2017). A modular platform for differentiation of human PSCs into all major ectodermal lineages. *Cell Stem Cell*, *21*, 399–410.e7. https://doi.org/10.1016/j.stem.2017.08.015.

Theveneau, E., & Linker, C. (2017). Leaders in collective migration: Are front cells really endowed with a particular set of skills? *F1000Res*, *6*, 1899. https://doi.org/10.12688/f1000research.11889.1.

Tomlinson, M. L., Rejzek, M., Fidock, M., Field, R. A., & Wheeler, G. N. (2009). Chemical genomics identifies compounds affecting Xenopus laevis pigment cell development. *Molecular BioSystems*, *5*, 376–384. https://doi.org/10.1039/b818695b.

Twigg, S. R. F., Forecki, J., Goos, J. A. C., Richardson, I. C. A., Hoogeboom, A. J. M., van den Ouweland, A. M. W., et al. (2015). Gain-of-function mutations in ZIC1 are associated with coronal craniosynostosis and learning disability. *American Journal of Human Genetics*, *97*, 378–388. https://doi.org/10.1016/j.ajhg.2015.07.007.

Ufartes, R., Schwenty-Lara, J., Freese, L., Neuhofer, C., Möller, J., Wehner, P., et al. (2018). Sema3a plays a role in the pathogenesis of CHARGE syndrome. *Human Molecular Genetics*, *27*, 1343–1352. https://doi.org/10.1093/hmg/ddy045.

Valluet, A., Druillennec, S., Barbotin, C., Dorard, C., Monsoro-Burq, A. H., Larcher, M., et al. (2012). B-Raf and C-Raf are required for melanocyte stem cell self-maintenance. *Cell Reports*, *2*, 774–780. https://doi.org/10.1016/j.celrep.2012.08.020.

Van Grunsven, L. A., Taelman, V., Michiels, C., Opdecamp, K., Huylebroeck, D., & Bellefroid, E. J. (2006). deltaEF1 and SIP1 are differentially expressed and have overlapping activities during Xenopus embryogenesis. *Developmental Dynamics: An Official Publication of the American Association of the Anatomists*, *235*, 1491–1500. https://doi.org/10.1002/dvdy.20727.

Vega-Lopez, G. A., Cerrizuela, S., Tribulo, C., & Aybar, M. J. (2018). Neurocristopathies: New insights 150 years after the neural crest discovery. *Developmental Biology*, *444*(Suppl. 1), S110–S143. https://doi.org/10.1016/j.ydbio.2018.05.013.

Verstappen, G., van Grunsven, L. A., Michiels, C., Van de Putte, T., Souopgui, J., Van Damme, J., et al. (2008). Atypical Mowat-Wilson patient confirms the importance of the novel association between ZFHX1B/SIP1 and NuRD corepressor complex. *Human Molecular Genetics*, *17*, 1175–1183. https://doi.org/10.1093/hmg/ddn007.

Wagner, R. W., Smith, J. E., Cooperman, B. S., & Nishikura, K. (1989). A double-stranded RNA unwinding activity introduces structural alterations by means of adenosine to inosine conversions in mammalian cells and Xenopus eggs. *Proceedings of the National Academy of Sciences of the United States of America*, *86*, 2647–2651. https://doi.org/10.1073/pnas.86.8.2647.

Wang, D., Chang, P. S., Wang, Z., Sutherland, L., Richardson, J. A., Small, E., et al. (2001). Activation of cardiac gene expression by myocardin, a transcriptional cofactor for serum response factor. *Cell*, *105*, 851–862. https://doi.org/10.1016/s0092-8674(01)00404-4.

Watt, K. E. N., & Trainor, P. A. (2014). Chapter 17—Neurocristopathies: The etiology and pathogenesis of disorders arising from defects in neural crest cell development. In P. A. Trainor (Ed.), *Neural crest cells* (pp. 361–394). Boston: Academic Press.

White, R. M., Cech, J., Ratanasirintrawoot, S., Lin, C. Y., Rahl, P. B., Burke, C. J., et al. (2011). DHODH modulates transcriptional elongation in the neural crest and melanoma. *Nature*, *471*, 518–522. https://doi.org/10.1038/nature09882.

Xiong, Y., Zhang, J., Shi, L., Ning, Y., Zhu, Y., Chen, S., et al. (2017). NOGO-B promotes EMT in lung fibrosis via MMP14 mediates free TGF-beta1 formation. *Oncotarget*, *8*, 71024–71037. https://doi.org/10.18632/oncotarget.20297.

Zalc, A., Rattenbach, R., Auradé, F., Cadot, B., & Relaix, F. (2015). Pax3 and Pax7 play essential safeguard functions against environmental stress-induced birth defects. *Developmental Cell*, *33*, 56–66. https://doi.org/10.1016/j.devcel.2015.02.006.

Printed in the United States
by Baker & Taylor Publisher Services